T0325919

Process Modelling and Simulation with Finite Element Methods

SERIES ON STABILITY, VIBRATION AND CONTROL OF SYSTEMS

Founder and Editor: Ardéshir Guran
Co-Editors: M. Cloud & W. B. Zimmerman

About the Series

Rapid developments in system dynamics and control, areas related to many other topics in applied mathematics, call for comprehensive presentations of current topics. This series contains textbooks, monographs, treatises, conference proceedings and a collection of thematically organized research or pedagogical articles addressing dynamical systems and control.

The material is ideal for a general scientific and engineering readership, and is also mathematically precise enough to be a useful reference for research specialists in mechanics and control, nonlinear dynamics, and in applied mathematics and physics.

Selected Volumes in Series A

SERIES ON STABILITY, VIBRATION AND CONTROL OF SYSTEMS

Series A   Volume 15

Founder & Editor: **Ardéshir Guran**

Co-Editors: **M. Cloud & W. B. Zimmerman**

Process Modelling and Simulation with Finite Element Methods

William B. J. Zimmerman
University of Sheffield, UK

World Scientific

NEW JERSEY • LONDON • SINGAPORE • BEIJING • SHANGHAI • HONG KONG • TAIPEI • CHENNAI

Published by

World Scientific Publishing Co. Pte. Ltd.

5 Toh Tuck Link, Singapore 596224

USA office: 27 Warren Street, Suite 401–402, Hackensack, NJ 07601

UK office: 57 Shelton Street, Covent Garden, London WC2H 9HE

British Library Cataloguing-in-Publication Data
A catalogue record for this book is available from the British Library.

ISBN-13 978-981-238-793-6
ISBN-10 981-238-793-5

Printed in Singapore

ABOUT THE AUTHOR

Dr William B.J. Zimmerman is a Reader in Chemical and Process Engineering. His research interests are in fluid dynamics and reaction engineering. He has previously created modules entitled Chemical Engineering Problem Solving with Mathematica, Modelling and Simulation in Chemical Processes, Numerical Analysis in Chemical Engineering, and FORTRAN programming. He has been modelling with finite element methods since 1986. He has authored over sixty scientific and scholarly works. He is a graduate of Princeton and Stanford Universities in Chemical Engineering, past Director of the M Sc in Environmental and Energy Engineering, originator of the M Sc in Process Fluid Dynamics, and a winner in US and UK national competitions of four prestigious fellowships:

2000-5 EPSRC Advanced Research Fellow

1994-99 Royal Academy of Engineering, Zeneca Young Academic Fellow.

1991-93 NATO postdoctoral fellow in science and engineering.

1988-91 National Science Foundation Research Fellow

FOREWORD

I would especially like to thank the Engineering and Physical Sciences Research Council of the United Kingdom for the award of an Advanced Research Fellowship on the topic of "models of helical mixing and reaction: a new approach to chemical reaction engineering." Without the flexibility of the fellowship, I doubt I would have felt adventuresome enough to afford the time to run so far – developing an intensive training module and writing a textbook on a topic that is relatively new to me and was not envisaged when I wrote the original research strategy for the fellowship in 1998. It has turned out to be integral to my plans for turbulence modeling, although this is not reflected in the book.

Johan Sundqvist and Ed Fontes of COMSOL have been most supportive of my two projects, lending considerable resources to helping me to iron out difficulties in modeling, contributing to the intensive modules, and providing critiques of the draft chapters. I would have thought the crew at COMSOL would have tired with the endless e-mails from my research group. With most packages I use, I have no intention of being "cutting edge," so the FAQs on the web site usually already have my queries – asked and answered. It is both novel and refreshing to have identified (and sometimes solved or worked around) new bugs or puzzles. To Ed and Johan, a wholehearted thanks for welcoming me to the FEMLAB developers community.

Many thanks to the team of collaborators and chapter co-authors who have encouraged this effort. Buddhi, Alex, Kiran, Jordan, Peter, George and Julia have always had a kind word and a willingness to brainstorm and contribute.

Finally, thanks to attendees at my intensive modules for spotting inconsistencies, patiently wading through "experimental" teaching material (guinea pigs who are so intelligent are a rare find!), and putting up with my sometimes convoluted explanations. Not to mention those awkward times as we uncovered clangers in the demonstrations. It has always been a conceit of mine that computer demonstrations should be realistic – bugs and all – since debugging is an integral programming skill that relies on intuition and experience. So thanks for sharing the experience!

CONTENTS

INTRODUCTION TO FEMLAB

W.B.J. ZIMMERMAN

Department of Chemical and Process Engineering, University of Sheffield,
Newcastle Street, Sheffield S1 3JD United Kingdom

E-mail: w.zimmerman@shef.ac.uk

FEMLAB is a relatively recent development in the MATLAB sphere. Perhaps a good fraction of the readers of this book were attracted by the title and the dust jacket description, so they might have little exposure to FEMLAB previously. To them, I would heartily recommend attending a FEMLAB seminar on their recurring academic roadshows. The experience of seeing FEMLAB in action is more illustrative than the printed word and screen captures shown here. This Introduction provides an overview of why I wrote the book and developed an intensive training module for FEMLAB modeling of chemical engineering applications – the unique features of FEMLAB that the reader will want to assess for her own modeling objectives. The FEMLAB User's Guide (available for download from the COMSOL web site) does a better job of familiarizing the reader with "What is FEMLAB?" than the brief introduction in this chapter to the FEMLAB graphical user interface (GUI). The point of the introduction to FEMLAB here is to describe how completely determined models are set up in FEMLAB, after which the methodology can be used in subsequent chapters without ambiguity. Nevertheless, I hope that this chapter whets your appetite for the cornucopia of modeling tools, along with an intellectual framework for using FEMLAB for modeling, that is described in this book.

0.1 Overview of the Book

Chapters 1-4 were taken as the text for the first intensive module "Chemical Engineering Modelling with FEMLAB." These chapters represent a personal odyssey with FEMLAB. It was not originally my intention to write a book about FEMLAB. For a long term project that I am still undertaking, I need a PDE engine that is readily customizable to additional terms and heterogeneous domains. Once I decided that FEMLAB could fill the bill, I needed to become an expert on it. One nefarious way of doing that is to declare a course on it, rope graduate students and other interested external parties into attending, and then study like mad to produce a coherent set of lectures and computer laboratories. I already had several templates for this, having taught undergraduate and postgraduate modules on numerical analysis, modeling, and simulation. So I adapted the storyline of those modules with FEMLAB models. Chapter One is the product of this adaptation. Chapter Two is an obvious outgrowth of my prior use of the PDE toolbox of MATLAB and a necessary explanation of finite element methods. Chapters 3-7 were far more deliberate attempts to exploit the powerful features of FEMLAB by systematically exploring models that illustrate the feature of the theme of each chapter. I searched through my own repertoire of PDE modeling and sought out contributions from colleagues that would illustrate the features. Chapters 8 and 9 are of a different type. These chapters would legitimately fit into the FEMLAB Model Library as case studies of modeling with FEMLAB, rather than organized along a particular programming theme. Nonetheless, the case studies highlight non-standard aspects of FEMLAB/MATLAB modeling, analysis, and postprocessing that are strikingly original.

Target audience

The book is aimed at graduate Chemical Engineers who use modelling tools and as a general introduction to FEMLAB for scientists and engineers.

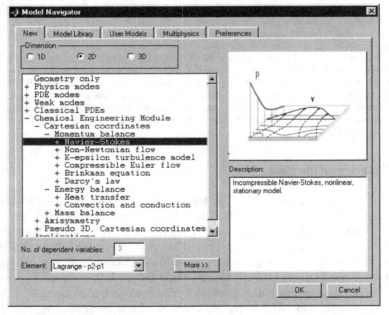

Figure 0.1 The pre-built application modes are arranged in a tree structure on the Model Navigator. Here is the Incompressible Navier-Stokes mode under the Chemical Engineering Module. The Model Navigator specifies that this mode is 2-D, has three dependent variables, and uses a mixed type of element Lagrange p2 for the velocities u and v, Lagrange p1 for the pressure. Using mixed order discretization schemes is quite common in finite element methods for numerical stability of the Navier-Stokes solvers. The SIMPLE scheme [1] pioneered the approach. The Model Navigator allows the user to specify pre-built application modes or to customize a generic PDE mode (coefficient, general, weak) to build up their own model.

Attitude

The attitude of this book is to demonstrate particular features of FEMLAB that make computational modelling easy to implement, and then emphasize those features that are advantages to modelling with FEMLAB. This will be illustrated with reference to Chemical Engineering Modelling, which has a special history and well known applications. The features, however, are generally applicable in the sciences and engineering.

Bias

The book is slanted toward applications in fluid dynamics, transport phenomena, and heterogeneous reaction, which reflect some of the research interests of the author that routinely involve mathematical modelling by PDEs and solution by numerical methods.

Modeling versus simulation

This book is about modelling and programming. The first four chapters, the core of the taught module, focus completely on modeling. The remaining chapters are slanted towards the use of FEMLAB for simulation. The distinction is that simulation has some stochastic and evolutionary elements. Simulations may have a PDE compute engine as an integral component, but generally involve much more "user defined programming." This book organizes case studies of modeling along the lines of a cookbook – here are some models that are important in chemical engineering applications that are computable in MATLAB/FEMLAB. What is lacking from this presentation style, however, are the philosophical and methodological aspects of modeling. This book is "How To", but not sufficiently "Why" and "How good?" are the models. There are two major classes of modeling activity – (1) rigorous physicochemical modeling, which takes the best understanding of physics and attempt to compute by numerical methods the exact value up to the limits of finite precision representation of numbers; (2) approximate modeling, which intends to approximate the true, rigorous dynamics with simpler relationships in order to estimate sizes of effects and features of the outcome, rather than exact, detailed accuracy. In this book, no attempt is made to systematically treat how to propose the equations and boundary conditions of modeling – decisions about modeling objectives and acceptable approximations are presumed to have already been taken rationally. Yet, in most modeling conundrums and trouble shooting, whether or not the model itself is sensible is a key question, and what level of approximation and inaccuracy are acceptable, are part and parcel of the modeling activity. Numerics and scientific/engineering judgement about what should be modelled and how should not be separated.

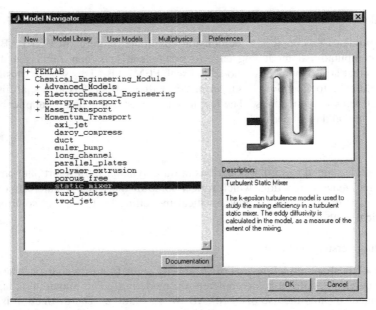

Figure 0.2 The Model Library contains already solved problems using existing application modes. The Model Library includes models created by COMSOL staff and donated by users. The growth in the content of the Model Library over the last twelve months has been phenomenal. Browsing the FEMLAB models and the Model Library documentation of them is an excellent way of generating modeling ideas. The Model Library is organized by subject matter. Here, the turbulent static mixer model is highlighted, under the tree structure with branch Chemical Engineering Module and sub-branch Momentum Transport. The k-ε model for turbulence is the simplest model of turbulence and the workhorse of most commercial CFD packages [2].

Why should I use FEMLAB for modelling?

1. FEMLAB has an integrated modelling environment.
2. FEMLAB takes a semi-analytic approach: You specify equations, FEMLAB symbolically assembles FEM matrices and organizes the bookkeeping.
3. FEMLAB is built on top of MATLAB, so user defined programming for the modelling, organizing the computation, or the post-processing has full functionality.
4. FEMLAB provides pre-built templates as Application Modes (see Figure 0.1) and in the Model Library for common modelling applications.
5. FEMLAB provides multiphysics modelling – linking well known "application modes" transparently.
6. FEMLAB innovated extended multiphysics – coupling between logically distinct domains and models that permits simultaneous solution. Examples: networks with different models for links and nodes, dispersed phases, multiple scales.

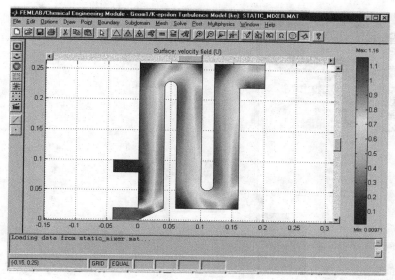

Figure 0.3 FEMLAB's postprocessing screen. Here the solution for the last executed run of the turbulent static mixer model is shown. FEMLAB's GUI provides pull down menus and toolbars to initiate all building blocks of model construction -- specifying analyzed geometries, meshing, specifying PDE equations and boundary conditions, analyzing and post processing the solutions found. Note that the status bar at the bottom shows the position of the cursor on the visualization window. The information window just above it echoes messages to the screen from the FEMLAB/MATLAB commands executed in FEMLAB's MATLAB workspace. The "Loading data from static_mixer.mat" message was the response to our request to load the model library entry for the turbulent static mixer.

As we will learn in Chapters Four and Seven, extended multiphysics is very similar to the linkages provided by process simulation tools common for integrated flowsheets of process plant such as HYSYS and Aspen, or which can be developed in MATLAB's Simulink environment. FEMLAB fully couples this functionality to a PDE engine that rivals CFD packages such as FLUENT and CFX or other commercial PDE engines such as ANSYS, but with competitive advantages listed above.

Modelling strategies in FEMLAB

This book is about how I think about modelling and simulation. Perhaps my thoughts will serve as a guide to help you with the modelling problem that drew you to FEMLAB. After posing myself the modelling problems in this book, I came up with a short list of guidelines for how to approach modelling with FEMLAB:

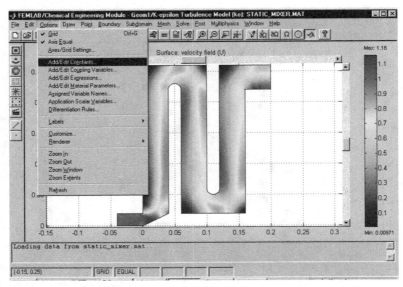

Figure 0.4 The Options Menu permits definition of many useful feature: constants, grids for drawing and visualization, and expressions used in entering the model equations are the most common uses.

1. Don't re-invent the wheel. Read the Model Library and User's Guide/Web pages.
2. Formulate a mathematical model. Compare with pre-built application modes.
3. Can it all be done in the FEMLAB GUI, or is the PDE engine only a subroutine?

FEMLAB as an integrated modelling environment

FEMLAB can be viewed two ways –

1. As an interactive, integrated GUI for setting up, solving, and post-processing a mathematical model – a PACKAGE.
2. As a set of MATLAB subroutines for setting up, solving, and post-processing a mathematical model – a PROGRAMMING LANGUAGE.

This book intends to show how to implement models built both ways in an efficient way. The FEMLAB GUI is so straightforward in setting up problems and trying "what if" scenarios that it must be the first port of call in "having a go." The great utility of a PACKAGE is that the barriers to entry are small, so the pay off is worth the investment of learning all the features of the tool.

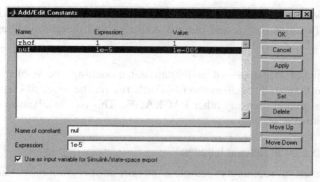

Figure 0.5 FEMLAB constants (rhof =1 and nuf=1e-5) defined for the turbulent static mixer model.

FEMLAB's GUI (Figure 0.3) allows sophisticated modularity in modelling

- Pre-built application modes provide templates for common calculations.
- The Model Library provides Case Studies.
- A model can be set up by systematically traducing the Menu bar from left to right.
  - The Model Navigator (Figure 0.1) accesses previously built application modes or existing models are loaded from the File Menu.
  - The Options menu (see Figure 0.4) provides definition space for constants (see Figure 0.5), variables, and expressions used in either setup, solution, or post-processing phases.
  - The Draw Menu (Figure 0.6) allows domain specifications in Draw Mode (Figure 0.7).
  - The Point Mode (Figure 0.8) provides entry for point constraints under Point Settings (Figure 0.9) dialogue box.
  - The Boundary Mode (Figure 0.10) provides entry for boundary constraints through the Boundary Settings (Figure 0.11) dialogue box.
  - The Subdomain Mode (Figure 0.12) permits PDE specifications (Figure 0.13) in the Subdomain Settings.
  - The Mesh Mode (Figure 0.14) shows the mesh and specifies it, which is generated by an elliptic mesh generator subject to constraints specified in the Mesh Parameters dialogue box. The Remesh button generates the mesh, or the triangle button on the Toolbar.
  - The Solve Menu specifies the type and parameters to be used in the solution scheme. The solution procedure is initiated by the Solve Button on the Solver Parameters dialogue box or the = button on the Toolbar. The solution is shown on the GUI main window with parameters defined in Post Mode. See Figure 0.3 again.
  - The Post Mode provides various graphical and computational processing.

— The **Multiphysics Menu** allows switching between "active" modes for the specifications menus and permits additions and deletions of "application modes"

The GUI makes the stages of computational modelling accessible in a much shorter time than traditional methods. Furthermore, the level of complexity in modelling is greater than any other PACKAGE. This has its advantages, as well as its own drawbacks.

<u>FEMLAB as a programming language</u>

I have learned a seemingly ceaseless stream of programming languages – BASIC, Assembler (asm for 8088 & Cray Assembly Language), FORTRAN, PASCAL, LISP, APL, C, Mathematica, C++, MATLAB.

Programming is hard. The languages are full of commands and syntax with intricate details that need to be mastered before complicated problems can be tackled.

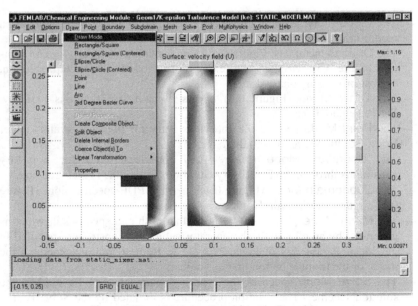

Figure 0.6 Draw Mode is selected from the Draw Menu.

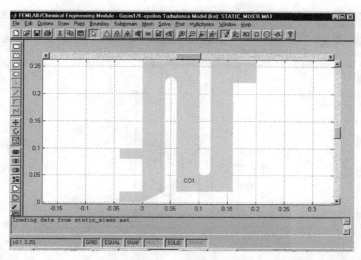

Figure 0.7 The single composite analyzed geometry (CO1) of the static mixer model. This geometry was drawn by geometry primitive commands (rectangles and arcs) and then merged together to form one contiguous domain.

Engineers usually put problem solving first, and skills and techniques are acquired as necessary to solve problems. Programming strategies should reflect this. My FORTRAN programming strategy is simple – I find the "off-the-shelf" subprograms that do the integral steps of what I want to achieve, and then build a program "shell" around it to read in parameters, set up storage, call the essential subprograms, and then "post process" (also often with canned routines) and then write out the data to files.

I treat FEMLAB/MATLAB programming the same.

The key is to get the FEMLAB GUI to do the work for you.

- The File Menu has the "Save Model m-file" and "Reset Mode m-file" options for you.
- Set up the "workhorse" of your model in the GUI, and then export the model m-file, which provides most of the "program body" needed to use FEMLAB as a programming language in MATLAB, thus providing all the subroutines and command syntax and logical structure, without the User needing to know the details.
- MATLAB m-files/ m-file functions can then be set up to provide data entry, storage set up, post-processing, and output. Complicated programmes can be built up modularly without the user specifying, or even knowing all the details. MATLAB programming expertise is needed, but crucially NOT FEMLAB programming expertise.

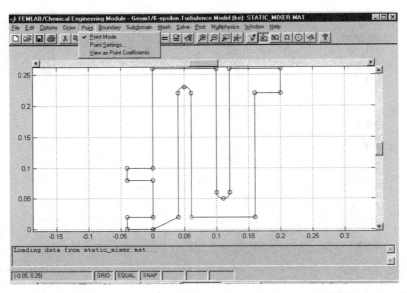

Figure 0.8 Point mode shows all the points (vertices and specifically identified points) distinguished in the geometry model by circles. The red circle is selected here.

The book provides a wealth of examples of "user defined programming" with MATLAB m-file scripts and m-file functions calling FEMLAB subprograms. In every case, however, I adapted models developed in the FEMLAB GUI and read out as model m-files. I have yet to write a MATLAB program around FEMLAB commands/functions "from scratch." The FEMLAB Reference Manual provides a complete description of all the commands, so I have tried to get functionality out of the MATLAB programming that is not achievable through the GUI alone. Perhaps this is a good juncture to point out that each FEMLAB GUI session has its own MATLAB workspace, separate from the one that launched it. So it is perfectly legitimate to write your own MATLAB m-file script and read it into the FEMLAB GUI. The MATLAB workspace will execute all the MATLAB commands, even those that are not possible to do through the GUI alone, and the GUI responds by showing the intermediate steps – drawing the geometry, meshing, solving, and postprocessing. If you are writing your own user-defined m-file script for FEMLAB/MATLAB, playing it back in the GUI shows you how far it gets before the program bugs (well, maybe you don't put them in your codes, but mine are usually infested to start) crash it. In this respect, MATLAB is a "macro" language for FEMLAB.

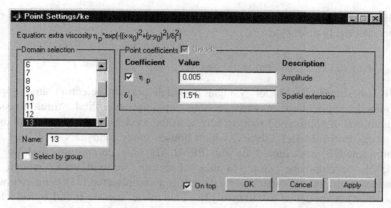

Figure 0.9 Point settings dialogue box. Here point 13 which is highlighted corresponds to the red circled vertex in Figure 0.8. The k-ε model has an extra point-wise viscosity associated with vertices as point coefficients in the FEM weak formulation. This page provides data entry.

Summary

FEMLAB has a powerful GUI that provides easy entry to try out "what if" scenarios and explore modelling methods/types without the investment of "programming time."

FEMLAB has unique modelling advantages in "multiphysics" and "extended multiphysics" which may make FEMLAB the only viable modelling tool for certain applications.

FEMLAB provides a method of automatically creating MATLAB m-file source code that reduces the programming effort for setting up more complicated models. Exporting solutions to MATLAB also makes post solution analysis more flexible.

FEMLAB/MATLAB programming provides automation opportunities, including running efficiently (least memory/processor overhead) as a background job.

0.2 An Example from the Model Library

Figures 0.1 – 0.14 run through the major data entry points for PDE models with an example of the turbulent static mixer from the Model Library, constructed on top of the k-ε turbulence model application mode. The figure captions tell the story, and the screen captures illustrate some of the key features of the FEMLAB GUI. This is the only time that the book will show FEMLAB GUI screen captures. From here on out, we will describe the information content for model specification in terms of the data entry needed for the dialogue boxes used in each model. This limitation to the printed word and graphical results of the models is a consequence of a desire to discuss many models, rather than to view menus and dialogue boxes with their content, limiting the number of models that can be effectively discussed.

0.2.1 k-ε model of a turbulent static mixer

Figure 0.1 shows the Model Navigator, which permits selection among pre-determined application modes, setting up user-specified "multiphysics" combinations of application modes, access to the models listed here by the user, and access to the Model Library, which houses many manyears of solved and explored models contributed by the FEMLAB user community and by the COMSOL development team.   In this subsection, we will walk through the "turbulent static mixer," which is modelled as a complicated 2-D geometry with the popular k-ε model of turbulence [2]. This model can be found by selecting the Model Library tab in Figure 0.1, which brings up the Model Library dialogue page, whose menu tree is traversed in Figure 0.2.   We arrive at the turbulent static mixer model with illustration and short blurb description. Selecting the OK button brings up the FEMLAB GUI with the geometry, model equations and boundary conditions completely specified. Figure 0.3 shows the postprocessing screen that the model was stored with in the file "static_mixer.mat." MAT files are the binary format for efficient disk storage of MATLAB variables. Those created by FEMLAB have the complete state of the FEMLAB environment saved. The postprocessing screen shows a color density plot of surface velocity of the last solution executed. Let's find out what situation that was, in terms of model equations and parameters.

The model is specified in a series of dialogue boxes. Traversing the pull down menus from left to right will show the pertinent specifications. Now pull down the **Options** menu, with the **Add/Edit Constants** choice highlighted as shown. The **Options** menu allows specification of a local database of constants, coupling variables, expressions, and differentiation rules, as well as specifying the display scales. Here, we only need to view the constants. Figure 0.5 shows the **Add/Edit Constants** dialogue box. We can see that rhof=1 and nuf=10^{-5} are specified. Note that these are pure numbers, i.e. no units are specified. It is up to the user to employ a consistent set of units for his models, or to specify the model in dimensionless form with dimensionless control parameters. This is not necessarily a trivial task.

The next pull down menu over is the **Draw** menu. Here we will only switch to **Draw** mode, which then takes over the display. Figure 0.7 shows the grey composite geometry CO1 that was constructed when the model was originally created. Note that the **Draw** toolbar has replaced the postprocessing toolbar on the left. We could use these tools to enter new geometric primitives. CO1 is a simply connected single domain, but we are not limited to either a single domain or to simply connected domains. FEMLAB accepts these graphic primitives, along with Boolean set theory operators (union and intersection) to construct analyzed geometries. Although the geometry specification can be done graphically in Draw mode, it can also be done through MATLAB functions, a power that is exploited in Chapter six on geometrical continuation.

Since we do not need to alter the geometry, we can move on to Point mode, shown in Figure 0.8. Here all the vertices required in specifying the analyzed geometry are shown as circles. You can add additional points within Point mode that you might need either for specifying the FEM model or for postprocessing. The FEM permits specification of a system of equations in weak form, which for a PDE system is equivalent to a conservation law in integral form. Weak terms that have no PDE equivalent may be added, like point sources and constraints. It may only be that postprocessing information is required at a particular point, so entering the point in Point mode will permit selection of a mesh to find the required solution more accurately.

Figure 0.9 shows the Point Settings dialogue box. The k-ε model uses pointwise contributions to the viscosity coefficient in weak form. These are all set at the vertices. Shown in Figure 0.9 is the contribution on vertex 13 (red circle in Draw mode). The upper left corner shows the specific expression "hard-wired" into the k-ε turbulence application mode for point viscosity contributions to the weak form. Here there are two coefficients that can be entered, η_p and δ_l and they have been preset to typical model values to the k-ε model.

Figure 0.10 moves us along to the Boundary mode, selected from the Boundary pull down menu as shown. All boundary segments are shown in the display, as well as the boundary sense. The boundary sense is the direction of increasing arc length of that particular boundary segment. FEMLAB does not try to coordinate boundary sense in adjacent boundary segments, as is clear from several reversals seen in the display here. If the user wants to specify a boundary condition that varies along a boundary, it can be done either with the independent variables defined when the model was created by the Model Navigator, say x and y for a typical 2-D geometry, or with the arc length s defined locally along the boundary, with positive sense matching the arrow shown here.

Figure 0.11 shows the Boundary Settings dialogue box. This application mode permits setting conditions on the mean field and/or on the turbulence quantities k (turbulent kinetic energy) or ε (dissipation rate). Since boundary 1 is an inflow boundary (or outflow, with opposite signs), the u,v,k, and ε terms are all specified, but not independently. Again, the upper left corner shows the equation being satisfied on boundary 1.

Figure 0.12 shows us how to select Subdomain mode. Here there is exactly one subdomain (highlighted in the display). Subdomain mode is where the PDE system is usually specified. For simple PDEs, it is the equation(s) that is specified in subdomain mode. In pre-built application modes, however, the form of the equations is "hard-wired" in, and only the coefficients are specified in subdomain mode.

Figure 0.13 shows the Subdomain Settings dialogue box for domain 1. The upper left corner shows the equation(s) that are hard-wired into this application mode. The entry boxes are for the coefficients in the equations, which can be specified as constants, expressions involving other dependent or independent variables, or even MATLAB m-file functions. The generality of "user defined programming" for just coefficients in pre-built application modes is impressive.

Figure 0.14 shows the Mesh mode with the mesh set up for the saved solution here. Mesh mode, Solve mode, and Post mode are the places where the solution methodology are specified. But up to this point, we have specified a complete FEM model *analytically*. Mesh, Solve, and Post modes are about numerics, and to demonstrate these well takes a whole chapter, and is done simply in Chapter one.

Just hitting the Solve button (=) on the toolbar, however, gives us the solution with this mesh and the default numerical solution settings. Post mode (Figure 0.3) shows the color density plot of the surface velocity U for the conditions specified.

I doubt we are any the wiser about turbulence from this tour, but we now know the steps necessary to specify a model analytically. In subsequent chapters, these steps are referred to, and they are equivalent to specifying a PDE or FEM model completely. The k-ε model and geometry specified here are both advanced models. Invariably, novice users wish to jump in at the deep end with the greatest model complexity all at once. In this book, we do precisely the opposite. The reductionist approach is adopted in Chapter one and two with surgical precision, where we introduce the basics with even simpler steps than envisaged by the creators of FEMLAB. Why? Because you do need to crawl before you can run, even if in other circumstances you are already a sprinter. The difficulty with complex computer packages is uncertainty on the part of the user about what the package does. So to remove the mystery, we start simple and build up capability with exact certainty about what we are asking FEMLAB to do.

0.2.2 Why the tour of k-ε model of a turbulent static mixer?

Clearly, since we learned rather little about turbulence from this tour of the turbulent static mixer entry in the Model Library, there is a different reason for the tour itself. The rationale for showing these features of FEMLAB is to give the non-FEMLAB initiated reader some flavor of how the FEMLAB GUI is laid out and how the data entry is organized. The actual intellectual content of models can be explained without the reader knowing the layout, but the reading experience would be more theoretically useful than practical. For this reason, if you are not already a FEMLAB user, I would recommend requesting a demonstration license for both FEMLAB and MATLAB. Mathworks [3] and COMSOL [4] will provide one month trial licenses for both products free of charge, with the software downloadable or available from CD-ROM shipped to you by request. The Users' Guide for FEMLAB is very good, and you might want to read it after this Introduction and before Chapter One. I read all the documentation that comes with FEMLAB cover to cover before designing and delivering my first intensive module on chemical engineering modeling with FEMLAB [5] and highly recommend it. Nevertheless, I felt there was something missing in the FEMLAB references, even though the Model Library and Chemical Engineering Module references have a wealth of fascinating case studies. I think it is the perspective of an expert user that is missing, but forgive my hubris in thinking it is my perspective!

By now, you must be thinking that this book is a thinly veiled sales pitch for FEMLAB/MATLAB. I would be dishonest if I did not make my preference for modeling with FEMLAB/MATLAB clear at the outset of this book. There are many packages for modeling available on the market, but FEMLAB is the first I have seen for general purpose modeling that is equation based in generating the PDE engine. Equations are the language of mathematical modeling and mathematical physics, and FEMLAB aims to speak the language of its target user community. So this book represents my personal odyssey in learning how to adapt FEMLAB to modeling of chemical engineering processes, especially but not exclusively PDE based. In the next section, I give a synopsis of the themes treated in each chapter. As an experienced programmer with nearly two decades of computational modeling and FEM experience, I could not have achieved these results in the six months spent writing this book by any other package in my arsenal, nor even by adapting research codes written by myself and other expert numerical analysts with which I am proficient. This is also the last endorsement for FEMLAB you will read in a book which only rarely makes use of other tools. Some readers might notice Mathematica, MATLAB and gnuplot graphics.

On the negative side, FEMLAB users are wont to complain that many interesting post-processing manipulations require MATLAB programming and exporting of results to the MATLAB workspace, the FEMLAB graphics are "quaint" and the FEMLAB error messages are obtuse and cryptic. Ferreting out errors in syntax is more difficult than with C/FORTRAN compilers, although MATLAB m-file scripts are generally more informative when they crash about the nature of the problem than the same m-file run in the GUI. In part this comes from the ability to interrogate the variables in the MATLAB workspace much as one uses a debugger to tease out post-crash information from C. Perhaps a future advance in FEMLAB will include access to the FEMLAB workspace. Modelling or conceptual errors, however, are notoriously difficult to identify.

We can lay those at the user's door morally, but since FEMLAB is not "idiot-proof", we are free to specify "badly conditioned or inconsistent" models. (Politely, wrong.) FEMLAB may never generate an error message at all. With just about every novice user who has sought my advice, I have shown them where they have specified an inconsistent boundary condition like 0=1 in General PDE mode. Yet, in many cases, FEMLAB generates output that is not superficially wrong, but certainly not satisfactory in the case of modeling, conceptual, or syntactical errors. At this point, the "tough love" approach is all that can be advised – there is no substitute for experience. This book encapsulates many of my experiences. I haven't tried to sugar-coat my chapters so that all models are magically perfect. Think of this as a cookbook that shows both good recipes and bad ones, but each labeled and the latter coming with a health warning. For instance, in Chapter Seven, I tried four attempts at modeling the population balance equations before the last came good. So you will learn from my mistakes that I own up to as well as from my triumphs. For better or worse, every modeling attempt I made during the six months of writing this book has been included. I will pat myself on the back for persistence, because in the end they all worked, but at many points I had my doubts and frustrations. I am pleased not to have cherry picked the models. Of course I have not shown every single computation nor "what if" line that I pursued in each model.

0.3 Chapter Synopsis

Chapter One treats the basics of numerical analysis with FEMLAB. No doubt many of the example models are artificial in that if you were handed the modeling problems in Chapter One, FEMLAB would not be the obvious choice of computational platform. The topics of root finding, numerical integration by marching, numerical integration of ordinary differential equations, and linear system analysis are universal to numerical analysis. They form the basis of my previous lecture courses in FORTRAN programming and chemical engineering problem solving with Mathematica. For pedagological purposes, Chapter One provides a firm basis for understanding what FEMLAB does. The common applications in chemical engineering that are treated as examples, flash distillation, tubular reactor design, diffusive-reactive systems, and heat conduction in solids, are understandable to the non-chemical engineer as well. Perhaps the single most important modeling feature introduced here, however, is the use of a conceptual 0-dimensional model. Consisting of a single element, the 0-D construct introduces a variable which is a scalar for which an ODE in time or an algebraic equation can be specified. This construct is important for describing equations or systems of equations that are mixed (partial)differential-algebraic, and is utilized with the extended multiphysics feature of FEMLAB in the more complicated models presented later.

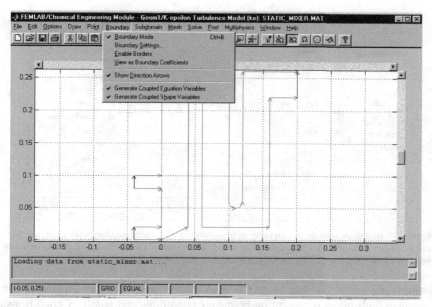

Figure 0.10 Boundary mode clarifies the boundary identifications and permits boundary data entry for the FEM model.

Chapter Two might be thought the normal point of departure for a textbook on finite element methods (FEM). In my opinion, FEMLAB is not so much a tool about FEM, but a modeling tool that happens to use FEM in its automated methodology. The key actions of FEMLAB that reduce the drudgery of modeling are (1) the translation of systems of equations in symbolic form to an algorithm that can be computed numerically, (2) the provision of a wide array of numerical solver, analysis, and post-processing tools at either the "touch of a button" or (3) through a powerful "scripting language" can be programmed in MATLAB as subroutines (function calls) and automated. So much of modeling of partial differential equations in the past has been devoted to the computer implementation of algorithms that the modeler did not get the chance to properly consider modeling alternatives. Who would consider a different modeling scheme if it meant spending three graduate student years building the tools before the scheme could be tested? FEMLAB is a paradigm shift for modelers – it frees them to ask those "what if" questions without the price of coding a new computer program. Nonetheless, FEMLAB uses FEM as the powerhouse of its PDE engine. Chapter Two gives an overview of how FEM is implemented in FEMLAB. For experienced FEM users, the takeaway message is that FEMLAB translates PDEs specified symbolically into the assembly of the FEM augmented

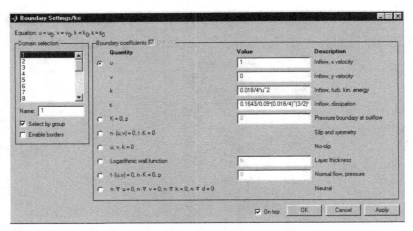

Figure 0.11 Boundary settings permit entry of boundary data for each boundary with a range of pre-built boundary conditions for the application mode. Here, not only is the inflow mean u,v-velocity specified for boundary 1, but the turbulence intensity k and energy dissipation rate ε as well.

stiffness matrix – the Jacobian, the load vector, and auxiliary equations for Lagrange multipliers representing boundary conditions and auxiliary conditions. Chapter Two illustrates these points about partial differential equations and the finite element method thorough treatment of canonical types of linear, second order PDEs: elliptic, parabolic, and hyperbolic and gives an overview of FEM, with particular emphasis on the treatment of boundary and auxiliary conditions by the method of Lagrange multipliers.

Chapter Three is about multiphysics modeling. What is it? How does FEMLAB do it so well? There are applications: thermoconvection, non-isothermal chemical reactors, heterogeneous reaction in a porous pellet. Furthermore, the workhorse methodology for nonlinear solving, parametric continuation, is explained. I won't steal the thunder of Chapter Three here by explaining multiphysics modeling in detail. Suffice to say that multiphysics modeling means the ability to treat many PDE equations simultaneously, and the provision of pre- built PDE equations that can be mixed and matched in the specification of a model so that the symbolic translation to a FEM assembly is transparent to the user.

Chapter Four is about extended multiphysics: the central role of coupling variables and the use of Lagrange multipliers. Example applications are: a heterogeneous reactor; reactor-separator-recycle; buffer tank modelling; and an immobilized cell bioreactor model.

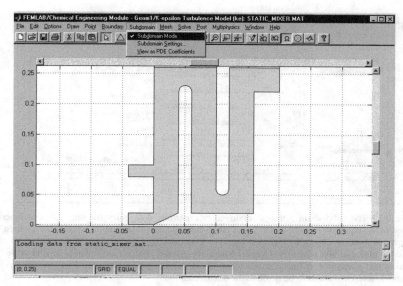

Figure 0.12 In the turbulent static mixer model, there is only one subdomain, exactly equivalent to the single composite geometry object specified in draw mode.

Chapter five starts the advanced concepts in modeling – nonlinear dynamics and simulation. Chapter six deals with geometric continuation, and Chapter seven treats integral equations and inverse problems. All three chapters are largely drawn from my own research portfolio, but there are also newly developed treatments or extended studies from previous works. Rather than systematically exploring the features of FEMLAB as in chapters 1-4, chapters 5-7 pose the question "Can FEMLAB be bent to solve the problems that interest me in stability theory (five), complex geometries and modulating domains (six) or inverse problems (seven), where I know the questions and desired forms of the answers, but can FEMLAB provide the solution tools? These chapters will have their own audiences for the direct questions they treat, but should provide many users with fertile proving grounds and a basketful of "tricks of the trade." Getting information into and out of the FEMLAB GUI is one of the weaknesses of the package. Many of my tricks are how to use the MATLAB interfaces to do intricate I/O.

Chapters eight through ten are purely about applications and are only co-authored by me. To a large degree, chapters 5-7 are about my applications and their generalizations, used to demonstrate FEMLAB functionality. Chatpers 8-10 are the applications of colleagues for which we thought FEMLAB and the concepts of chapters 1-4 should be exploitable. My co-authors of these chapters have other agendas and that is evident in the narrative voice adopted in these chapters.

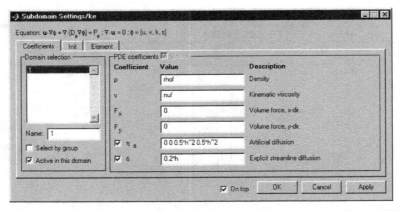

Figure 0.13 The subdomain settings dialogue box permits data entry for the PDE coefficients defined in the equation line above the select tabs. Here the coefficient tab is selected for domain 1, the same domain as shown in Figure 0.12. The constants rhof and nuf defined in the Add/Edit Constants dialogue box under the Options Menu are entered here, as are formula for the artificial diffusion and streamline diffusion coefficients. Note that the production term P is not specifically defined, so reference needs to be made to the documentation.

Chapter eight is about the level set method for modeling two phase flows that are dominated by interfacial dynamics and transport. The subject matter was mastered and modelled in record time for one of my doctoral students. The simulations are a reflection of the need for researchers to be able to run numerical experiments in complex systems dynamics to augment understanding of laboratory experiments. Such *"in silico"* experiments are more flexible than laboratory experiments, provide a much greater wealth of detailed knowledge, but at the expense of modeling errors of all varieties.

Chapter nine focuses on electrokinetic flow modeling in microfluidic applications. A substantial fraction of FEMLAB users are numbered in the microfluidics community, especially with biotech end-uses. Rather early on, we targeted FEMLAB as a potentially useful modeling tool for microfluidic reactor networks for the "chemical-factory-on-a-chip" community. The extended multiphysics capabilities of FEMLAB for designing such factories are an explosive growth area which should benefit the community. Microfluidics 2003 [6] was sponsored by COMSOL for just this reason.

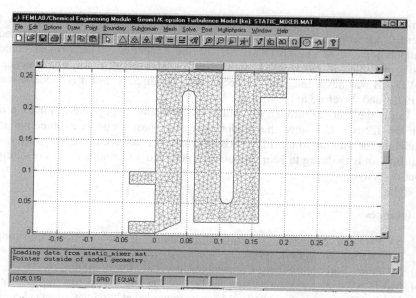

Figure 0.14 Mesh mode shows the existing mesh and permits specification of mesh parameters for the elliptic mesh generator routine.

The appendix, a MATLAB/FEMLAB primer for vector calculus, is a compromise between the recurrent suggestion of students taking the module for more MATLAB instruction and my desire for the students to grasp vector calculus more intuitively. I am actually a late convert to MATLAB, with apologies to Cleve Moler, its creator. I was one of the graduate students gifted with the beta test edition of MATLAB 1.0 while he was developing it. At the time, computational power was expensive and there was a bias against interpreted environments for scientific computing. To programmers, the same matrix utilities were available as library subroutines, and the final product, a compiled executable, was more efficient. MATLAB has come a long way since version 1.0beta, and the number of man years and breadth of applications in the toolboxes, as well as judicious use of compilation within the environment, simply invalidates my early prejudices. I cannot access programming libraries with anywhere near the functionality of the MATLAB toolboxes. The GUIs for the toolboxes make manmonths of programming effort evaporate at the touch of a button (OK, the click of a mouse). And if speed is still an issue, the MATLAB C compiler is available. Or just my favorite trick of running MATLAB as a background job (no GUIs to clutter the memory) is usually sufficient for big jobs. So to get the most functionality out of FEMLAB, MATLAB programming ability is valuable. But anything other than a primer is outside the scope of this book. I presume a modest MATLAB familiarity of the reader which is readily achieved from over-the-counter books. So to add more MATLAB support, I decided to write a short primer about vector calculus representations and computations in MATLAB/FEMLAB for the appendix. This project could easily get out of hand, so I apologize for abridging it for convenience. MATLAB was never intended for vector calculus directly, but it is fundamental to PDEs and therefore to FEMLAB.

Enjoy the journey through this book. As it is an odyssey, the destination is not the focus. Certainly the reader, however, has a concrete objective in modeling for wanting to use FEMLAB. Perhaps somewhere in this odyssey you will find tools to bring to bear on your problem and will find useful in reaching your objective.

References

[1] SV Patankar, <u>Numerical heat transfer and fluid flow</u>. Hemisphere Publishing corporation, New York, 1980.
[2] SB Pope, <u>Turbulent Flows</u>, Cambridge University Press, 2000.
[3] MATLAB demonstration version can be found at http://www.mathworks.com
[4] FEMLAB demonstration CD-ROM can be requested at http://www.comsol.com
[5] WBJ Zimmerman, http://eyrie.shef.ac.uk/femlab
[6] WBJ Zimmerman, http://eyrie.shef.ac.uk/fluidics

Chapter 1

FEMLAB AND THE BASICS OF NUMERICAL ANALYSIS

W.B.J. ZIMMERMAN

*Department of Chemical and Process Engineering, University of Sheffield,
Newcastle Street, Sheffield S1 3JD United Kingdom*

E-mail: w.zimmerman@shef.ac.uk

In this chapter, several key elements of numerical analysis are profiled in FEMLAB with 0-D and 1-D models. These elements are root finding, numerical integration by marching, numerical integration of ordinary differential equations, and linear system analysis. These methods underly nearly all problem solving techniques by numerical analysis for chemical engineering applications. The use of these methods in FEMLAB is illustrated with reference to some common applications in chemical engineering: flash distillation, tubular reactor design, diffusive-reactive systems, and heat conduction in solids.

1.1 Introduction

This chapter is rather busy, as it must accomplish several different goals. Primarily, it is intended to introduce key features of how FEMLAB works. Secondarily, it is to illustrate how these key features can be used to analyse simple enough chemical engineering problems that 0-D and 1-D spatial or spatial-temporal systems can describe them. The chapter is also intended to whet your interest to investigate modeling and simulation with FEMLAB by presenting at least a glimpse of the power of the FEMLAB and MATLAB tools when applied to chemical engineering analysis.

Because FEMLAB is not intended to be a general tool for problem solving, some of these goals are achieved in a roundabout fashion. The author has previously taught courses in chemical engineering problem solving by numerical analysis using FORTRAN, *Mathematica*™, and MATLAB™, and used all the examples implemented here with those tools. Furthermore, the most extensive compilations of chemical engineering problem solving by numerical analysis have been done in POLYMATH [1], which only seems to be used by the chemical engineering community through the CACHE program.

The upshot is that for the examples in this chapter, FEMLAB is probably not the package of first choice for the analysis. From the author's experience either MATLAB or *Mathematica* is preferable, with less overhead in setting up the calculations. Nevertheless, even though FEMLAB was not necessarily envisaged to solve such problems, that its numerical analysis tools are general enough to do so is important information that will benefit the reader in later chapters, where very clearly FEMLAB is the first choice package for the analysis – 2-D and 3-D spatial-temporal systems with multiphysics.

1.2 Method 1: Root Finding

Typically, courses in numerical analysis go into great detail in the description of the algorithm classes used for root finding. From experience, there are only two algorithms that are really useful – the bisection method and Newton's method. Instead of presenting all the methods, here we will consider why root finding is one of the most useful numerical analysis tools. Finding roots in linear systems is fairly easy. Nonlinear systems are the challenge, and nearly all interesting dynamics stem from nonlinear systems. The interest in root finding in nonlinear systems results from its utility in describing inverse functions. Why? Because with most nonlinear functions, the "forward direction", $y=f(u)$, is well described, but the inverse function of $u=f^{-1}(y)$ may be analytically indescribable, multi-valued (non-unique), or even non-existent. But if it exists, then the numerical description of an inverse function is identical to a root finding problem – find u such that $F(u)=0$ is equivalent to $F(u)=f(u)-y=0$. Since the goal of most analysis is to find a solution of a set of constraints on a system, this is equivalent to inverting the set of constraints. FEMLAB has a core function for solving nonlinear systems, femnlin, and in this section its use to solve 0-D root finding problems will be illustrated.

femnlin uses Newton's method which with only one variable u uses the first derivative F'(u) which is used iteratively to drive toward the root. The method takes a local estimate of the slope of the function and projects to the root. The slope can be computed either analytically (Newton-Raphson Method) or numerically (the secant method). If the slope can be computed either way, you can use Taylor's theorem to project to the root. The basic idea is to use a Taylor expansion about the current guess u_0:

$$f(u) = f(u_0) + (u - u_0) f'(u_0) + \ldots \qquad (1.1)$$

which can be re-arranged to estimate the root as

$$u = u_0 - \frac{f(u_0)}{f'(u_0)} \qquad (1.2)$$

This methodology is readily extendable to a multiple dimension solution space, i.e. u is a vector of unknowns, and division by $f'(u_0)$ represents multiplication by the inverse of the Jacobian of f. The next subsection illustrates root finding in FEMLAB.

1.2.1 Root finding: A simple application of the FEMLAB nonlinear solver

As implied in the previous section, root finding is a "0-D" activity, at least in terms of the spatial-temporal dependence of the solution vector of unknowns, u, which can be a multi-dimensional vector. FEMLAB does not have a "0-D" application mode, so we must improvise in 1-D. This has the undesirable feature that we will unnecessarily solve the problem redundantly at several points in space. Given the small size of the problem, the efficiency of FEMLAB coding, and the speed of modern microprocessors, this causes no guilt whatsoever!

Start up MATLAB and type FEMLAB in the command window. After several splash screens, you should be facing the Model Navigator window.

Model Navigator
• Select 1-D dimension
• Select PDE modes → Coefficient form
• Element: Lagrange – linear
• More >>
• OK

This application mode gives us one dependent variable u, but in a 1-D space with coordinate x. Now we are in a position to set up our domain. Pull down the Draw menu and select Specify Geometry.

Draw Mode
• Name: interval
• Start: 0
• Stop: 1
• Apply
• OK

Now for the boundary conditions. Since we wish to emulate 0-D (no spatial variation) then Neumann boundary conditions (no slope at either boundary) are appropriate. Pull down the Boundary menu and select Boundary Settings.

Boundary Mode / Boundary Settings
• Select domains 1 and 2 (hold down ctrl key)
• Select Neumann boundary conditions
• Apply
• OK

Subdomain mode specifies the equation to be satisfied in each subdomain. Pull down the Subdomain menu and select Subdomain settings. Notice the equation in the upper left given in vector notation. In 1-D, this equation can be simplified to

$$d_a \frac{\partial u}{\partial t} - \frac{\partial}{\partial x}\left(c\frac{\partial u}{\partial x} + \alpha u\gamma \right) + au + \beta\frac{\partial u}{\partial x} = f \qquad (1.3)$$

Clearly, $\alpha\gamma$ and β are redundant with the simplification to 1-D. Since we want to find roots in 0-D, however, all the coefficients on the LHS of (1.3) can be set to zero. Let's solve for the roots of the polynomial equation $u^3 + u^2 - 4u + 2 = 0$.

Subdomain Mode / Subdomain Settings

- Select domains 1
- Set c=0; a=4; f=u^3+u^2+2; d$_a$=0
- Apply
- Select the init tab; set u(t0)=-2
- Apply / OK

By rearranging the polynomial, we can readily see that $a=4$ and $f = u^3 + u^2 + 2$.

One last step – discretizing the domain with elements. Since we do not wish to replicate our effort, we will mesh the interval with exactly one element, the closest we can get to 0-D! Pull down the Mesh menu and select the Parameters option.

Mesh Mode

- Set Max element size, general = 1
- Select Remesh
- OK

The report window now declares "Initialized mesh consists of 2 nodes and 1 elements."

Now to find the root nearest to the initial guess of –2. If you are wondering why a=4 was set, rather than all of the dependence put into f, it is so that the finite element discretization of the RHS of (1.3) does not result in a singular stiffness matrix. Now pull down the Solve menu and select the Parameters option. This pops up the Solver Parameters dialog window.

Solver Parameters

- General tab: select stationary nonlinear solver type.
- Jacobian: select Numeric option
- Solve
- Cancel

Note in proof: I got into the habit of using the coefficient mode and numeric Jacobian as it mirrors my style of FEM coding – the coefficient mode does not include all the potential nonlinearity in the coefficients in assembling the stiffness matrix, but by computing the Jacobian numerically, it is all included in the iterative scheme. For small problems, you will see no performance degradation. On larger problems, it is wiser to use the PDE general mode and the exact Jacobian option, which assembles the full nonlinear contributions to the Jacobian analytically. If I had to write this chapter again, all the coefficient modes would disappear.

During the Solve step, the report window shows a runout of several columns per iteration; particularly important is the error estimate ErrEst, which for iteration 4 is about 10^{-7} which is smaller in magnitude than the default tolerance set at 10^{-6}. in the **Nonlinear** tab of the **Solver Parameters** dialog box. Click anywhere on the grid and the report window will "Value of u(u) at 0.456: -2.73205." The analytically determined root nearest to this is $-1-\sqrt{3}$, showing the numerical solution in good agreement. According to the structure of the quadratic formula of algebra, clearly another root is $-1+\sqrt{3}$, and by inspection, the third root is 1. Returning to the subdomain settings, set the initial guess to u(t0)=-0.5 and FEMLAB converges to u=0.732051, again a good approximation. u(t0)=1.2 as an initial guess converges to u=1.

This exercise clarifies two features of nonlinear solvers and problems – (i) nonlinear problems can have multiple solutions; (ii) the initial guess is key to convergence to a particular solution. With Newton's method, it is usually the case that convergence is to the nearest solution, but overshoots in highly nonlinear problems may override this stereotype. These features persist in higher dimensional solution spaces and with spatial-temporal dependence.

The MATLAB model m-file rootfinder.m contains all the MATLAB source code with FEMLAB extensions to reproduce the current state of the FEMLAB GUI. This file is available from the website http://eyrie.shef.ac.uk/femlab. Just pull down the file menu, select **Open model m-file**, and use the Open file dialog window to locate it. You can rapidly place your nonlinear function in the **Subdomain settings**, specify an initial guess, and use the stationary nonlinear solver to converge to a solution. But what if your function does not have a linear component to put on the LHS of (1.3)? For instance, $\tanh(u) - u^2 + 5 = 0$ results in a singular stiffness matrix when FEMLAB assembles the LHS of (1.3). The suggestion is to set the coefficient of the second derivative of u, c=1 in the **Subdomain settings**. Coupled with the Neumann boundary conditions, this artificial diffusion cannot change the fact that the solution must be constant over the single element, yet it prevents the stiffness matrix from becoming singular.

Root Finding in General Mode

The difficulty with a singular stiffness matrix assembly for $\tanh(u) - u^2 + 5 = 0$ can be averted by using General Mode, which solves

$$d_a \frac{\partial u}{\partial t} + \frac{\partial \Gamma}{\partial x} = F \qquad (1.4)$$

where $\Gamma(u, ux)$ is in principle the same functionality as the coefficient form (1.3), but is treated differently by the Solver routines. In Coefficient Mode, the coefficients are treated as independent of u unless the numerical Jacobian is used, which brings out some of the nonlinear dependency – iteration does the rest. The exact Jacobian in General Mode differentiates both Γ and F with respect to u symbolically in assembling the stiffness matrix. Typically, General Mode requires fewer iterations for convergence than Coefficient Mode with the numerical Jacobian. The use of the exact Jacobian below does not require any special treatment to avoid a singular stiffness matrix in the treatment of the linear terms as the coefficient mode did. In general, General Mode is more robust at solving nonlinear problems than Coefficient Mode. It is my opinion that Coefficient Mode is a "legacy" feature of FEMLAB – the PDE Toolbox of MATLAB, in many ways a precursor to FEMLAB, uses coefficient representations extensively. Further, the coefficient formulation with numerical Jacobian is a long standing FEM methodology, so for benchmarking against other codes, it is a useful formulation.

Here's the recipe for General Mode – a minor modification of what we just did.

Model Navigator	1-D geom., PDE modes, general form (nonlin stat)
Options	Set Axes/Grid to [0,1]
Draw	Name: Interval; Start = 0; Stop =1
Boundary Mode/ Boundary Settings	Set both endpoints (domains) to Neumann BCs
Subdomain Mode/ Subdomain Settings	Set Γ = 0; da = 0; F = u^3+u^2-4*u+2
Mesh mode	Set Max element size, general = 1; Remesh
Solve	Use default settings (nonlinear solver, **exact** Jacobian)
Post Process	After five iterations, the solution is found. Click on the graph to read out u=0.732051. Play with the initial conditions to find the other two roots.

Although setting up this template (rtfindgen.m) for root finding of simple functions of one variable was rather involved, and in fact MATLAB has a simpler procedure for root finding using the built-in function fzero and inline declarations of functions, the FEMLAB GUI still provides many options and flexibility to root finding that may not be available in other standard packages. The next subsection applies our newly constructed nonlinear root finding scheme to a common chemical engineering application, flash distillation, which clarifies a few more features of the FEMLAB GUI.

1.2.2 Root finding: Application to flash distillation

Chemical thermodynamics harbors many common applications of root finding, since the constraints of chemical equilibrium and mass conservation are frequently sufficient, along with constitutive models like equations of state, to provide the same number of constraints as unknowns in the problem. In this subsection, we will take flash distillation as an example of simple root finding for one degree of freedom of the system, which is conveniently taken as the phase fraction ϕ.

A liquid hydrocarbon mixture undergoes a flash to 3.4 bar and 65°C. The composition of the liquid feed stream and the 'K' value of each component for the flash condition are given in the table. We want to determine composition of the vapour and liquid product streams in a flash distillation process and the fraction of feed leaving the flash as liquid. Table 1.1 gives the initial composition of the batch.

Table 1.1 Charge to the flash unit

Component	X_I	K_i at 65°C and 3.4 bar
Ethane	0.0079	16.2
Propane	0.1281	5.2
i-Butane	0.0849	2.6
n-Butane	0.2690	1.98
i-Pentane	0.0589	0.91
n-Pentane	0.1361	0.72
Hexane	0.3151	0.28

Flash at 3.4 bar and 65°C

A material balance for component i gives the relation

$$X_i = (1-\phi)y_i + \phi x_i \tag{1.5}$$

where X_i is the mole fraction in the feed (liquid), x_i is the mole fraction in the liquid product stream, y_i is the mole fraction in the vapour product, and f is the ratio of liquid product to feed molar flow rate. The definition of the equilibrium coefficient is $K_i = y_i / x_i$. Using this to eliminate x_i from the balance relation results in a single equation between y_i and X_i:

$$y_i = \frac{X_i}{1-\phi\left(1-\dfrac{1}{K_i}\right)} \tag{1.6}$$

Since the y_i must sum to 1, we have a nonlinear equation for ϕ:

$$f(\phi) = 1 - \sum_{i=1}^{n} \frac{X_i}{1-\phi\left(1-\dfrac{1}{K_i}\right)} = 0 \tag{1.7}$$

where n is the number of components. This function $f(\phi)$ can be solved for the root(s) ϕ, which allows back-substitution to find all the mole fractions in the product stream. The Newton-Raphson method requires the derivative $f'(\phi_k)$ at the current estimate to determine the improved estimate, and FEMLAB will compute this analytically as an option. It is fairly straightforward to arrive at the Newton-Raphson iterate as

$$f'(\phi) = \sum_{i=1}^{n} \frac{X_i \left(1 - \dfrac{1}{K_i}\right)}{\left[1 - \phi\left(1 - \dfrac{1}{K_i}\right)\right]^2} \qquad (1.8)$$

Now onto the FEMLAB solution for root finding. As an exercise, we will set up the solution using the general PDE mode. We could just load rootfinder.m or rtfindgen.m and customize it, but of course becoming familiar with FEMLAB's features is an important goal.

Start up FEMLAB and await the Model Navigator window. If you already have a FEMLAB session started, save your workspace as a model MAT-file or the commands as a model m-file, and the pull down the file menu and select New.

Model Navigator
- Select 1-D dimension
- Select PDE modes → General
- Element: Lagrange – linear
- More >>
- OK

This application mode gives us one dependent variable u and one space coordinate x. Next, set up the domain. Pull down the Draw menu and select Specify Geometry.

Draw Mode
- Name: interval
- Start: 0
- Stop: 1
- Apply / OK

Now for something new. We must enter our data. Pull down the options menu and select Add/Edit constants. The Add/Edit constants dialog box appears. Now enter our fourteen pieces of data:

Add/Edit Constants

- Name of constant: X1
- Expression: 0.0079
- Apply
- Name of constant: K1
- Expression: 16.2
- Continue with the rest of Table 1.1
- OK

Now onto the Neumann boundary conditions. Pull down the **Boundary** menu and select **Boundary Settings**.

Boundary Mode

- Select domains 1 and 2 (hold down ctrl key)
- Select Neumann boundary conditions
- Apply / OK

Next Subdomain mode. Pull down the **Subdomain** menu and select **Subdomain mode**. Before setting the equations, it is useful to define some intermediate variables to make the data entry more concise. Pull down the **options** menu and select **Add/Edit expressions**. By experience, you should be in **Subdomain mode** to add expressions for the first time.

Add/Edit Expressions

- Variable name: t1
- Variable type: subdomain
- Add
- Repeat to create t2 through t7
- Now select variable t1 and click on the definition tab.
- Select level: subdomain 1
- Enter expression: -X1/(1-u*(1-1/K1))
- Apply
- Now select the variables tab, select t2, and then click on the definition tab, and enter the similar expression for t1, substitute index 2 where appropriate.
- Continue with indices 3 through 7
- OK

Now pull down the **Subdomain** menu and select **Subdomain settings**. Default values of $d_a=1$, $\Gamma=-ux$, and $F=1$ are specified in the data entry locations. Because we select Neumann BCs for spatial dependency, we can take the default setting for $\Gamma=-ux$ without contradiction. As you will see, the solution is "spatially flat" – pseudo-0D.

Subdomain Mode / Settings
- Select domains 1
- Set $F=1+t1+t2+t3+t4+t5+t6+t7$; $d_a=0$
- Apply
- Select the init tab; set $u(t0)=0.5$
- Apply / OK

Pull down the Mesh menu and select the Parameters option to set up our single element.

Mesh Mode
- Set Max element size, general = 1
- Select Remesh
- OK

The report window now declares "Initialized mesh consists of 2 nodes and 1 elements."

Now pull down the **Solve** menu and select the **Parameters** option. This pops up the **Solver Parameters** dialog window.

Solver Parameters
- General tab: select stationary nonlinear solver type.
- Jacobian: retain default Exact option
- Solve
- Cancel

After three iterations, we find that $\phi=0.458509$ solves for the phase fraction at equilibrium. For your own information, resolve with the initial guesses $\phi=0$ and $\phi=1$. How many roots would you expect to this function? If you wish to avoid all of the data entry, then you could just load the MATLAB model m-file flash.m that came with the distribution.

Exercises:

1.1 Find the root of the equation $f(u) = ue^u - 1 = 0$. This function is transcendental, which means that it has no analytic solution in the rational numbers. If you use Coefficient Mode, put c=1 to aid convergence.

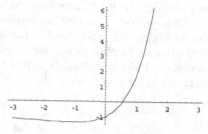

Figure 1.1 u exp(u) −1.

1.2 Find the roots of the equation $f(u) = u^3 - 3u^2 + \frac{5}{2}u - \frac{1}{2} = 0$. As this function is a cubic polynomial, there is an analytic solution in the irrational numbers, $u = 1$, $u = 1 - \frac{1}{\sqrt{2}}$, $u = 1 + \frac{1}{\sqrt{2}}$.

1.3 Method 2: Numerical Integration by Marching

Numerical integration is the mainstay of numerical analysis. The first duty of scientific computing before there were digital computers were to fill the handbooks with tables of special functions, nearly all of which were solutions to special classes of ordinary differential equations. And the computational methodology? One-dimensional numerical integration.

There are two classes of 1-D integration: initial value problems (IVP) and boundary value problems (BVP). The latter will be considered in the next section. The easiest to integrate are IVPs, as if all the initial conditions are all specified at a point, it is straightforward to step along by small increments according to the local first derivative. Clearly, if the ODE is first order, i.e.

$$\frac{dy}{dt} = f(t), \qquad y|_{t+\Delta t} = y|_t + \Delta t f(t) \qquad (1.9)$$

The second statement in (1.9) is true exactly in the limit of $\Delta t \to 0$. It is termed the Euler method and is the most straight-forward way of integrating a first order ODE. In one dimension, you simply step forward according to the local value of the derivative of f at the point (x_n, y_n), where n refers to the n-th discretization step of the interval upon which you are integrating. Thus,

$$y_{n+1} = y_n + hf(x_n, y_n)$$
$$x_{n+1} = x_n + h \qquad (1.10)$$

This assumes that the derivative does not change over the step of size h, which is only actually true for a linear function. For any function with curvature, this is a lousy assumption. Consider, for instance, how far wrong we go with a large step size in Figure 1.2. So clearly, one important point in improving on Euler's Method is to be able to use big steps, since it requires small steps for good accuracy. Euler's method is called "first order" accurate, as the error only decreases as the first power of h.

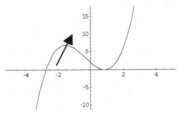

Figure 1.2 Curvature effects are lost in the Euler method.

Runge-Kutta methods

So if we want to use big step sizes, we need a "higher order method", one that reduces the error faster as step size decreases. A k-th order method has error which diminishes as h^k. Given that it is curvature that we know we are neglecting, we can estimate the curvature of the graph $y(x)$ by evaluating the slope $f(x)$ at several intermediate points between x_n and x_{n+1}. Second order accuracy is obtained by using the initial derivative to estimate a point halfway across the interval, then using the midpoint derivative across the full width of the interval.

$$k_1 = hf\left(x_n, y_n\right)$$

$$k_2 = hf\left(x_n + \frac{1}{2}h, y_n + \frac{1}{2}k_1\right) \tag{1.11}$$

$$y_{n+1} = y_n + k_2 + O\left(h^3\right)$$

The upshot is that by making two function evaluations, we have saved a whole order in accuracy. So, for instance, with a first order method, N calculations gives us an error $O(1/N)$, but for a second order method, 2N calculations gives us error $O(1/4N^2)$. It would take N^2 calculations to do so well with a first order method.

Higher order Runge-Kutta methods

Can we do better? Clearly, we can use a three midpoint method to achieve third order accuracy, a four midpoint method for fourth order accuracy, etc. When should we stop? Well, there is more programming work for higher order methods, so our time is a consideration. But intrinsically, functions may not be very smooth in their k-th derivative that we are estimating. It is possible that in increasing the *"accuracy* of the approximation", the round-off error of higher derivative terms so estimated becomes appreciable. If that is the case, with each successive step, the error may grow rapidly. This implies that higher order methods are less *stable* than lower order methods. The common choice for integrating ODEs is to use a fourth order Runge-Kutta method. This is fairly compact to programme, gives good accuracy, and typically has good stability character.

<u>Other methods</u>

There are two other famous problems in numerical integration that need particular programming attention:

Numerical Instability. Suppose your integration diverges to be very far from known test-cases, even with a high order accuracy method. Then it is likely that your method is numerically unstable. You can cut down your step size and eventually achieve numerical stability. However, this means a longer calculation. If you are computing a great many such integrations and the slowness really bothers you, try a semi-implicit method like predictor-corrector schemes.

Stiff Systems. Stiff systems usually have two widely disparate length or time scales on which physical mechanisms occur. Stiff systems may have "numerical instability" of the explosive sort mentioned above, or they may have non-physical oscillations. Try the book of Gear [2] for a recipe to treat stiff systems.

1.3.1 Numerical integration: A simple example

Higher order ODEs are treatable by marching methods by reduction of order. Suppose you have an ode:

$$\frac{d^2 y}{dx^2} + q(x)\frac{dy}{dx} = r(x) \qquad (1.12)$$

Unless $q(x)$ and $r(x)$ are constants, then you are out of luck with most textbook analytic methods for finding a solution. There are special cases of $q(x)$ and $r(x)$ that lead to analytic solutions, but these days you are better off computing the numerical solution in nearly all cases anyway. Why? Because you need to plot the graph of the solution $y(x)$ to make sense of it, so you will need to harness some computing horse power for the graphics. How? First let's reduce the order of the second order system above to two first order systems:

$$\frac{dy}{dx} = z(x)$$

$$\frac{dz}{dx} = r(x) - q(x)z(x)$$

Each of these ODEs can be numerically integrated by time marching methods as in (1.10) or (1.11), *simultaneously*. A simple example is

$$\frac{d^2u}{dt^2} + u = 0 \qquad\qquad (1.13)$$

Reduction of order yields two first order ODEs:

$$\frac{du_1}{dt} = -u_2$$

$$\frac{du_2}{dt} = u_1 \qquad\qquad (1.14)$$

Taking the initial condition to be $u_1=1$ and $u_2=0$, we can now set up a 0-D spatial system to integrate this coupled set of ODEs.

Start up FEMLAB and await the Model Navigator window. If you already have a FEMLAB session started, save your workspace as a model MAT-file or the commands as a model m-file, and the pull down the file menu and select New.

Model Navigator
- Select 1-D dimension
- Select PDE modes → Coefficient+time dependent
- Select 2 dependent variables
- Element: Lagrange - linear
- More >>
- OK

This application mode gives us two dependent variables u_1 and u_2 and one space coordinate x. Next, set up the domain.

Pull down the Draw menu and select Specify Geometry.

Draw Mode
- Name: interval
- Start: 0
- Stop: 1
- Apply / OK

Now onto the Neumann boundary conditions. Pull down the **Boundary** menu and select **Boundary Settings**.

Boundary Mode

- Select domains 1 and 2 (hold down ctrl key)
- Select Neumann boundary conditions
- Apply / OK

Now pull down the **Subdomain** menu and select **Subdomain settings**. Notice the equation in the upper left given in vector notation.

Subdomain Mode

- Select domains 1
- Set $f_1=0$; $f_2=0$; $a_{12}=1$; $a_{21}=-1$; $c_1=1$; $c_2=1$;
- Apply
- Select the init tab; set $u_1(t0)=1$.
- Apply / OK

Pull down the Mesh menu and select the Parameters option to set up our single element.

Mesh Mode

- Set Max element size, general = 1
- Select Remesh
- OK

Now pull down the **Solve** menu and select the **Parameters** option. This pops up the **Solver Parameters** dialog window.

Solver Parameters

- Time-stepping tab: set output times
  `linspace(0,2*pi,50)`
- Jacobian: numeric
- Solve
- Cancel

`linspace(0,2*pi,50)` is the MATLAB command to create a vector of length 50 which uniformly goes from 0 to 2π. Click anywhere on the graph and you learn that $u_1(t=2\pi)=1.02475$. Given that the analytic solution is $u_1(t=2\pi)=1$, this is rather inaccurate (2%). The ode15s solver is a stiff solver with low to medium accuracy. ode45 has the best accuracy of the suite of solvers, but is for non-stiff systems. It gives $u_1(t=6.29)=0.99994$, which is pretty good. Greater accuracy, however, comes from decreasing the output time interval. If the output times are set to linspace (0, 2*pi, 500), i.e. tenfold smaller time-step, than $u_1(t=2\pi)=0.999939$ with ode15s.

These two figures (1.3 & 1.4) clarify that FEMLAB can reproduce the numerical integration of the cosine and sine functions with high fidelity if given a small enough time step. Although we think of sine and cosine as "analytic functions," when tabulated this way, it is clear that the distinction between analytic functions and those that require numerical integration is specious – they are no more analytic than Bessel functions, elliptic functions, etc.

1.3.2 Numerical integration: Tubular reactor design

In this subsection, a coupled set of first order nonlinear ODEs are solved simultaneously for the design of a tubular reactor undergoing a homogeneous chemical reaction. Typically, the key element in the design of a tubular reactor is the estimate of the length of the reactor.

A tubular reactor is used to dehydrate gaseous ethyl alcohol at 2 bar and 150°C. The formula for this chemical reaction is

$$C_2H_5OH \rightarrow C_2H_4+H_2O$$

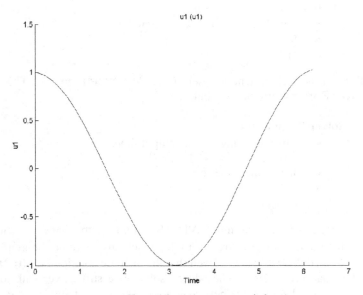

Figure 1.3 u1(t) over one period.

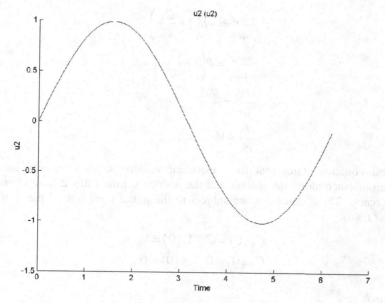

Figure 1.4 u2(t) over one period.

Some experiments on this reaction have suggested the reaction rate expression at 2 bar pressure and 150°C, where C_A is the concentration of ethyl alcohol (mol/litre) and R is the rate of consumption of ethyl alcohol (mol/s/m³):

$$R = \frac{52.7C_A^2}{1 + \dfrac{0.013}{C_A}}$$

The reactor is to have a 0.05m diameter and the alcohol inlet flowrate is to be 10g/s. The objective is to determine the reactor length to achieve various degrees of alcohol conversion. We wish to determine reactor length for the outlet alcohol mole fractions 0.5, 0.4, 0.3, 0.2, and 0.1.

Chemical Engineering Design Theory

Assuming small heat of reaction, plug flow and ideal gas behaviour, it can be shown that the reacting flow is described by four ordinary differential equations in terms of the dependent variables C_A, C_W (the water concentration), V (the velocity) and x (the distance along the reactor from the inlet):

$$\frac{dC_A}{dt} = -R\left(1 + \frac{C_A}{C}\right)$$

$$\frac{dC_W}{dt} = R\left(1 - \frac{C_W}{C}\right)$$

$$\frac{dV}{dt} = \frac{RV}{C}$$ (1.15)

$$\frac{dx}{dt} = V$$

The last equation states that the superficial velocity creates an equivalence between distance along the reactor and the residence time t that a fluid element has to react. These equations are subject to the initial condition of the flow at the inlet $(t=0)$:

$$C_A(0) = C \quad V(0) = V_0$$
$$C_W(0) = 0 \quad x(0) = 0$$ (1.16)

Approach

Clearly from the initial condition and stoichiometry, $C_W = C_E$ (the concentration of ethyl alcohol, and the value of C is constant as temperature and pressure are assumed constant. C can be found from the ideal gas law, with

$$C = \frac{p}{T\left(8314 \dfrac{J}{kmolK}\right)}$$ (1.17)

And the initial flow velocity V_0 can be determined from the flowrate given, the inlet density (the molecular weight of ethyl alcohol is 46 kg/kmol), and the tube cross-sectional area. The equations will need to be integrated numerically in space-time t until the required alcohol mole fractions have been reached. Use either simple Euler or Runge-Kutta numerical integration.

You may note that it is possible to solve for C_A without recourse to the other variables, but C_W, V, and x depend explicitly on t. But since the requirement is to find positions x where specific mole fractions occur, it is best to solve for all four variables simultaneously.

Partial Results

A resolved numerical solution gives

$$\frac{C_A}{C} = 0.1$$

$$t = 5.65225 \tag{1.18}$$

$$x = 18.5435$$

with a profile for C_A/C as in Figure 1.5.

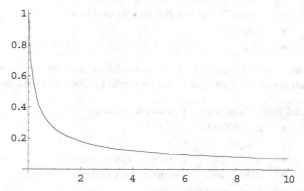

Figure 1.5 Profile of normalized alcohol concentration vs. space time t.

FEMLAB Implementation

We wish to create our pseudo-0D simulation environment yet again, this time with four dependent variables. Start up FEMLAB and await the Model Navigator window.

Model Navigator
• Select 1-D dimension
• Select PDE modes → Coefficient+time dependent
• Select 4 dependent variables
• Element: Lagrange - linear
• More >>
• OK

This application mode gives us four dependent variables u_1 u_2 u_3 u_4 and one space coordinate x. Next, set up the domain.

Pull down the Draw menu and select Specify Geometry.

Draw Mode

- Name: interval
- Start: 0
- Stop: 1
- Apply / OK

Now onto the Neumann boundary conditions. Pull down the Boundary menu and select Boundary Settings.

Boundary Mode

- Select domains 1 and 2 (hold down ctrl key)
- Select Neumann boundary conditions
- Apply
- OK

Since there are several parameters, it is useful to specify them with named constants. Pull down the Options menu and select Add/Edit constants.

Add/Edit Constants: Fill out as below

- $P = 200000.$ (kPa)
- $T = 423.$ (K)
- $R = 8314.$ (J/mol K, gas constant)
- $MM = 46.$ (atomic mass units)
- Flowrate = 0.01 (m/s)
- Dia = 0.05   (m)
- C=P/RT
- area=Pi*Dia^2/4
- rho=MM*C
- vel=Flowrate/rho/area

Now pull down the Subdomain menu and select Subdomain mode. Next pull down the Options menu and select Add/Edit Expressions. Create an expression rate= $52.7*u_1{\wedge}2/(1+0.013/ u_1)$. Next enter Subdomain settings.

Subdomain Mode

- Select domains 1
- Set $c_{11}=1$; $c_{22}=1$; $c_{33}=1$; $c_{44}=1$
- Set $f_1 = $ -rate*(1+ u_1/C);
  f_2= rate*(1- u_2/C);
  f_3= rate* u_3/C;
  f_4= u_3/C;
- Select the init tab; set $u_1(t0)$=C; $u_3(t0)$=vel
- Apply / OK

Pull down the Mesh menu and select the Parameters option to set up our single element.

Mesh Mode
• Set Max element size, general = 1
• Select Remesh
• OK

Now pull down the Solve menu and select the Parameters option. This pops up the Solver Parameters dialog window.

Solver Parameters
• Time-stepping tab: set output times 0:0.1:10.
• Jacobian: numeric
• Solve
• Cancel

Try plotting point plots of u_1, u_2, u_3 and u_4 for the whole range of times. How good is the qualitative agreement with Figure 1.5? Does it agree numerically with the fully resolved solution?

Exercises:

1.3 Find the value of $y'(x=1)$ from the system of equations below. Plot y' for x between 0 and 3.

$$y'' + y' + y^2 = 0$$
$$y(x=0) = 1$$
$$y'(x=0) = 0$$

1.4 Linear systems of ODEs result from first order reversible reaction systems in a continuously stirred tank reactor. For instance, consider the isomerization reactions

$$A \leftrightarrow B \leftrightarrow C$$

with forward reaction rates k_1 and k_3, respectively, as written; reverse reaction rates k_2 and k_4, as written. First order kinetics leads to the following system of ODEs:

$$\frac{dC_A}{dt} = -k_1 C_A + k_2 C_B$$

$$\frac{dC_B}{dt} = k_1 C_A - k_2 C_B - k_3 C_B + k_4 C_C$$

$$\frac{dC_C}{dt} = k_3 C_B - k_4 C_C$$

It may surprise you, but because the above system is linear, it has a general, analytic solution. Though general, it lends little insight into the dynamics of the system. Plot the graph of concentrations versus time for the initial value problem. Start with pure $C_A=1$ with parametric values $k_1= 1$ Hz, $k_2 =0$ Hz, $k_3=2$ Hz, $k_4=3$ Hz and plot the graph versus time of concentrations.

1.4 Method 3: Numerical Integration of Ordinary Differential Equations

In the previous section, numerical integration was treated by marching methods, commonly referred to as "time-stepping," although in the reactor design application, it was clearly spatial integration. In marching methods, the unknowns are found *sequentially*. The other common method for integration is to approximate the ODE and solve *simultaneously* for the unknown dependent variables at the grid points. With marching methods, all solutions must be initial value problems (IVP). The number of initial conditions must match the order of the ODE system. But for second order and higher systems, a second type of boundary condition is possible – the boundary value problem (BVP), where in 1-D, there are conditions at the initial and final points of the domain. Hence, these are two point boundary value problems. Marching methods can laboriously treat BVPs by shooting – artificially prescribing an IVP and guessing the initial conditions that satisfy the actual BVP by trial and error. In higher dimensional PDEs, a BVP specifies conditions on the boundaries of the domain.

One of the major advantages of the finite element method is that it naturally solves two-point BVPs. As an example, the reaction and diffusion equation in 1-D is

$$\frac{\mathcal{D}}{L^2}\frac{\partial^2 u}{\partial x^2} = R(u) \tag{1.19}$$

where u is the concentration of the species, $\mathcal{D}$ is the diffusivity, L is the length of the domain, $R(u)$ is the disappearance rate by reaction, and x is the dimensionless spatial coordinate. If the unknown function $u(x)$ is approximated by discrete values $u_j = u(x_j)$ at the grid points $x=x_j=j\ \Delta x$, then with central differences, the system of equations becomes

$$\sum_{j=1}^{N} M_{ij} u_j = \frac{L^2 \overline{\Delta x}^2}{\mathcal{D}} R_i \qquad (1.20)$$

where M_{ij} is a tridiagonal matrix with the diagonal element -2, and 1 on the super and subdiagonals:

$$M = \begin{bmatrix} -2 & 1 & 0 & 0 & \cdots \\ 1 & -2 & 1 & 0 & \cdots \\ 0 & 1 & \ddots & \ddots & \ddots \\ 0 & 0 & \ddots & \ddots & \ddots \\ \vdots & \ddots & \ddots & \ddots & \ddots \end{bmatrix} \qquad (1.21)$$

and $R_j=R(u_j)$. This system can be solved by iteration for u''_i by matrix inversion, where n refers to the n-th guess:

$$u_i^n = \frac{L^2 \overline{\Delta x}^2}{\mathcal{D}} \sum_{j=1}^{N} M_{ij}^{-1} R_j \qquad (1.22)$$

and $R_j=R(u_i^{n-1})$. For either IVP or BVP, the appropriate rows of the matrix M in (1.21) can be altered to accommodate the boundary conditions. As written, (1.21) supposes $u=0$ at both $x=0$ and $x=1$. This is a Dirichlet type boundary condition, and is the natural boundary condition for finite difference methods – natural because it occurs if no effort is made to overwrite rows of (1.21) with specified boundary conditions.

We will now illustrate the solution of (1.19) with FEMLAB on a small 1-D domain with first order reaction $R(u)=k\ u$ and representative values for the resulting dimensionless parameter, the Damkohler number:

$$Da = \frac{kL^2}{\mathcal{D}} = \frac{\left(10^{-3} s^{-1}\right)\left(10^{-3} m^{-1}\right)^2}{1.2 \times 10^{-9}\ m^2 \big/ s} = 0.833 \qquad (1.23)$$

and with boundary conditions $u=1$ at $x=0$ and no flux at $u=1$.

First launch FEMLAB and enter the Model Navigator:

Model Navigator
- Select 1-D dimension
- Select PDE modes → Coefficient
- Element: Lagrange – linear
- More >>
- OK

This application mode gives us one dependent variable u, but in a 1-D space with coordinate x. Now we are in a position to set up our domain. Pull down the Draw menu and select Specify Geometry.

Draw Mode
- Name: interval
- Start: 0
- Stop: 1
- Apply
- OK

Now for the boundary conditions. Pull down the Boundary menu and select Boundary Settings.

Boundary Mode
- Select domain 1
- Check Dirichlet and set h=1; r=1
- Select domain 2
- Select Neumann boundary conditions
- Apply
- OK
---

h and r are the two handles on Dirichlet BCs in Coefficient Mode. If you want to set u to a given value U_0 on a boundary, then it is accomplished with setting h=1 and r= U_0. Now pull down the Subdomain menu and select Subdomain mode. Select Subdomain settings.

Subdomain Mode
- Select domains 1
- Set c=-1; f=0.833*u
- Select the init tab; set u(t_0)=1-x
- Apply
- OK

Click on the triangle symbol to mesh (15 elements) and the triangle with the embedded upside-down triangle to refine the mesh (30 elements).

Now pull down the Solve menu and select the Parameters option. This pops up the Solver Parameters dialog window.

Solver Parameters
• Select stationary linear
• Solve
• Cancel

You should get a graph with the information as in Figure 1.6. Clearly the desired boundary conditions are met: $u=1$ at $x=0$ and the slope vanishes at $x=1$. But did FEMLAB solve the problem we thought we posed?

Now resolve with the stationary nonlinear solver. First note that FEMLAB takes thirteen iterations to converge. Do you notice that the final value has dropped from 0.86 to 0.69? One might wonder why there is a difference. The linear solver only evaluates $R(u)$ once at the initial condition u(t0)=1-x and thus only needs one iteration of (1.22). The nonlinear solver evaluates $R(u)$ for each iteration at the old estimate for u. Thus, the nonlinear solver might "forget" the initial guess completely after a number of iterations as it homes in on a converged solution.

Let's test this explanation. Changing the initial condition should change the stationary linear solution. Return now to Subdomain settings and try the initial condition $u(t0)=1$. What final value do you get for $u(x=1)$? Now try the stationary nonlinear solver. Do you get the same solution as with the other initial condition?

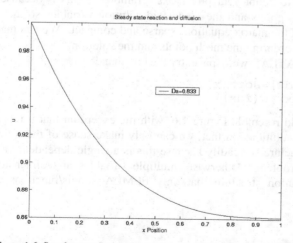

Figure 1.6 Steady state first order reaction and diffusion with Da=0.833.

This example should illustrate the importance of selecting the right solver for your equations. If there is any dependence of f on the dependent variables, then the stationary nonlinear solver should be used. The linear solver is faster, but it also presumes that the coefficients of the PDE do not depend on the dependent variable u (else the problem would be nonlinear). When in doubt, use the nonlinear solver. After all, (1.19) with $R(u)=k\ u$, is a *linear* problem, but FEMLAB only finds the correct steady state solution with the nonlinear solver! The slow convergence rate is also the consequence of the form of the model – general mode with the exact Jacobian solver option for the nonlinear solver converges in two iterations to the correct profile.

We argued that (1.20) is the finite difference matrix equation for this problem, yet later applied the argument that (1.22) should describe the FEMLAB finite element problem. Because we used Lagrange linear elements, in this special case the finite element and finite difference matrix operators coincide, up to the boundary conditions. To see this, we will take a foray into the MATLAB representation of FEMLAB problems.

Pull down the File Menu and select **Export FEM structure as 'fem'.** This puts the current solution as a MATLAB data structure in the MATLAB workspace. We can then manipulate it using the built-in MATLAB functions and commands, as well as the special function set of FEMLAB.

In your MATLAB workspace, try the commands

```
>>x=fem.xmesh.p{1};
>>u=fem.sol.u;
>>plot(x,u)
```

This should pop up a MATLAB Figure plotting the solution u versus the array of mesh points. No doubt your plot looks scrambled. This is because FEMLAB stores the mesh points and the associated solution variables so as to make the specification of the matrix equations sparse and compact. We can make sense of the solution by ordering the mesh points and the solution:

In your MATLAB workspace, try the commands

```
>> [xx,idx]=sort(x);
>>plot(xx,u(idx))
```

This plot should resemble Figure 1.6, with the exception that it represents your last FEMLAB solution. In fact, we can only make sense of the solution format of the **fem** structure so readily because this is a single dependent variable, one-dimensional problem. Otherwise, multiple variables and dimensions leave a mesh and solution structure that only FEMLAB tools/functions can readily decode.

The final MATLAB manipulation we will consider here is interrogation of the finite element matrix. The `fem` structure does not hold the finite element stiffness matrix, but rather contains the information necessary for FEMLAB functions to construct it. This activity is a vital part of the finite element method, and the FEMLAB function that does it is called assemble. Type in the command below:

```
>>[K,L,M,N]=assemble(fem);
>>K/30
```

You should now see a MATLAB sparse representation of a matrix, all of the elements of which are 1, -2, and 1, arranged on different diagonals. This is the stiffness matrix of the finite element method, and up to the ordering of the unknowns, is equivalent to (1.21). If you return to the Subdomain Settings, element tab, and select Lagrange quadratic elements, and repeat the solution, exporting FEM and assemble K as above, you will note that although sparse, the matrix is distinctly different from the Lagrange linear elements.

Exercise

1.5 The coefficient form has a pde term α u. Repeat the implementation of the reaction-diffusion example, but this time entering $\alpha = 0.833$ and *f=0* for the subdomain settings. Now compare the stationary linear and nonlinear solver solutions. Can you explain why this formulation leads to this result? What effect does this formulation of the problem have on the stiffness matrix *K*. Can you think of a difficulty that might occur if the Da is chosen so that the diagonal element is nearly zero in magnitude, i.e. Da $\Delta x^2 = 2$?

1.5 Method 4: Linear Systems Analysis

Central to MATLAB, and hence to FEMLAB, is linear systems analysis. In this section, we will briefly review the concepts of linear operator theory – typically lumped as "matrix equations" in undergraduate engineering mathematics modules. The good news is that it is not necessary to do any matrix manipulations yourself. That was the *raison d'etre* for MATLAB: to serve as a user interface to libraries of subroutines for engineering matrix computations. Much of the history of scientific computing is encapsulated in efficient and sparse methods for matrix computations. An excellent guide to matrix computations, but surely for experts, is the book of Golub and Van Loan [3]. However, at the introductory level to MATLAB, a good and readable survey can be found in the up-to-date book by Hanselman and Littlefield [4].

Briefly, the standard matrix equations look like this:

$$a_{11}x_1 + a_{12}x_2 + a_{13}x_3 + \ldots + a_{1N}x_N = b_1$$
$$a_{21}x_1 + a_{22}x_2 + a_{23}x_3 + \ldots + a_{2N}x_N = b_2$$
$$a_{31}x_1 + a_{32}x_2 + a_{33}x_3 + \ldots + a_{3N}x_N = b_3 \qquad (1.24)$$
$$\vdots$$
$$a_{M1}x_1 + a_{M2}x_2 + a_{M3}x_3 + \ldots + a_{MN}x_N = b_M$$

Here there are N unknowns x_j which are related by M equations. The coefficients a_{ij} are known numbers, as are the constant terms on the right hand side, b_i. In engineering, models are frequently derived that satisfy such linear systems of equations. Mass and energy balances, for instance, commonly generate such sets of linear equations.

Solvability

When N=M, there are as many constraints as there are unknowns, so there is a good chance of solving the system for a unique solution set of x_j's. There can fail to be a unique solution if one or more of the equations is a linear combination of the others (row degeneracy) or if all the equations contain only certain combinations of the variables (column degeneracy). For square matrices, row and column degeneracy are equivalent. A set of degenerate equations are termed **singular**. Numerically, however, at least two additional things can go wrong:

- While not exactly linear combinations of each other, some of the equations may be so close to linearly dependent that within round-off errors on the computer they are.
- Accumulated round-off errors in the solution process can swamp the true solution. This frequently occurs for large N. The procedure does not fail, but the computed solution does not satisfy the original equations all that well.

Guidelines for Linear Systems

There is no "typical" linear system of equations, but a rough idea is that round-off error becomes appreciable:

- N as large as 20-50 can be solved by normal methods in single precision without recourse to specialist correction of the two numerical pathologies.
- N as large as several hundred can be solved by double precision.
- N as large as several thousand can be solved when the coefficients are sparse (i.e. most are zero) by methods that take advantage of sparseness. MATLAB has a special data type for sparse matrices, and a suite of functions that exploit the sparseness.

However, in engineering and physical sciences, there are problems that by their very nature are singular or nearly singular. You might find difficulty with N=10. Singular value decomposition is a technique which can sometimes treat singular problems by projecting onto non-singular ones.

Common Tasks in Numerical Linear Algebra

Equation (1.24) can be succinctly written as a matrix equation (cf. equation 1.20).

$$\mathbf{A} \cdot \mathbf{x} = \mathbf{b} \tag{1.25}$$

- Solution for the unknown vector **x**, where **A** is a square matrix of coefficients, and **b** is a known vector.
- Solution with more than one **b** vector with the matrix **A** held constant.
- Calculation of the matrix $\mathbf{A}^{-1}$, which is the inverse of a square matrix **A**.
- Calculation of the determinant of a square matrix **A**.
- If M<N, or if M=N but the equations are degenerate, then there are effectively fewer equations than unknowns—an underdetermined system. In this case, either there can be no solution, or there is more than one solution vector **x**. The solution space consists of a particular solution $\mathbf{x_p}$ plus any linear combination of typically N-M vectors called the nullspace of **A**. The task of finding this solution space is called *singular value decomposition*.
- If M>N, there is, in general, no solution vector x to (1.24). This overdetermined system happens frequently, and the best compromise solution that comes closest to satisfying the equations is sought. Usually, the closeness is "least-squares" difference between the right and left hand sides of (1.24).

Matrix Computations in MATLAB

Matrix inversion is easily entered using the inv(matrix) command. Solution of matrix equations is represented by the matrix division \ operator as here:

```
>> A=[ 3 -1 0; -1 6 -2; 0 -2 10];
>> B=[1; 5; 26];
>> X=A\B
X =
    1.0000
    2.0000
    3.0000
```

Determinants

Determinants are used in stability theory and in assessing the degree of singularity of a matrix. Why do you need to know the determinant? Most of the time, you want to know when a determinant is zero. However, when the determinant is zero, or numerically close to zero, it is numerically difficult to compute due to "round-off" swamping effects mentioned earlier. This is yet another application for singular value decomposition.

MATLAB computes determinants by the simple function det(A). Either enter by hand the matrix below at the MATLAB command line, or cut and paste from the file matrix2.dat:

```
>> A=[0.45, -0.244111, -0.0193373, 0.323972, -0.118829;
    -0.244111, 0.684036, -0.103427, 0.205569, 0.00292382;
    -0.0193373, -0.103427, 0.8295, 0.0189674, -0.011169;
    0.323972, 0.205569, 0.0189674, 0.659479, 0.197388;
    -0.118829, 0.00292382, -0.011169, 0.197388, 0.776985]
```

The determinant is found from

```
>> det(A)
   ans =
   -1.9682e-008
```

Principal Axis Theorem: Eigenvalues and Eigenvectors

MATLAB has built-in functions for computing the eigenvalues and eigenvectors of a matrix:

```
>> eig(A)
ans =
   -0.0000
    0.7000
    0.8000
    0.9000
    1.0000
```

The eig() function can also return the eigenvectors as the columns of the matrix V when called as below:

```
>> [V,D]=eig(A)
V =
   -0.6836   -0.0000   -0.5469   -0.4785   -0.0684
   -0.4181    0.6162    0.1831    0.4530   -0.4547
   -0.0837    0.4003    0.6189   -0.6232    0.2479
    0.5409    0.2582   -0.2415   -0.4042   -0.6474
   -0.2416   -0.6272    0.4755   -0.1190   -0.5550
D =
   -0.0000        0        0        0        0
        0   0.7000        0        0        0
        0        0   0.8000        0        0
        0        0        0   0.9000        0
        0        0        0        0   1.0000
```

The `eigs()` function is a variant of `eig()` which computes a specific number of eigenvalues/eigenvector pairs for sparse matrices. Its use will be demonstrated in the next subsection in conjuction with FEMLAB.

The matrix A has a determinant that is little different from zero and a single eigenvalue that is effectively zero. The eigenvector associated with it is effectively the null space of A – the direction that gets mapped to zero:

```
>> A*V(:,1)
ans =
  1.0e-007 *
    0.2669
    0.1633
    0.0327
   -0.2112
    0.0943
```

All the other eigenvectors can be verified by the property that they map onto themselves, scaled by the eigenvalue, for instance:

```
>> A*V(:,2)./ V(:,2)
ans =
    0.7000
    0.7000
    0.7000
    0.7000
    0.7000
```

In MATLAB, the ./ division operator is element-by-element division. The colon above refers to the whole of the column.

Because the system is nearly singular, we should not be surprised that the solutions to any matrix equation involving it are poorly conditioned. For instance,

```
>> B=[0; 1; 0; 1; 0];
>> A\B

ans =
  1.0e+006 *
    2.1487
    1.3142
    0.2631
   -1.7001
    0.7593
```

Since the elements of A are of order one, the forcing vector B is of order one, one would expect the solution to (1.25) to be order one, not order one million. For chemical engineers, this is like being told that a mass balance involves input flow rates of about 1 kg/hr, constraints on mass balances with appreciable fractions in the splitters (order one), and that the solution mass flow rates are about one million kg/hr for internal streams. Not likely. Yet this is the solution proposed by a nearly singular matrix.

Singular Value Decomposition (SVD)

SVD offers a better solution in many respects. All matrices have a unique decomposition, similar to the principal axis theorem for eigenvalues and eigenvectors

$$A = U \cdot diag \cdot V^T \tag{1.26}$$

where U and V are square real and orthogonal. *diag* is a diagonal matrix which contains the singular values. In terms of U, V, and *diag*, the system (1.25) is readily solved

$$A^{-1} = V \cdot \left[1 / diag_j\right] \cdot U^T \tag{1.27}$$

U and V being orthogonal means that their transposes are also their inverses. The inverse of a diagonal matrix is just the reciprocal of the diagonal elements. So the only time we have a problem solving the system is when one or more of the singular values (diag$_j$), relative to the largest, is close to zero. It follows that (1/ diag$_j$) is a very large number, which distorts our numerical solution, sending it off to infinity along a direction which is spurious. A good approximation is to throw these spurious directions away completely by setting (1/ diag$_j$) for the offending singular values to zero! The vector,

$$x = V \cdot \left[1 / diag_j\right] \cdot U^T b \tag{1.28}$$

with this substitution for nearly zero elements, should be the smallest in magnitude to approximately satisfy the equations.

In the case of our example matrix A, the MATLAB command svd() gives the singular values if called with one output, and the three matrices U, diag, V if called with three:

```
>> [U,D,V] =svd(A)

U =
     -0.0684     -0.4785      0.5469      0.0000     -0.6836
     -0.4547      0.4530     -0.1831     -0.6162     -0.4181
      0.2479     -0.6232     -0.6189     -0.4003     -0.0837
     -0.6474     -0.4042      0.2415     -0.2582      0.5409
     -0.5550     -0.1190     -0.4755      0.6272     -0.2416
```

```
  D =
        1.0000          0          0          0          0
             0     0.9000          0          0          0
             0          0     0.8000          0          0
             0          0          0     0.7000          0
             0          0          0          0     0.0000
  V =
       -0.0684    -0.4785     0.5469     0.0000     0.6836
       -0.4547     0.4530    -0.1831    -0.6162     0.4181
        0.2479    -0.6232    -0.6189    -0.4003     0.0837
       -0.6474    -0.4042     0.2415    -0.2582    -0.5409
       -0.5550    -0.1190    -0.4755     0.6272     0.2416
```

The SVD prescription for solution with smallest magnitude is implemented as follows:

```
>> ss=[1. 1./0.9 1./0.8 1./0.7 0];
>>dinv=diag(ss);
>> V*dinv*U'*B
ans =
     0.0893
     1.2820
     0.1479
     1.0317
    -0.2130
```

This is a far more physically acceptable solution, for instance, for internal mass flow rates in the hypothetical mass balance discussed above.

This excursion into linear systems theory is important for modeling with FEMLAB because finite element methods are matrix based. When the generalized stiffness matrix becomes nearly singular, FEMLAB may not be providing a satisfactory solution. These matrix computations and their sparse implementations in MATLAB can readily serve as diagnostics for the health of the FEMLAB solution. They also provide an insight into the natural dynamics of the system through the eigen analysis of the operator. These ideas will be made concrete with an example computed as a FEMLAB model in the next subsection.

1.5.1 Heat transfer in a nonuniform medium

The steady state heat transfer equation is commonly met in engineering studies as the simplest PDE that is analytically solvable: Laplace's equation. Nevertheless, series solutions for complicated geometries may be intractable. The author has recently shown that some series so derived are purely asymptotic and poorly convergent [5]. Consequently, numerical solutions are likely to be better behaved than series expansions. Furthermore, any variation on the processes of heat transfer may destroy the analytic structure. In this section, we will consider the typical one-dimensional heat transfer problem in a slab of non-uniform conductivity and a distributed source that is differentially heated on the ends:

$$-\frac{d}{dx}\left(k\frac{dT}{dx}\right)=f(x)$$

$$T\mid_{x=0}=1 \qquad T\mid_{x=1}=0$$

(1.29)

Launch FEMLAB and enter the Model Navigator:

Model Navigator
- Select 1-D dimension
- Select Physics modes → Heat transfer→Linear stationary
- Element: Lagrange - linear
- More >>
- OK

This application mode gives us one dependent variable u, but in a 1-D space with coordinate x. Now we are in a position to set up our domain. Pull down the Draw menu and select Specify Geometry.

Draw Mode
- Name: interval
- Start: 0
- Stop: 1
- Apply / OK

Now for the boundary conditions. Pull down the Boundary menu and select Boundary Settings.

Boundary Mode
- Select domain 1
- Set T=1
- Select domain 2
- Set T=0
- Apply / OK

Now pull down the Subdomain menu and select Subdomain mode. Select Subdomain settings.

Subdomain Mode
• Select domains 1
• Set k=-1; Q=x*(1-x)
• Select the init tab; set T(t_0)=1-x
• Apply / OK

Click on the triangle symbol to mesh (15 elements) and the triange with the embedded upside-down triangle to remesh (30 elements).

Now click on the equals sign = on the toolbar to solve. The solution should be found fairly quickly with a nearly linear profile with almost a slope of −1. Verify that $T|_{x=0.499} = 0.474742$. This problem has an analytic solution with $T|_{x=0.499} = 0.474958$:

$$T = 1 - \frac{13}{12}x + \frac{x^3}{6} - \frac{x^4}{12} \qquad (1.30)$$

Now try k=-(1-x/2). There is also an analytic solution in this case, but in the complex numbers requiring logarithms in the complex plane and a branch cut. The analytic solution gives $T|_{x=0.499} = 0.550622$. How good is your solution?

Now for the linear systems theory. Pull down the File menu and select Export FEM structure as 'fem.' You can now manipulate the solution in MATLAB. As in the last section, you can assemble the stiffness matrix:

```
>>[K,L,M,N]=assemble(fem);
```

As K is sparse, you can find the smallest six eigenvalues in magnitude with

```
>> dd=eigs(K,6,0);
```

and the eigenvectors with

```
>>[V,D]=eigs(K,6,0);
```

Note that K has one zero eigenvalue, and that all its eigenvalues are negative otherwise. This should not worry you, as FEMLAB implements its boundary conditions through the block matrix N and auxiliary forcing vector M. It could replace rows of K and elements in L to approximate boundary conditions, but the structure of boundary conditions in FEMLAB allows for more general types of boundary conditions when augmenting the matrix equations with N and M. The fact that K is singular as a block matrix is a consequence of the natural boundary conditions for finite element methods being Neumann conditions (no flux). There are an infinity of solutions to the pure Neumann boundary conditions, as an arbitrary value can be added to any solution and it is still a solution. That K is singular naturally tripped up the author when he first used finite element methods as an undergraduate. Purely Neumann boundary conditions are badly behaved in FEMLAB. For instance, if you change the boundary conditions in our example to purely Neumann conditions, you should find that the steady state solution is 10^{11} in size. Yet MATLAB can solve such a problem by SVD or by the principal axis theorem. Since the matrix K is negative-semi-definite, all its eigenvalues are real. So pseudo-inversion to eliminate the zero eigenvalue of K follows from

```
>> ss=1./dd;
>> ss(1)=0.;
>> dinv=diag(ss);
>> uneumann=V*dinv*V'*L
```

Finally, interpreting this solution must be done remembering that the structure of a FEMLAB mesh is not monotonic. These commands plot the solution:

```
>>[xs,idx]=sort(fem.xmesh.p {1});
>>plot(xs,fem.sol.u(idx));
```

Similarly, the approximate Neumann solution found from the projection onto the first five eigenvectors with smallest magnitude non-zero eigenvalues is found from

```
>>plot(xs,uneumann(idx));
```

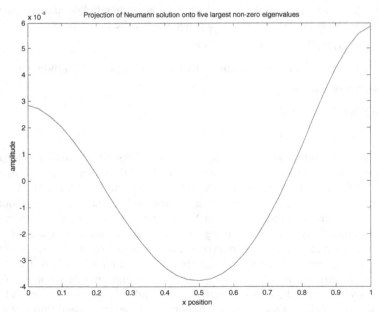

Figure 1.7 Projection solution for the purely Neumann solution to the non-uniform conductivity and distributed source heat transfer problem (1.29).

Furthermore, the eigenvectors can be interpreted the same way:

```
>>col2=V(:,2);
>>plot(xs,col2(idx));
>>col3=V(:,3);
>>plot(xs,col3(idx));
```

It should be noted that in the case of the Neumann solution, any constant value can be added to the solution and it will remain a solution. The eigenvectors are not normalized, so they can be multipled by any number and still be eigenvectors. Figures 1.8 and 1.9 show the two eigenpairs with largest eigenvalues in magnitude.

1.6 Summary

This chapter has illustrated that FEMLAB is constructed upon four standard methods in numerical analysis: root finding, numerical integration by marching methods, numerical integration for BVPs, and linear systems theory. These tools are conducive to solving many common problems that arise in chemical engineering applications in 0-D and 1-D spaces. In the next chapter, we will begin to see applications where FEMLAB solutions have value added over the standard solution techniques in 2-D.

References

1. Cutlip M.B. and Shacham M., Problem Solving in Chemical Engineering with Numerical Methods, Prentice-Hall, Upper Saddle River, NJ, 1999.
2. Gear W.C., Numerical Initial Value Problems in Ordinary Differential Equations, 1971.
3. Golub G.H. and Van Loan C.F., Matrix Computations, 3rd ed. - Baltimore; London: Johns Hopkins University Press, 1996.
4. Hanselman D. and Littlefield B.,Mastering MATLAB 6: A comprehensive tutorial and reference, Prentice Hall, Saddle River NJ, 2001.
5. Zimmerman W.B., "On the resistance of a spherical particle settling in a tube of viscous fluid" Submitted to SIAM J. Applied Maths, 2002.

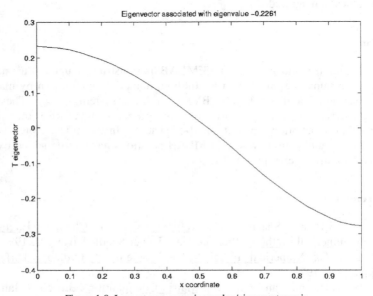

Figure 1.8 Largest non-zero eigenvalue/eigenvector pair.

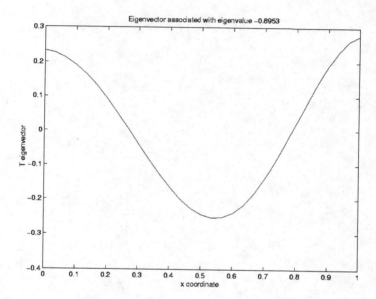

Figure 1.9 Second largest eigenvalue/eigenvector pair.

Chapter 2

PARTIAL DIFFERENTIAL EQUATIONS AND THE FINITE ELEMENT METHOD

W.B.J. ZIMMERMAN and B.N. HEWAKANDAMBY

Department of Chemical and Process Engineering, University of Sheffield,
Newcastle Street, Sheffield S1 3JD United Kingdom

E-mail: w.zimmerman@shef.ac.uk

Partial differential equations (PDEs) arise naturally in science and engineering from complex balance equations. Commonplace PDEs are derived from conservation laws for transport of mass, momentum, species and energy. Because these conservation laws are integral equations over the domain, the PDEs that arise from the continuum hypothesis have a structure that is readily represented by the finite element method as an approximation. In this chapter, the three different classes of differential equations that arise in spatial-temporal systems – elliptic, parabolic, and hyperbolic – are defined and representative cases are treated by FEMLAB computations. An overview of the finite element method is given, but greater depth of detail will await later chapters where the applications particularly exploit features of finite element methods that intrinsically permit elegant and accurate computation.

2.1 Introduction

Partial differential equations are usually found in science and engineering applications as the local, infinitesimal constraint imposed by conservation laws that are typically expressed as integral equations. The whole class of transport phenomena due to conservation of mass, momentum, species and energy lead to PDEs in the continuum approximation. Chemical engineers are well acquainted with shell balances in transport phenomena studies for heat, mass and momentum transfer.

In contrast to the previous chapter, where 0-D and 1-D spatial systems were treated by FEMLAB with example applications in chemical engineering, the chemical engineering curriculum is not overflowing with 2-D and 3-D example computations of the solutions to PDEs. A rare example is found in [1]. In fact, historically, many of the common chemical engineering models and design formula are simplifications of higher spatial dimension dynamics that are treated phenomologically. Resistance coefficients in fluid dynamics, mass transfer and heat transfer coefficients, Thiele moduli in heterogeneous catalysis, McCabe-Thiele diagrams for distillation column design, and many more common techniques are convenient semi-empiricisms that mask an underlying transport or non-equilibrium thermodynamics higher spatial dimension system, possibly expressible as a PDE system, but traditionally thought too difficult to solve given the complexity of the fundamental physical chemistry. These simpler methodologies are still preferable for quick estimates desired for preliminary design calculations, but may be insufficient for detailed design, retrofit, or process analysis and optimization purposes. For fundamental science, these methods are still migrating from chemical engineers to biotechnologist or material scientists in the first approach to multidisciplinary work. Nevertheless, computational fluid dynamics (CFD) has forever changed the paradigm for what is considered the state-of-the-art in transport modeling. Phenomenological methods may still have a niche, and a particular important one in interpreting distributed system models, yet CFD has a unique role for visualization and quantification of transport phenomena.

FEMLAB is not a "commercial CFD code", but it will do some CFD. There are several general purpose CFD packages available, with their own advantages in supporting certain applications. By CFD, most process engineers would envisage support for many turbulence and combustion models. FEMLAB, however, has a different niche in the area of multiphysics. In addition to the traditional transport phenomena that CFD treats, FEMLAB includes application modes for electrodynamics, magnetodynamics, and structural mechanics, permitting simultaneous treatment of these and transport phenomena. But its greatest strengths are actually least trumpeted – first, the ease of "user defined programming", which is the ability to implement the user's own model or parametric variation of coefficients, boundary conditions, initial conditions and to link to simultaneous physics, even on other domains; second, that it is built on MATLAB so that all the programming functionality needed to set up greater complexity of models or simulations is available, treating FEMLAB as a convenient suite of subroutines for high-level finite element programming and analysis. In the last chapter, we saw some of the power of user defined programming and analysis. In this chapter, we introduce FEMLAB's core strength of finite element modeling of higher dimensional PDE systems. The greater functionality of multiphysics, extended physics, and treatment on non-PDE constraints will be left for later chapters.

Partial Differential Equations

PDEs are classified according to their order, boundary condition type, and degree of linearity (yes, no or quasi). Amazingly, most PDEs encountered in science and engineering are second order, i.e. the highest derivative term is a second partial derivative. Is this a coincidence? Lip service is usually paid at this point to variational principles underlying most of physics. Yet, recently Frieden [2] has demonstrated that all known laws of physics can be derived from the principle of minimum Fisher information, which naturally introduces a second order operator of a field quantity as the highest order term in the associated law of physics – from the wavefunction in Schrodinger's equation to classical electrodynamics. Thus, classification and solution of second order spatial temporal systems in 2-D and 3-D are of wide applicability and importance in the sciences and engineering. For this reason, and that finite element methods (FEM) are intrinsically well-suited to treating second order systems, FEM are techniques with wide applicability.

In this chapter, we focus on second order systems in 2-D and 3-D. There are three canonical exemplar systems that are nearly uniformly treated in the standard textbooks. We shall not disappoint. They are:

$$\text{Laplace's equation (elliptic): } \frac{\partial^2 u}{\partial x^2} + \frac{\partial^2 u}{\partial y^2} = 0 \tag{2.1}$$

$$\text{Diffusion equation (parabolic): } \frac{\partial u}{\partial t} = \frac{\partial^2 u}{\partial x^2} \tag{2.2}$$

$$\text{Wave equation (hyperbolic): } \frac{\partial^2 u}{\partial t^2} = \frac{\partial^2 u}{\partial x^2} \tag{2.3}$$

The terms elliptic, parabolic and hyperbolic are traditional guides to the features of a PDE system from characterization of the linear terms by reference to the general linear, second-order partial differential equations in one dependent and two independent variables:

$$a\frac{\partial^2 u}{\partial x^2} + 2b\frac{\partial^2 u}{\partial x \partial y} + c\frac{\partial^2 u}{\partial y^2} + d\frac{\partial u}{\partial x} + e\frac{\partial u}{\partial y} + fu + g = 0 \tag{2.4}$$

where the coefficients are functions of the independent variables x and y only, or constant. The three canonical forms are determined by the following criterion:

elliptic: $\qquad\qquad\qquad b^2 - ac < 0$ $\qquad\qquad\qquad$ (2.5a)

parabolic: $\qquad\qquad\qquad b^2 - ac = 0$ $\qquad\qquad\qquad$ (2.5b)

hyperbolic: $\qquad\qquad\qquad b^2 - ac > 0$ $\qquad\qquad\qquad$ (2.5c)

These classifications serve as a rough guide to the information flow in the domain. For instance, in elliptic equations, information from the boundaries is propagated instantaneously to all interior points. Thus, elliptic equations are termed "non-local", meaning that information from far away influences the given position, versus "local", where only information from nearby influences the field variable. In parabolic systems, information "diffuses", i.e. it spreads out in all directions. In hyperbolic systems, information "propagates", i.e. there is a demarcation between regions that have already received the information, regions that will receive the information, and possibly regions that will never receive the information. If the system is linear or quasi-linear (i.e. some coefficient depends on the dependent variable or a lower order partial derivative than that it multiplies), this classification system and the intuition about how information is transported serves as a robust guide to second order systems. For nonlinear systems, however, nonlinearity can destroy the information transport structure. In nonlinear systems, information may be "bound", i.e. never transferred, beyond given attractors, or it may be created from noise (one view) or lost (a different view) by forgetting initial conditions in a given window in time.

2.1.1 Poisson's equation: An elliptic PDE

A modest variant on Laplace's equation is the Poisson equation:

$$\nabla^2 u = f\left(x\right) \tag{2.6}$$

We saw this equation in 1-D form in (1.19) which described heat transfer in a non-uniform medium with a distributed source. Here, the thermal conductivity is uniform. In order to give a different spin on (2.6), one should note that it is the equation satisfied by the streamfunction with an imposed vorticity profile:

$$\nabla^2 \psi = -\omega\left(x\right) \tag{2.7}$$

There are two common types of vortices that are easy to characterize – the Rankine vortex, where vorticity is constant within a region, and the point-source vortex, where vorticity falls off rapidly and thus is idealized as point vortex. One might be curious about the streamlines generated by these two vortex types.

Start up FEMLAB and enter the Model Navigator:

Model Navigator
•   Select 2-D dimension
•   Select Classical PDEs → Poisson's Equation
•   Element: Lagrange - quadratic
•   More >>
•   OK

This application mode gives us one dependent variable u, but in a 1-D space with coordinate x. Now we are in a position to set up our domain. Pull down the Draw menu and select Circle/Ellipse.

Draw Mode
• Draw circle and edit it to achieve a unit radius centered at the origin
• Select Axes/Settings
• Set y in the range –1.2 to 1.2
• Apply
• OK

Now for the boundary conditions. Pull down the **Boundary** menu and select **Boundary Settings**. Note that there are four segments of the boundary, even though we thought we drew a contiguous circular domain.

Boundary Mode
• Select domain 1-4
• Check Dirichlet and set h=1, r=0 (u=0)
• Apply
• OK

Now pull down the **Subdomain** menu and select **Subdomain mode**. Select **Subdomain settings**.

Subdomain Mode
• Select domains 1
• Set c=1; f=1
• Apply
• OK

Click on the triangle symbol to mesh (620 elements).

Now click on the equals sign = on the toolbar to solve.

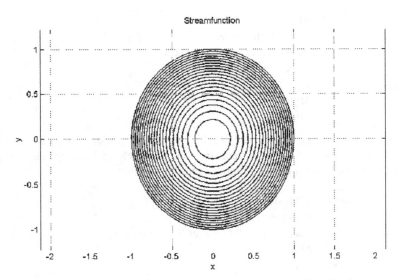

Figure 2.1 Streamlines for constant vorticity in a circular cavity with slip.

The streamlines are viewed by a contour plot. Figure 2.1 was generated by the "Copy to Figure" option on the Edit Menu.

Clearly, the streamlines are concentric circles.   As the boundary is a streamline (ψ=0) and the maximum occurs in the center (ψ=0.25), the volumetric flow rate induced by the constant vorticity is 0.25. Refining the mesh yields 2480 elements, but no apparent change in the solution (still concentric circles).

Now, for the other case: point source at the origin. We could conceivably approximate a point source by drawing a small circle centered at the origin as a second domain, and in that domain, have f=1/πR^2, where R is the smaller circle's radius, and f=0 outside. Then f integrates to unity, and in the limit $R \rightarrow 0$, f approaches the Dirac delta function. However, the limit can be approached more elegantly with the power of finite elements and weak terms.

Pull down the Draw menu and select Point.

Draw Mode
•   Click on the origin
•   Edit the points coordinates to be (0,0)
•   Apply / OK

Pull down the Point menu and select View as Coefficients.

Point Mode/Coefficient View
• Click on the origin
• Select point 3
• Select the weak tab
• Enter u_test
• Apply / OK

Click on the triangle on the toolbar to re-mesh (592 elements).

Now pull down the Solve menu and select the Parameters option.

Solver Parameters
• Select stationary linear
• Solution form: weak
• Solve
• Cancel
• OK

You should get a graph with the information as in Figure 2.2. In particular, one should note that the streamlines are not so uniformly spaced in Figure 2.1, and that the higher contours at the origin are clearly not circular. Refining the mesh to 2368 elements does not visually improve the smoothness of the circular contours, however, the maximum streamfunction increases from 0.807 to 0.918. Improvement comes from adapting the mesh. Now pull down the Menu menu and select the Parameters option.

Mesh Parameters
• Select more>>
• Max element size near vertices: 3 0.001
• Remesh
• OK

The 1428 elements are now packed in much more closely about the origin. Max element size near vertex 3 is set to 0.001 by this specification. The data entry is a MATLAB vector, where a space delimits vector elements 3 and 0.001. We can add more vertex/size pairs as desired to constrain the mesh generation.

Figure 2.2 has a maximum streamfunction of 1.56. Remeshing to 5712 elements achieves maximum streamfunction of 1.67. Remeshing again to 22848 elements results in 1.78. Although it is not clear that grid convergence is ever achieved, the qualitative arrangement of streamlines has converged as the swirling falls off with distance from the source rapidly.

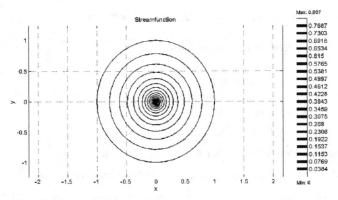

Figure 2.2 Streamlines for a point vortex at the center of a circular cavity.

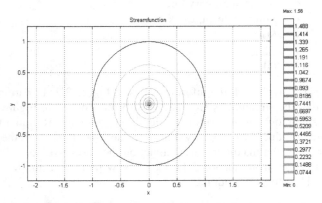

Figure 2.3 Streamlines for a point vortex at the center of a circular cavity, adapted mesh.

Exercise 2.1

Solve for the streamlines when the vorticity falls off exponentially with radius, i.e.

$$\omega = \omega_0 \exp\left(-\sqrt{x^2 + y^2}\right).$$

2.1.2 The diffusion equation: A parabolic PDE

The 1-D unsteady diffusion equation (2.2) can refer equally well to any of three common transport phenomena:

$$\text{Mass diffusion: } \frac{\partial c}{\partial t} = D \frac{\partial^2 c}{\partial x^2} \tag{2.8}$$

$$\text{Thermal conduction: } \frac{\partial T}{\partial t} = \frac{k}{\rho c_p} \frac{\partial^2 T}{\partial x^2} \tag{2.9}$$

$$\text{2-D Vorticity transport: } \frac{\partial \omega}{\partial t} = v \frac{\partial^2 \omega}{\partial x^2} \tag{2.10}$$

where c,T, and ω are concentration, temperature, and the z-component of vorticity in a 2-D flow, respectively, and their corresponding diffusivities are D, α, and v. This equation is thoroughly studied in the undergraduate curriculum. It has solutions by Fourier and Laplace transforms, and similarity solutions for initial and boundary conditions that collapse on the variable $\eta = \dfrac{x}{\sqrt{Dt}}$. That doesn't leave much room for finite element methods – just another technique for a tired old problem, right? Wrong. FEMLAB can still give this problem a boost which is not commonly considered. FEMLAB solutions are well suited to non-constant coefficients, i.e. transport properties that depend on the field variable. For instance, for suitably low pressures and high temperatures, a gas must satisfy the ideal gas law:

$$\rho = \frac{nM}{V} = \frac{PM}{RT} \tag{2.11}$$

where R is the gas constant and M is the relative molecular mass of the species. Under these conditions, it is rare to find a gas that has a constant heat capacity. For instance, over a range of temperatures, the heat capacity of CO_2 gas is well approximated by a quadratic in temperature, (in MJ/kg-mol°C), with T in °C:

$$c_p = 36.11 + 0.04233\,T - 2.887 \times 10^{-5} T^2 \tag{2.12}$$

It follows that

$$\frac{k}{\rho c_p} = \frac{kR}{36.11PM} f(T)$$

$$f(T) = \frac{(T+273)}{1.+ 1.172 \times 10^{-3}\,T - 7.995 \times 10^{-7} T^2} \tag{2.13}$$

Suitable scalings for time and position,

$$\tau = t \frac{kR}{36.11PML^2} \qquad X = \frac{x}{L} \tag{2.14}$$

substituted into (2.9) yields this simple form of the equation:

$$\frac{\partial T}{\partial \tau} = f(T) \frac{\partial^2 T}{\partial X^2} \tag{2.15}$$

Now, let's consider conduction across a stagnant CO_2 gas layer of length L where horizontal boundaries are held constant at 400°C and 500°C.

Start up FEMLAB and enter the Model Navigator:

Model Navigator
• Select 1-D dimension
• Select Classical PDEs → Heat Equation
• Element: Lagrange – quadratic
• More >>
• OK

This application mode gives us one dependent variable u, but in a 1-D space with coordinate x. Now we are in a position to set up our domain. Pull down the Draw menu and select Specify Geometry.

Draw Mode
• Name: interval
• Start: 0
• Stop: 1
• Apply
• OK

Now for the boundary conditions. Pull down the Boundary menu and select Boundary Settings.

Boundary Mode
• Select domain 1
• Check Dirichlet and h=1; r=500 (T=500)
• Select domain 2
• Select Dirichlet and h=1; r=400 (T=400)
• Apply/OK

Pull down the options menu and select **Add/Edit** constants. The Add/Edit constants dialog box appears.

Add/Edit Constants

- Name of constant: a1
- Expression: 1.172E-3
- Name of constant: a2
- Expression: 7.995E-7
- Name of constant: F400
- Expression: 421.5
- Apply
- OK

Pull down the Subdomain menu and select Subdomain mode. Before setting the equations, it is useful to define some intermediate variables to make the data entry more concise. Pull down the options menu and select **Add/Edit** expressions.

Add/Edit Expressions

- Variable name: FofT
- Variable type: subdomain
- Add
- Now click on the definition tab.
- Enter expression: (u+273)/(1+a1*u+a2*u^2)/F400
- Apply
- OK

Now pull down the Subdomain menu, select Coefficient View, and select Subdomain settings.

Subdomain Mode

- Select domain 1
- Set f=0; c=1; d_a=1/FofT
- Apply
- Select the init tab; set u(t0)=500-100*x
- Apply
- OK

Click on the mesh triangle on the toolbar, and then refine the mesh twice to 60 elements. Click on the = button on the toolbar to solve (stationary linear solver). The steady state solution is unchanged from the initial condition, since for long enough times, the accumulation term is immaterial. It certainly is for the

stationary solver. So we can only influence the transient solution with a temperature dependent diffusivity. So change the initial condition to u(t0)=400, i.e., the left boundary jumps to u=500 to define time τ=0.
elements."

Now pull down the Solve menu and select the Parameters option. This pops up the Solver Parameters dialog window.

Solver Parameters

- General tab: select time dependent
- Jacobian: numeric
- Time-stepping tab. Take output times 0:0.01:0.2
- Solve
- Cancel / OK

Figures 2.4 and 2.5 show the rate of advance of the diffusive front.

In particular, since diffusivity increases with temperature, we find that the profile reaches steady state more rapidly than with constant diffusivity. The self-similarity with $\eta = \dfrac{x}{\sqrt{Dt}}$ is not apparent in Figure 2.4, as the higher temperatures home in on the steady-state linear profile faster than the lower temperatures. Figure 2.5 shows the rise in temperature to the steady state value at the midpoint of the domain, which has the expected s-shape, but again rises faster than expected at short times.

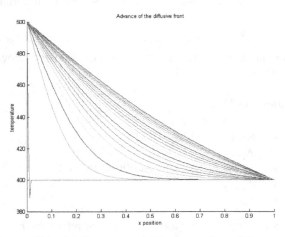

Figure 2.4 Temperature profiles from τ=0 to τ=0.2.

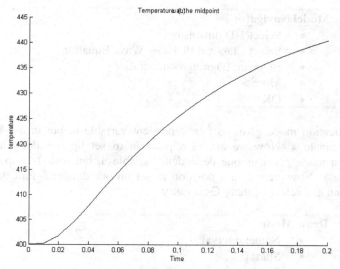

Figure 2.5 Temperature at position x=0.5.

Exercise 2.2

Solve for the same plots as Figure 2.4 and 2.5 with constant diffusivity f(T)=1. How do the profiles differ?

2.1.3 The wave equation: A hyperbolic PDE

The 1-D wave equation (2.3) has also been studied to death. Nor does it particularly turn up in chemical engineering applications. The obvious place is the study of sound waves, which receives little attention in the chemical engineering curriculum. So the major reason for including it here is completeness. Does this mean that the wave equation is unimportant in the chemical and process industries? Probably not. For instance, reactors are known to exhibit chemical waves, waves on interfaces in condensers, swirl atomizers, and distillation columns effect mass transfer, and acoustics, power ultrasound, and sonochemistry are receiving much attention on the research front. It just so happens that chemical engineers are taught little about waves, and thus it is difficult to find classical textbook analyses of chemical engineering unit operations in which waves play any role.

In this subsection, we will attempt to make the demonstration of wave dynamics slightly more interesting by the use of periodic boundary conditions and animation. Start up FEMLAB and enter the Model Navigator:

Model Navigator

- Select 1-D dimension
- Select Classical PDEs → Wave Equation
- Element: Lagrange - quadratic
- More >>
- OK

This application mode gives us one dependent variable u, but in a 1-D space with coordinate x. Now we are in a position to set up our domain. This application mode gives us one dependent variable u, but in a 1-D space with coordinate x. Now we are in a position to set up our domain. Pull down the Draw menu and select Specify Geometry.

Draw Mode

- Name: interval
- Start: 0
- Stop: 1
- Apply / OK

Pull down the Mesh menu and select Parameters. By creating symmetry boundaries, the endpoints become equivalent.

Mesh Mode

- Symmetry boundaries: 1 2
- Apply
- Remesh
- Apply
- OK

Now refine the mesh three times (120 elements). Now for the boundary conditions. Pull down the Boundary menu and select Boundary Settings.

Boundary Mode

- Select domain 1 and 2 (hold down CTRL key)
- Check Dirichlet
- Apply
- OK

Pull down the Subdomain menu and select Subdomain settings.

Subdomain Mode
- Select domain 1
- Set f=0; c=1; d_a=1
- Apply
- Select the init tab; set u(t0)= sech((x-0.5)*10)^2
- Set u_t(t0)= x*(1-x)/10.
- Apply
- OK

Now pull down the Solve menu and select the Parameters option. This pops up the Solver Parameters dialog window.

Solver Parameters
- General tab: select time dependent
- Jacobian: numeric
- Time-stepping tab. Take output times 0:0.05:2
- Solve
- Cancel
- OK

Now generate an animation from the Post Menu, Plot Parameters: Animate Tab. The final state at t=2 should look like Figure 2.6.

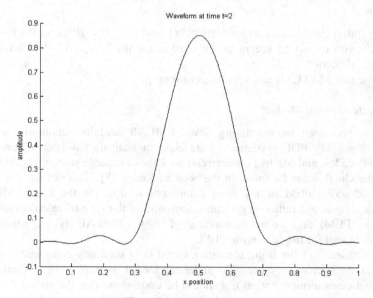

Figure 2.6 Waveform at t=2.

Note that at all times, these interesting wave dynamics remain "fixed" at the boundaries, since the Dirichlet boundary condition ensures zero boundary amplitude. So how do we combat this? Let's establish periodic boundary conditions, which are effected by a little knowledge of how FEMLAB keeps its books. When two boundaries are equivalent, FEMLAB adds the Dirichlet conditions on both boundaries to the constraint:

$$h_1 u(0) + h_2 u(0) = r_1 + r_2 \qquad (2.16)$$

So the modest change to implement periodic boundaries is to set $h_2 = -1$.

Pull down the **Boundary** menu and select **Boundary Settings**.

Boundary Mode
• Select domain 2
• Check Dirichlet h=-1; r=0
• Apply
• OK

Repeat the solution procedure. How does the final state (t=2) compare with the state without periodic boundary conditions (Dirichlet)? Did you notice any difference in the wave dynamics during the animation sequence?

Exercise 2.3

Try the initial conditions u(t0)=sin(10*pi*x) and u_t(t0)= -10*pi*cos(10*pi*x). What do you expect to see in the animation for u(t)? u_t(t)? Did anything unexpected occur?
  Note that MATLAB has a built-in constant pi.

The Finite Element Method

By now, you must be wondering how FEMLAB actually accomplishes this magic of solving PDE systems. Finite element analysis has been around for several decades, and has had commercial packages available since the 1980s. A good introduction can be found in the book of Reddy [3]. It is not the intention here to describe FEM in any great detail, nor to describe the full FEMLAB implementation, but rather to give an impression of the type of calculations that occur in FEM, and an understanding of why FEMLAB is a particularly convenient tool for implementing FEM.

  The essence of the finite element method is to state any constraints on the field variables in *weak form*. To understand what a weak form is (and why mathematicians termed it weak), it should be understood that the *strong form* of a system of constraints is the partial differential equation system and appropriate boundary conditions. Why is it strong? Because the field variables are required to be continuous and have continuous partial derivatives up through the order of the equation. That is a strong requirement. The weak form places a weaker restriction on the functions that could satisfy the constraints – discontinuities must be integrable.

To see the equivalence between a PDE and its weak form, consider a stationary PDE for a single dependent variable u in three spatial dimensions in a domain Ω, e.g. the general form:

$$\nabla \cdot \Gamma(u) = F(u) \qquad (2.17)$$

Let's suppose that v, called a test function, is any arbitrary function defined on the domain Ω and restricted to a class of functions $v \in V$. Multiplying (2.17) by v and integrating over the domain results in

$$\int_{\Omega} v\nabla \cdot \Gamma(u)dx = \int_{\Omega} vF(u)dx \qquad (2.18)$$

where dx is the volume element. Upon applying the divergence theorem, we achieve

$$\int_{\partial\Omega} v\Gamma \cdot ndS - \int_{\Omega} \nabla v \cdot \Gamma(u)dx = \int_{\Omega} vF(u)dx \qquad (2.19)$$

When the PDE is constrained by Neumann boundary conditions, the boundary term on (2.19) vanishes. This is one of the reasons that FEM have Neumann natural boundary conditions. Recall in Chapter 1 we observed that finite difference methods have natural Dirichlet conditions. This results in the condition on the volume integral:

$$\int_{\Omega} \left[vF(u) - \nabla v \cdot \Gamma(u) \right] dx = 0 \qquad (2.20)$$

This must hold for every $v \in V$. Now for the magic: finite elements and basis functions. Let's suppose that u is decomposed onto a series of basis functions:

$$u(x) = \sum_{i} u_i \phi_i(x) \qquad (2.21)$$

For instance, if the ϕ_i are sines and cosines with the fundamental and progressive harmonics, then (2.21) is a Fourier series. Instead, in FEM, the basis functions are chosen to be functions that only have support within a single element, i.e. they are zero in every element but one.

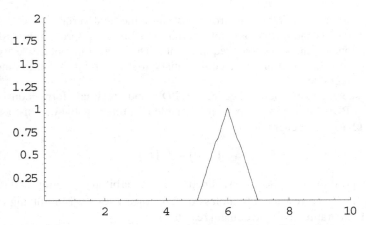

Figure 2.7 Two piecewise linear basis functions in 1-D on adjacent elements.

Figure 2.7 gives an example of two Lagrange linear basis functions in 1-D. Clearly, any function u(x) can be approximated to arbitrary accuracy with piecewise linear basis functions and sufficiently small elements.   The basis functions can be taken to be higher order, in which case more than one unknown u_i is needed per element. So the number of unknowns rises with the order of the basis functions.   The number of basis function rises with order as well.   For Lagrange linear basis functions, the representation of the function in any element is through two basis functions:

$$\phi = \xi \qquad \phi = 1 - \xi \tag{2.22}$$

where ξ is the local coordinate in the element. So for N elements, there are 2N basis functions ϕ_i. Lagrange quadratic elements have three basis functions:

$$\phi = (1 - \xi)(1 - 2\xi) \qquad \phi = 4\xi(1 - \xi) \qquad \phi = \xi(2\xi - 1) \tag{2.23}$$

Thus, for Lagrange quadratic elements there are 3N basis functions for N elements.

Recall that (2.20) must be satisfied for all $v \in V$, which we now take to be the function space of all functions that are linear combinations of the basis functions ϕ_i, i.e.

$$v(x) = \sum_i v_i \phi_i(x) \tag{2.24}$$

But because v enters (2.20) linearly, it suffices to show that if (2.20) is satisfied for each of the basis functions ϕ_i playing the role of v, then it is satisfied for all linear combinations of the basis functions (2.24), and thus for all $v \in V$.

Thus, the condition (2.24) is equivalent to a system of (k+1)N equations [(2.20) for each ϕ_i] in (k+1)N unknowns (the u_i), where k is the order of the element (k=1 linear, k=2 quadratic, etc.).

Then there is the reason why FEM with FEMLAB has such utility. FEMLAB automates the assembly of the (k+1)N equations (2.20). First, we note that Γ(u) and F(u) are general, potentially nonlinear, functions of u. So, in general, the solution is not achievable in closed form. In Chapter One, we showed that FEMLAB has a built-in nonlinear solver for 0-D problems, i.e. f(u)=0, where u was a single unknown value. The nonlinear solver was based on Newton's Method. The N-dimensional analogue of Newton's Method for the vector equation

$$L(U) = 0 \qquad (2.25)$$

where U is the vector of unknowns u_i and $L(U)$ is the system of equations found by substituting the basis functions ϕ_i for v in (2.20), is

$$K(U_0)(U - U_0) = L(U_0) \qquad (2.26)$$

where $K(U_0)$ is called the stiffness matrix and $L(U_0)$ is called the load vector. The stiffness matrix is the negative Jacobian of L:

$$K(U_0) = -\frac{\partial L}{\partial U}(U_0) \qquad (2.27)$$

So (2.26) is now a linear equation for U given the previous approximate solution U_0. Thus, if U_0 were close enough to a solution, the linear equation (2.26) should find an improved approximate solution U, and this procedure can be iterated until a solution is found to acceptable accuracy. Clearly, the nonlinear solver by Newton's Method is central to FEMLAB's PDE solver. Yet FEMLAB automates all of the steps involved in generating the finite element analysis of a PDE. It symbolically forms the Jacobian of the nonlinear operator $L(U)$ if it can. If it cannot, it numerically assembles the Jacobian. If the PDE were itself linear, this is not too cumbersome. Yet assembling the stiffness matrix is a Herculean task – it was common that the finite element analysis, both meshing the elements and assembling the stiffness matrix was the central feature of many doctoral studies in the sciences and engineering not too long ago. For new combinations of PDEs, or even variations on the coefficients (quasi-linear rather than constant, for instance, as in §2.1.2), the bookkeeping to organize the assembly of the stiffness matrix is a daunting task. Furthermore, the sparse solver methods for (2.26) and time-integration required substantial programming effort to coordinate for a single problem. Yet FEMLAB has done it as a set of subroutines (MATLAB functions) that coordinate multiple PDE systems (application modes) seamlessly.

Before moving in to explain the implementation of boundary conditions we shall work through a simple ODE to highlight the concepts we discussed so far. Following the worked example should guide you through the weak formulation and finding approximate numerical solutions to a given second order ODE.

Exercise 2.4: A worked example of finite element calculations in detail

To elaborate the concepts described above a simple ODE is solved using the variational principles that forms the core to FEM. The problem is simple. Solve the second order ODE

$$\frac{d^2u}{dx^2} + 4u = 8x^2 \tag{2.28}$$

subjected to boundary conditions

$$u(0) = u(\pi/4) = 0$$

using the weak formulation.

This simple second order ODE has an analytic solution (Prove it!)

$$u(x) = \cos(2x) + \left(1 - \frac{\pi^2}{8}\right)\sin(2x) + 2x^2 - 1. \tag{2.29}$$

Since we know the analytic solution, a comparison would give the error of the approximate solution we find.

The first step for the weak formulation is to assume a weight function and a trial function. Take $U(x)$ as the trial function and $\phi(x)$ as the weight function. We discuss the exact forms of these functions later. The trial function $U(x)$ forms a solution to the ODE. Therefore, if we substitute $U(x)$ in (2.28), the resulting equation gives the residual:

$$R = \left|\frac{d^2U}{dx^2} + 4U - 8x^2\right| \tag{2.30}$$

Subsequent steps really amount to the minimization of this residual. The minimization process starts by evaluating the *weighted residual*. To evaluate the weighted residual, multiply (2.30) by $\phi(x)$ and integrate over the domain (i.e. $0 \le x \le \pi/4$).

$$R(x) = \int_0^{\pi/4}\left(\phi\frac{d^2U}{dx^2} + 4\phi U - 8\phi x^2\right)dx \tag{2.31}$$

Using integration by parts, one can simplify above equation to obtain

$$R(x) = \int_0^{\pi/4} \left(4\phi U - \frac{d\phi}{dx} \frac{dU}{dx} \right) dx - 8 \int_0^{\pi/4} \phi x^2 dx \qquad (2.32)$$

To advance further we need to make some crucial assumptions. Since we are free to assign any function to $U(x)$ and $\phi(x)$ as far as they agree with the boundary conditions, we assume $U(x) = \phi(x)$. This is known as Galerkin's method. If $U(x) \neq \phi(x)$, then it gives the Rayleigh-Ritz formulation. We have to select an algebraic function of x to satisfy the boundary conditions $u(0) = u(\pi/4) = 0$.

$$\varphi(x) = U(x) = \phi(x) = c_1\varphi_1 + c_2\varphi_2 + \cdots + c_N\varphi_N = \sum_{i=0}^{N} c_i\varphi_i \qquad (2.33)$$

We assume N functions as follows.

$$\varphi_1 = x(\frac{\pi}{4} - x) \qquad \varphi_2 = x^2(\frac{\pi}{4} - x) \qquad \cdots \qquad \varphi_N = x^N(\frac{\pi}{4} - x)$$

Therefore

$$\varphi(x) = \sum_{i=1}^{N} c_i x^i (\frac{\pi}{4} - x) \qquad (2.34)$$

This selection satisfies the boundary conditions regardless the number of terms included in the series. Since $U(x) = \phi(x)$, the weighted residual becomes

$$R(x) = \int_0^{\pi/4} \frac{1}{2} \left(4\varphi^2 - \left(\frac{d\varphi}{dx} \right)^2 \right) dx - 8 \int_0^{\pi/4} \phi x^2 dx \qquad (2.35)$$

By substituting (2.34) in to (2.35) and evaluating the integral we obtain an expression for R independent of x. However, that expression contains N unknowns, c_i. We have to evaluate values of the c_i so that the weighted residual is minimum.

From (2.34)

$$\frac{d\varphi}{dx} = \sum c_i \left(\frac{\pi}{4} i x^{i-1} - (i+1)x^i \right) \qquad (2.36)$$

Therefore we have

$$
R(c_i) = 2\sum c_i c_j \left(\frac{\pi}{4}\right)^{i+j+3} \left(\frac{1}{i+j+1} - \frac{2}{i+j+2} + \frac{1}{i+j+3}\right)
$$
$$
- \frac{1}{2}\sum c_j c_j \left(\frac{\pi}{4}\right)^{i+j+1} \left(\frac{ij}{i+j-1} - \frac{2ij+i+j}{i+j} + \frac{(i+1)(j+1)}{i+j+1}\right)
$$
$$
- 8\sum c_i \left(\frac{\pi}{4}\right)^{i+4} \left(\frac{1}{i+3} - \frac{1}{i+4}\right) \qquad i, j = 1,2,3,\cdots, N
$$

(2.37)

Then we minimize the residual by taking derivatives of R w.r.t c_i. For predetermined number N, this results in N algebraic equations that have to be solved simultaneously.

$$
\frac{dR(c_i)}{dc_i} = 0
$$

(2.38)

For $N = 2$ there are only two unknowns; c_1 and c_2. It produces two linear equations.

$$
0.122c_1 + 0.048c_2 = -0.120
$$
$$
0.048c_1 + 0.033c_2 = -0.063
$$

(2.39a,b)

In matrix form

$$
\begin{bmatrix} 0.122 & 0.048 \\ 0.048 & 0.033 \end{bmatrix} \begin{Bmatrix} c_1 \\ c_2 \end{Bmatrix} = \begin{Bmatrix} -0.120 \\ -0.063 \end{Bmatrix}
$$

(2.40)

It resembles the general form

$$
[K]\{x\} = \{L\}
$$

(2.41)

where K is the Jacobian (stiffness matrix), and x is the vector of unknowns. L is the forcing vector (load vector).

Solution to (2.40) gives us

$$
c_1 = -0.554579
$$
$$
c_2 = -1.112560
$$

Therefore the solution to (2.28) is

$$u(x) = -0.554579x(\frac{\pi}{4} - x) - 1.11256x^2(\frac{\pi}{4} - x) \qquad (2.42)$$

If we go one step further by assuming $N = 3$, then we get 3 algebraic equations with three unknowns; c_1, c_2 and c_3. The resulting matrix equation is

$$\begin{bmatrix} 0.1216 & 0.0477 & 0.0228 \\ 0.0477 & 0.0328 & 0.0200 \\ 0.0228 & 0.0200 & 0.0139 \end{bmatrix} \begin{Bmatrix} c_1 \\ c_2 \\ c_3 \end{Bmatrix} = \begin{Bmatrix} -0.120 \\ -0.630 \\ -0.351 \end{Bmatrix} \qquad (2.43)$$

The solution for (2.43) is

$$c_1 = -0.588 \qquad c_2 = -0.838 \qquad c_3 = -0.349$$

Therefore the new solution for (2.28) becomes

$$u(x) = -0.588x(\frac{\pi}{4} - x) - 0.838x^2(\frac{\pi}{4} - x) - 0.349x^3(\frac{\pi}{4} - x) \qquad (2.44)$$

Figure 2.8 shows the plots of (2.42) and (2.44) together with the analytic solution (2.29). As we can clearly see, the approximate algebraic solutions can achieve good agreement with the exact solution if more terms of the series are included.

If you worked the example, by now you have a clear idea of the weak formulation of a solution. In next section we discuss the implementation of the boundary conditions in FEMLAB.

2.1.4 Boundary conditions

As described for the canonical case above, one should note that the stiffness matrix K is equivalent to Neumann boundary conditions. As we saw in Chapter One, pure Neumann conditions lead to a singular stiffness matrix, which FEMLAB could not directly treat, since it resulted in the addition of an arbitrary and large constant to the solution found by projection methods on to the eigensystem of the stiffness matrix. One of the vagaries of FEM is the treatment of boundary conditions.

We could propose to treat boundary conditions much as is done with finite difference methods. The appropriate lines of the matrix equation are replaced by direct constraints on the unknowns u_l so that the order of the matrix is preserved. This has the unpleasant effect of breaking the sparsity of the stiffness

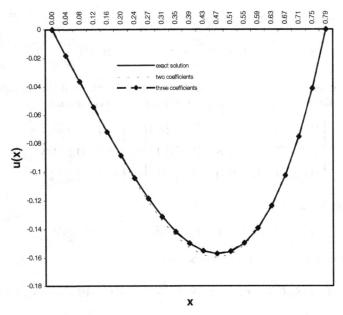

Figure 2.8 Plot of solutions to (2.4.1). Algebraic function with 3 components of the series is in good agreement with the analytic solution.

matrix with regard to some boundary conditions, and thus artificially requiring full matrix solvers that are much less accurate and inefficient by comparison to sparse solvers for the same matrix equation. FEM has an elegant solution using Lagrange multipliers – a well known method for dealing with equality constraints in optimization problems.   Suppose in addition to the PDE constraints, we have a series of boundary conditions that are to be satisfied in weak form for all $v \in V$. By applying the basis function expansion and writing the boundary integrals for each basis function, by analogy to the PDE constraints (2.25), we arrive at a vector equation for the boundary constraints:

$$M(U) = 0 \qquad (2.45)$$

This constraint residual equation, as it is known, need not be N equations. Usually it is just a handful of equations in N unknowns, as not all basis vectors taken as test functions v contribute a boundary constraint. The linearized version of (2.45) reads similarly to (2.26):

$$N(U_0)(U - U_0) = M(U_0) \qquad (2.46)$$

where N is the negative Jacobian of M:

$$N(U_0) = -\frac{\partial M}{\partial U}(U_0) \qquad (2.47)$$

Now for the clever trick. The stiffness matrix equation is augmented by a vector of unknowns Λ, called the Lagrange multipliers, which multiply N^T, where the superscript T means transpose:

$$K(U_0)(U - U_0) + N(U_0)^T \Lambda = L(U_0) \qquad (2.48)$$

Why is this clever? Well, if the constraint (2.45) is satisfied, then there is a unique set of Lagrange multipliers satisfying (2.48). (2.46) and (2.48) permit the simultaneous solution of more than just boundary conditions, however. Any constraint – internal pointwise, subdomain integral, edge or boundary that can be expressed in weak form can be treated by the Lagrange multiplier method.

Lagrange Multipliers

So how do Lagrange multipliers ensure that $M(U)=0$ is satisfied? By a variational principle. With $\Lambda=0$ (2.48) is equivalent to the minimum principle for

$$\min_U \left\{ \frac{1}{2} U^T K U - U^T L \right\} \qquad (2.49)$$

If we wish to ensure the constraint (2.45) is satisfied simultaneously, then we add a weighted penalty to (2.49) for the extent to which $M(U)=0$ is not satisfied. The weights are called Lagrange multiplers Λ.

$$\min_U \left\{ \frac{1}{2} U^T K U - U^T L + \Lambda \cdot M \right\} \qquad (2.50)$$

Now we use the linearization of $M(U)=M(U_0)+N(U_0)(U-U_0)$ to simplify (2.50). Note that constant terms do not contribute to the minimization.

$$\min_U \left\{ \frac{1}{2} U^T K U - U^T L + \Lambda \cdot NU \right\} \qquad (2.51)$$

The minimization (2.51) is then equivalent to (2.48) , i.e. the solution to (2.48) minimizes (2.51). In the parlance of FEM, (2.51) is the "minimization in the energy", i.e. weighted by the stiffness matrix K. It should not be confused with the least squares minimization, which by analogy with (2.50) is

$$\min_U \left\{ \frac{1}{2} \|L(U)\|^2 + \Lambda \cdot M \right\} \qquad (2.52)$$

Linearization of L and M leads, after much re-arrangement and neglect of constant terms, to the condition

$$\min_{U}\left\{\frac{1}{2}U^{T}K^{T}K\left(U-U_{0}\right)-U^{T}K^{T}L+U^{T}N^{T}\cdot\Lambda\right\} \qquad (2.53)$$

Thus, the solution U to (2.53) by a theorem in linear algebra, is the solution to the *normal equations* [4]:

$$K^{T}K\left(U-U_{0}\right)+K^{T}\left(K^{T}\right)^{-1}N^{T}\cdot\Lambda=K^{T}L \qquad (2.54)$$

which is the least squares solution to

$$K\left(U-U_{0}\right)+\left(K^{T}\right)^{-1}N^{T}\cdot\Lambda=L \qquad (2.55)$$

So the solution to (2.55) ensures that the constraint (2.45) *M(U)=0* is satisfied in the least squared error sense (2.52), whereas the constraint (2.48) satisfies (2.45) in the sense of lowest energy. FEMLAB uses (2.48) rather than (2.55) for simplicity, rather than (2.55) for greatest accuracy. The distinction is important as the least squared error minimization (2.52) is defined for any general nonlinear operator L(U), but the stiffness energy (2.50) is only sensible for K that is symmetric and positive definite. If this is the case, then the Lagrange multipliers in (2.48) are merely a convenience, not a guarantee that the constraint (2.45) is satisfied in any approximate sense. (2.55) is a stronger condition, yet at the price of extra matrix manipulations. (2.52) is open to the criticism that M(U) is not constrained to be a penalty, so a stronger condition is to explicitly consider each constraint as a penalty individually [5]

$$\min_{U}\left\{\frac{1}{2}\left\|L\left(U\right)\right\|^{2}+\sum_{j}\Lambda_{j}\left\|M_{j}\left(U\right)\right\|^{2}\right\} \qquad (2.56)$$

This technique does not render its solution so succinctly as a matrix equation, as the constraint term involves N,M, K and Λ.

Weak terms

So if you were wondering how we treated the point source of vorticity in (2.1.1), it was by a weak term, which merely evaluated the integral

$$\int_{\Omega}v\delta\left(x\right)dx=v\left.\right|_{x=0} \qquad (2.57)$$

and made the appropriate contributions to the stiffness matrix and load vector in (2.49).

2.1.5 Basic elements

Fundamental to the FEM is the concept that any domain can be implemented as a collection of smaller subdomains of preferred shape. These subdomains are called *finite elements*. Corners of an element are called nodes at which the solutions to field variables are computed. There can be nodes in between corner points that are commonly called *edge nodes*. In FEMLAB, when you generate the mesh, it subdivides the computational domain in to a selected form of elements and form of nodes accordingly. One can find more than a hundred types of elements in use. If you are a beginner, it is natural to be puzzled over the type of elements that should be used and the number of elements to be used.

The discretization process proscribes the type and the number of elements. The number of elements is directly connected with the accuracy of the solution. The higher the number of elements used, the lesser will be the error. However, having a large number of elements would be computationally expensive, demanding a large chunk of RAM and an extended runtime.

Defining an unnecessary number of elements is a very common practice. There is no formula that allows you to choose optimally exact number of elements. It is only by experience that you would be able to decide the right amount of elements to pack in a domain. Though the accuracy increases with the number of elements N, there will be a certain number N_c beyond which the sensitivity of accuracy becomes negligible.

Figure 2.9 shows the normalized error against the number of elements N. The number of elements doubles in each iteration. One can see that the last three points do not make any considerable improvement on the accuracy. However one can perform a few short runs to find out the appropriate number of elements to be used. There are instances where one is interested in a certain region of the domain rather than the whole domain. For instance, take the case of the flow past a cylinder. The boundary layer behavior around the cylinder is to be investigated. In such cases one can and should pack more elements around the cylinder having a lower density of elements in the far field. This way one can attain the accuracy required without increasing the number of elements. FEMLAB allows such grid stretching. Another point worth mentioning at this juncture is the skewness of the elements in stretched grids. Skewed elements make the formation of the Jacobian impossible. Therefore great care should be taken in using stretched meshes.

The type of the elements to be used depends on the problem that has to be solved. The dimensionality of the domain defines the dimensionality of the elements. The most simple is the 1-D element that represents a line segment between two nodes at each end. The most fundamental element in 2-D is a triangle where as in 3-D it is a tetrahedron. Table 2.1 shows some of the basic elements in use with respect to the dimensionality. Though we used straight

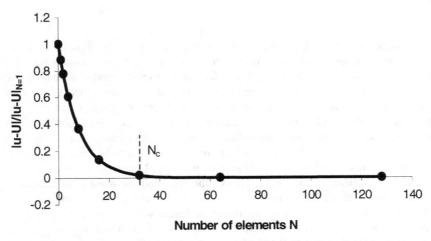

Figure 2.9   Normalized error vs Number of elements. The increase of N results in improving the accuracy but beyond certain number N_c, the effect become negligible.

lines, curvilinear segments between nodes would provide a more general form of the elements.

Though the geometry of the element is an important factor, the element types are categorized according to the interpolation polynomials used with them. According to this categorization there are three types of elements: Simplex, Complex and Multiplex. If the polynomials used have linear terms and constants with nodes at the corners, then the elements are called simplex. Complex elements use higher order polynomials (quadratic, cubic, quintic, etc.) with edge and internal nodes together with corner nodes. Multiplex elements have their sides in parallel with the coordinate axes and use higher order polynomials.

As mentioned above, complex elements use higher order polynomials. The combinations of polynomials and nodal configurations can be determined using the Pascal triangle, Pascal tetrahedron or Pascal hypercubes. Table 2.2 shows the Pascal triangle with polynomials up to fifth order. The polynomials that are selected should be complete: i.e. it should contain all terms up to the highest order. For example $c_1+c_2x+c_3x^2$ is complete while $c_1+c_2x^2$ is not since it does not contain the first order term. For any 2-D element, by taking all the terms above a selected horizontal line, one can easily obtain the complete polynomial up to the required order. In general there should be a node for each term of the polynomial. For example a cubic element should have four nodes along each side. But there can be more complex combinations. Table 2.3 shows the linear, quadratic and cubic elements for a 2-D triangular element.

Dimensionality	Shape
1-D	•———————•
2-D	Triangle　　　　　Rectangle　　　　　Quadrilateral
3-D	Tetrahedral　　Regular Hexahedral　　Irregular Hexahedral

Table 2.1 Basic elements.

$$
\begin{array}{ll}
1 & \text{Constant} \\
x \quad y & \text{Linear} \\
x^2 \quad xy \quad y^2 & \text{Quadratic} \\
x^3 \quad x^2y \quad xy^2 \quad y^3 & \text{Cubic} \\
x^4 \quad xy^3 \quad x^2y^2 \quad x^3y \quad y^4 & \text{Quartic} \\
x^5 \quad x^4y \quad x^3y^2 \quad x^2y^3 \quad xy^4 \quad y^5 & \text{Quintic}
\end{array}
$$

Table 2.2 Pascal Triangle for 2-D elements.

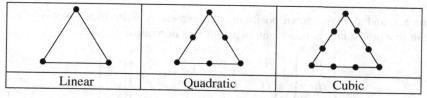

Linear	Quadratic	Cubic

Table 2.3 Types of complex elements with basis functions up to third order.

FEMLAB provides you with two major types of elements irrespective to the geometry, namely Lagrange and Hermite elements. These elements are so named because of the type of the interpolation polynomials used in them. As the name suggests, the Lagrange polynomials are used as the basis functions in Lagrange elements. Suppose a field variable $u(x)$ is expressed using Lagrange polynomials L_n over a 1-D element. Then,

$$\mathbf{u}(x) = L_1(x)u_1 + L_2(x)u_2 + \cdots + L_N(x)u_N \tag{2.58}$$

where u_n are the unknown coefficients. The $L_n(x)$ is given by

$$L_n = \prod_{M=1, M \neq N}^{n} \frac{x - x_M}{x_N - x_M} \tag{2.59a}$$

$$= \frac{(x - x_1)(x - x_2) \cdots (x - x_{N-1})(x - x_{N+1}) \cdots (x - x_n)}{(x_N - x_1)(x_N - x_2) \cdots (x_N - x_{N-1})(x_N - x_{N+1}) \cdots (x_N - x_n)} \tag{2.59b}$$

The expansion generates the polynomials of desired order. Lagrange elements are the most commonly used type in CFD. They provide the value of the variable at nodes.

Hermite elements use the Hermite polynomials to interpolate the values of the field variables. The main difference between Lagrange and Hermite elements is the degrees of freedom (DOF) available. In the case of Lagrange elements DOF are the values of the function at nodes (This consists values of all variables at the node). However in Hermite elements other than the function values at nodes, the first derivatives of the variables at corner points are available. Again, suppose a variable u(x) to be determine over 1-D elements. Since the values at n th and $n+1$th nodes, $u(x_n)$ and $u(x_{n+1})$ and the derivatives $du/dx\big|_i$ and $du/dx\big|_{i+1}$ are known a polynomial of four unknowns should be used to approximate $u(x)$.

$$u(x) = \alpha_1 + \alpha_2 x + \alpha_3 x^2 + \alpha_3 x^3 \tag{2.60}$$

Since x_i and x_{i+1} are known positions, one can easily write four equations: two with function values at nodes and two with first derivatives at nodes.

$$\begin{Bmatrix} u_i \\ du_i \\ u_{i+1} \\ du_{i+1} \end{Bmatrix} = \begin{bmatrix} 1 & x_i & x_i^2 & x_i^3 \\ 0 & 1 & 2x_i & 3x_i^2 \\ 1 & x_{i+1} & x_{i+1}^2 & x_i^3 \\ 0 & 1 & 2x_{i+1} & 3x_{i+1}^2 \end{bmatrix} \begin{Bmatrix} \alpha_1 \\ \alpha_2 \\ \alpha_3 \\ \alpha_4 \end{Bmatrix} \tag{2.61}$$

where du denotes the derivative of u w.r.t. x at the nodes. By matrix inversion α_n can be expressed in terms of nodal values and values of derivatives. The resulting equation is

$$u(x) = u_i \varphi_1(x) + du_i \varphi_2(x) + u_{i+1} \varphi_3(x) + du_{i+1} \varphi_4(x) \tag{2.62}$$

where $\varphi_i(x)$ are the Hermite interpolation functions (cubic functions in this case, also known as cubic splines).

If you observed closely, you could see that the cubic interpolation function in Lagrange elements needs four nodes whereas in Hermite elements only two nodes are required. Hermite elements are commonly used in solving load and stress distributions of trusses.

Further to these two types, FEMLAB offers curved mesh elements and Argyris elements. The curved mesh elements are provided to facilitate the approximation of true boundaries with higher accuracy. The Argyris elements are fifth order Hermite elements using nodal values as well as derivatives up to second order. It also uses the normal components of ∇u at the midpoints of the sides. Argyris elements require determination of 21 constants of a quintic polynomial. In addition to predefined elements, FEMLAB allows user defined elements. We advise interested readers to consult FEMLAB manuals for detailed description on how to define a new class of elements.

Here we provided a sufficient description of basic elements to make a beginner comfortable with the jargon and using the elements with some understanding. With FEMLAB, one does not need to become an expert in meshing techniques and development of elements. As you have already seen, once you generate the domain over which the differential equations are to be solved, meshing is just a click of a button away. However, if a reader is interested in understanding the basic concepts, development of elements and meshing techniques, refer to [6] and references there in.

Exercise:

To explain the formulation and solution methods involved in FEM, a worked example is considered. In 1.5.1, heat transfer in a nonuniform medium was treated. Here we consider similar problem with more simplifications. Instead of a variable heat transfer coefficient, we consider a constant coefficient and consider $f(x)$ to be uniform over the length of the domain. The length of the domain L=1. Figure 2.10 shows the physical system. We assume the heat flow across any cross section (shown as A in Fig. 2.10) normal to the centre axis to be uniform. Therefore the temperature varies only along the axis: hence reduced dimensionality.

The heat transfer in this problem is fully described by the equation (1.29). With the uniform cross section A=1, and constant source $Q = f(x)$ in (1.29). The equation becomes:

$$\frac{d}{dx}\left(k\frac{dT}{dx}\right) + Q = 0; \quad 0 \le x \le 1 \qquad (2.63)$$

The data for the problem are:

$k = 3.3\text{J/°Cm s}$

$Q = 10\text{J/s m}$

$T|_{x=0} = 1$

$q|_{x=1} = 1.25\text{J/m}^2 \text{ s}$

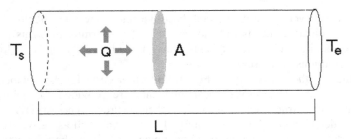

Figure 2.10 Axial heat transfer along an insulated rod. Each end has temperatures Ts and Te. The length of the rod is L and cross sectional area A=1. Heat is generated within the rod at a constant rate of Q.

Step 1: Variational Formulation

This PDE is the strong form of the equation for heat conduction within a cylinder. The first step in FEM is to derive the weak form of the equations. To derive the weak form, equation (2.63) is multiplied by a weight function and integrated over the domain.

$$\int_0^1 w\left[\frac{d}{dx}\left(k\frac{dT}{dx}\right)+Q\right]dx=0 \qquad (2.64)$$

Integrating by parts (using the divergence theorem in 1-D) we obtain

$$\int_0^1\left(\frac{dw}{dk}k\frac{dT}{dx}\right)dx=\left[wk\frac{dT}{dx}\right]_0^1+\int_0^1 wQdx \qquad (2.65)$$

From heat transfer theory, Fourier's law gives the heat flux across a unit cross section is given by Fourier's law $q=-k\dfrac{dT}{dx}$. Therefore,

$$\int_0^1\left(\frac{dw}{dk}k\frac{dT}{dx}\right)dx=-[wq]_0^1+\int_0^1 wQdx \qquad (2.66)$$

From earlier sections, we know that the polynomial basis functions have to be used to approximate the unknowns w and T. Selection of these polynomials is the second step of the FEM procedure.

Step 2: Discretization and Choice of Polynomials

It is obvious that we are going to use 1-D elements. We can have simplex elements for simplicity i.e. linear polynomials to approximate the unknowns.

Figure 2.11 shows the descretization of the cylinder. It uses nodes to divide the length into equal segments. The higher the number of segments, the higher will be the accuracy. However in this example, to demonstrate the FEM formulation we use fewer segments with nodes at each end. The polynomial defined piecewise varies linearly between any two nodes. In general, suppose the temperature $T(x)$ is approximated as

$$T = a + bx \tag{2.67}$$

Now, suppose we have divided the length of the cylinder to N-1 elements with N nodes as shown in Figure 2.11 (d). A random element extending from x_i to x_j is considered. The temperatures at nodes are assumed to be T_i and T_j. From (2.67) we can write the temperatures at nodes i and j.

$$T_i = a + bx_i \tag{2.68a}$$

$$T_j = a + bx_j \tag{2.68b}$$

Solving for a and b gives

$$a = \frac{1}{l}\left(T_i x_j - T_j x_i\right) \tag{2.69a}$$

$$b = \frac{1}{l}\left(T_j - T_i\right) \tag{2.69b}$$

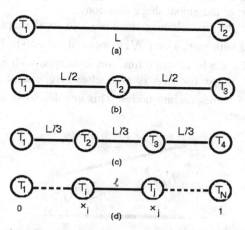

Figure 2.11 Discretization of the cylinder. (a) is the schematic representation of the cylinder. (b) and (c) shows two and three element discretization. In calculation we use (c). (d) shows the general case where the cylinder is divided to N elements. (c) is used to derive the general form of shape functions.

where $l = x_j - x_i$ = length of the element. Since we know a and b (2.67) can be rewritten as

$$T^e = \frac{1}{l}(x_j - x)T_i + \frac{1}{l}(x - x_i)T_j \tag{2.70}$$

The equation (2.70) is the linear approximation function for the element. It describes the temperature variation at any point within the element (hence the notation T^e). Instead of a and b, we now have temperature values at the nodes T_i and T_j as unknowns.

Let $N_i = \frac{1}{l}(x_j - x)$ and $N_j = \frac{1}{l}(x - x_i)$. Then (2.70) can be rewritten as

$$T^e = N_i T_i + N_j T_j \tag{2.71}$$

N_i and N_j are known as the shape functions.

$$N_i = 1 \quad \text{at} \quad x = x_i \quad \text{and} \quad N_i = 0 \quad \text{at} \quad x = x_j$$
$$N_j = 1 \quad \text{at} \quad x = x_j \quad \text{and} \quad N_j = 0 \quad \text{at} \quad x = x_i$$

The temperature distribution along the element is determined by these two functions and end values. Figure 2.12 shows the profiles of N_i, N_j and the resulting temperature T^e. One can generate T^e's for all elements. These element shape functions can be used to formulate the global shape functions. Figure 2.13 shows the definition of the global shape functions.

If we consider the first element there are two local shape functions: N_1^1 which is associated with node 1 and N_2^1 associated with node 2. For the second element again we have a local shape function associated with node 2 defined as N_2^2. Each global shape function is zero elsewhere except in the elements associated with the corresponding nodes. This enables us to define the global temperature variation.

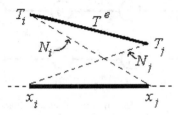

Figure 2.12 Profiles of shape functions N_i and N_j and temperature profile T^e along the element constructed using shape functions.

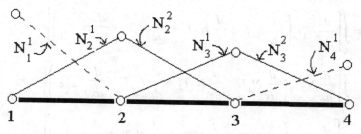

Figure 2.13 Global shape functions.

$$T = \sum T^e = N_1 T_1 + \cdots + N_i T_i + \cdots N_N T_N = \sum_{j=1}^{N} N_j T_j \qquad (2.72)$$

This completes the specification of the basis functions. Now we can return to the variational formulation again.

Step3: Assembling the Element Equations to Form the Global Problem

In step 2 we derived the approximation function for T. Galerkin's formulation assumes the weight function to be same as the approximation for the unknown variables. Therefore we have $w = T$. With this, we can substitute T and w in (2.66).

$$\underbrace{\int_0^1 \left[k \frac{d}{dx} \left(\sum N_i T_i \right) \frac{d}{dx} \left(\sum N_j T_j \right) \right] dx}_{I_k} = \underbrace{-\left[q \sum N_i T_i \right]_0^1}_{I_b} + \underbrace{\int_0^1 q \sum N_i T_i dx}_{I_s} \quad (2.73)$$

For clarity we consider terms of (2.73) separately. The equation (2.73) actually gives the error of the approximate solution (refer to exercise 2.4). To minimise the error, equation (2.73) should be differentiated w.r.t T_i which are the coefficients of the polynomials. The minimization process converts I_k into the stiffness matrix **K**. For four nodes (i.e. three elements as in figure 2.11c) we can expand I_k as below. In what follows we indicate the limits as x_i and x_j. This is to reduce the complications that arising in evaluating the terms. The intervals in the integrals depend on global shape functions.

$$I_k(T) = \int_{x_i}^{x_j}\left[\frac{d}{dx}N_1T_1\left(\frac{d}{dx}N_1T_1 + \frac{d}{dx}N_2T_2 + \frac{d}{dx}N_3T_3 + \frac{d}{dx}N_4T_4\right)\right]dx +$$

$$\int_{x_i}^{x_j}\left[\frac{d}{dx}N_2T_2\left(\frac{d}{dx}N_1T_1 + \frac{d}{dx}N_2T_2 + \frac{d}{dx}N_3T_3 + \frac{d}{dx}N_4T_4\right)\right]dx +$$

$$\int_{x_i}^{x_j}\left[\frac{d}{dx}N_3T_3\left(\frac{d}{dx}N_1T_1 + \frac{d}{dx}N_2T_2 + \frac{d}{dx}N_3T_3 + \frac{d}{dx}N_4T_4\right)\right]dx +$$

$$\int_{x_i}^{x_j}\left[\frac{d}{dx}N_4T_4\left(\frac{d}{dx}N_1T_1 + \frac{d}{dx}N_2T_2 + \frac{d}{dx}N_3T_3 + \frac{d}{dx}N_4T_4\right)\right]dx$$

(2.74)

In the minimization process we differentiate $I_k(T)$ w.r.t. each T_i and set each derivative to zero. This procedure generates a number of equations equal to number of nodes. For instance the differentiation of (2.74) gives

$$\frac{d}{dT_1}I_k(T) = \int_{x_i}^{x_j}\frac{dN_1}{dx}\frac{dN_1}{dx}T_1dx + \int_{x_i}^{x_j}\frac{dN_1}{dx}\frac{dN_2}{dx}T_2dx$$

$$+ \int_{x_i}^{x_j}\frac{dN_1}{dx}\frac{dN_3}{dx}T_3dx + \int_{x_i}^{x_j}\frac{dN_1}{dx}\frac{dN_4}{dx}T_4dx$$

Likewise, there will be three more equations. In matrix form it gives the stiffness matrix.

$$\mathbf{K} = \begin{bmatrix} \int_{x_i}^{x_j}k\frac{dN_1}{dx}\frac{dN_1}{dx}dx & \int_{x_i}^{x_j}k\frac{dN_1}{dx}\frac{dN_2}{dx}dx & \int_{x_i}^{x_j}k\frac{dN_1}{dx}\frac{dN_3}{dx}dx & \int_{x_i}^{x_j}k\frac{dN_1}{dx}\frac{dN_4}{dx}dx \\ \int_{x_i}^{x_j}k\frac{dN_2}{dx}\frac{dN_1}{dx}dx & \int_{x_i}^{x_j}k\frac{dN_2}{dx}\frac{dN_2}{dx}dx & \int_{x_i}^{x_j}k\frac{dN_2}{dx}\frac{dN_3}{dx}dx & \int_{x_i}^{x_j}k\frac{dN_2}{dx}\frac{dN_4}{dx}dx \\ \int_{x_i}^{x_j}k\frac{dN_3}{dx}\frac{dN_1}{dx}dx & \int_{x_i}^{x_j}k\frac{dN_3}{dx}\frac{dN_2}{dx}dx & \int_{x_i}^{x_j}k\frac{dN_3}{dx}\frac{dN_3}{dx}dx & \int_{x_i}^{x_j}k\frac{dN_3}{dx}\frac{dN_4}{dx}dx \\ \int_{x_i}^{x_j}k\frac{dN_4}{dx}\frac{dN_1}{dx}dx & \int_{x_i}^{x_j}k\frac{dN_4}{dx}\frac{dN_2}{dx}dx & \int_{x_i}^{x_j}k\frac{dN_4}{dx}\frac{dN_3}{dx}dx & \int_{x_i}^{x_j}k\frac{dN_4}{dx}\frac{dN_4}{dx}dx \end{bmatrix}$$

(2.75)

K is a symmetric matrix and Galerkin's method forces this symmetry. $\mathbf{I}_b$ and $\mathbf{I}_s$ in (2.73) give rise to two 1×4 matrices:

$$\mathbf{f}_b = -\begin{bmatrix} qN_1\big|_0^1 \\ qN_2\big|_0^1 \\ qN_3\big|_0^1 \\ qN_4\big|_0^1 \end{bmatrix} \text{ and } \mathbf{f}_s = \begin{bmatrix} \int_0^1 QN_1 dx \\ \int_0^1 QN_2 dx \\ \int_0^1 QN_3 dx \\ \int_0^1 QN_4 dx \end{bmatrix} \tag{2.76}$$

The compact equation is $[\mathbf{K}]\{\mathbf{x}\} = \{\mathbf{L}\}$ where $\mathbf{F} = \mathbf{f}_b + \mathbf{f}_s$. The column matrix $\mathbf{f}_b$ contains the boundary terms and $\mathbf{f}_s$ contain the source terms. $\mathbf{x}$ is the vector of unknowns (nodal temperatures in our case). Components in $\mathbf{L}$, $\mathbf{f}_b$ and $\mathbf{f}_s$ have to be evaluated elementwise.

Step 4: Numerical Manipulation

As we formulated the global problem in step 3, the rest is down to matrix manipulation to evaluate the unknowns. As the first step we have to evaluate the components K_{mn} of stiffness matrix **K**. K_{1n} corresponds to node 1. Therefore N_1^1 and N_2^1 are the only non-zero global shape functions.

$$K_{11} = \int_0^{0.33} k \frac{dN_1^1}{dx} \frac{dN_1^1}{dx} dx \qquad\qquad K_{12} = \int_0^{0.33} k \frac{dN_1^1}{dx} \frac{dN_2^1}{dx} dx$$

$$= \int_0^{0.33} 3.3 \left(-\frac{1}{0.33}\right)\left(-\frac{1}{0.33}\right) dx = 10 \qquad = \int_0^{0.33} 3.3 \left(-\frac{1}{0.33}\right)\left(\frac{1}{0.33}\right) dx = -10$$

$$K_{13} = K_{14} = 0$$

In evaluating terms in the second row we immediately make use of the symmetry of the matrix.

$$K_{21} = K_{12} = -10$$

Upon evaluating the K_{22} we run into a problem -- which shape function to use N_2^1 or N_2^2? The solution is simple. Since those two functions are defined over two elements, we have to integrate relevant function over the appropriate element considering the limits from 0 to 0.66 (or more generally $2l$).

$$K_{22} = \int_0^{0.66} k \frac{dN_2}{dx} \frac{dN_2}{dx} dx = \int_0^{0.33} k \frac{dN_2^1}{dx} \frac{dN_2^1}{dx} dx + \int_{0.33}^{0.66} k \frac{dN_2^2}{dx} \frac{dN_2^2}{dx} dx$$

$$= \int_0^{0.33} 3.3 \left(\frac{1}{0.33} \right) \left(\frac{1}{0.33} \right) dx + \int_{0.33}^{0.66} 3.3 \left(-\frac{1}{0.33} \right) \left(-\frac{1}{0.33} \right) dx = 20$$

K_{23} involves functions defined only over the second element. Hence N_2^2 and N_3^1 are to be considered.

$$K_{23} = \int_{0.33}^{0.66} k \frac{dN_2^2}{dx} \frac{dN_3^1}{dx} dx$$

$$= \int_{0.33}^{0.66} 3.3 \left(-\frac{1}{0.33} \right) \left(\frac{1}{0.33} \right) dx = -10$$

$K_{24} = 0$ since N_4 does not share node 2. $K_{31} = 0$ according to the same line of reasoning. $K_{32} = K_{23} = -10$ and K_{34} is to be evaluated in the same manner we evaluated K_{23}. Again $K_{41} = K_{42} = 0$ as the shape functions do not share the node in question. $K_{43} = K_{34}$ by symmetry. K_{33} and K_{44} are to be evaluated in the same way we evaluated the K_{22}, considering the relevant shape function over the relevant domain. The completed stiffness matrix is given below.

$$\mathbf{K} = \begin{bmatrix} 10 & -10 & 0 & 0 \\ -10 & 20 & -10 & 0 \\ 0 & -10 & 20 & -10 \\ 0 & 0 & -10 & 10 \end{bmatrix} \tag{2.77}$$

This is the famous tridiagonal matrix in FEM. In this case it is only 4×4 since we have only four nodes. With full modeling, one would get a huge, sparse matrix of few thousands of components, yet still banded.

Next step is to evaluate the components in $\mathbf{f}_b$ and $\mathbf{f}_s$. In evaluating the terms in $\mathbf{f}_b$ it is important to identify only N_1 and N_4 remain nonzero at x=0 and x=1. In fact $N_1 = N_4 = 1$ at x=0 and x=1. All other shape functions are zero as far as start and end points are concerned. Therefore

$$f_b = -\begin{bmatrix} qN_1\big|_0^1 \\ qN_2\big|_0^1 \\ qN_3\big|_0^1 \\ qN_4\big|_0^1 \end{bmatrix} = \begin{bmatrix} q_{x=0} \\ 0 \\ 0 \\ -1.25 \end{bmatrix} \tag{2.78}$$

In computing the terms in f_s, the integrals are to be evaluated taking into consideration where the global shape functions are defined.

$$f_s = \begin{bmatrix} \displaystyle\int_0^1 QN_1 dx \\ \displaystyle\int_0^1 QN_2 dx \\ \displaystyle\int_0^1 QN_3 dx \\ \displaystyle\int_0^1 QN_4 dx \end{bmatrix} = \begin{bmatrix} \displaystyle\int_0^{0.33} QN_1^1 dx \\ \displaystyle\int_0^{0.33} QN_2^1 dx + \int_{0.33}^{0.66} QN_2^2 dx \\ \displaystyle\int_{0.33}^{0.66} QN_3^1 dx + \int_{0.66}^{1.0} QN_3^2 dx \\ \displaystyle\int_{0.66}^{1.0} QN_4^1 dx \end{bmatrix} = \begin{bmatrix} 1.65 \\ 3.3 \\ 3.3 \\ 1.65 \end{bmatrix} \tag{2.79}$$

By putting together (2.77), (2.78) and (2.79), the complete matrix equation is obtained.

$$\begin{bmatrix} 10 & -10 & 0 & 0 \\ -10 & 20 & -10 & 0 \\ 0 & -10 & 20 & -10 \\ 0 & 0 & -10 & 10 \end{bmatrix} \begin{bmatrix} T_1 \\ T_2 \\ T_3 \\ T_4 \end{bmatrix} = \begin{bmatrix} q_{x=0} \\ 0 \\ 0 \\ -1.25 \end{bmatrix} + \begin{bmatrix} 1.65 \\ 3.3 \\ 3.3 \\ 1.65 \end{bmatrix} \tag{2.80}$$

From here onward, matrix manipulation become the main focus. $q_{x=0}$ is to be evaluated once temperatures are estimated. This is possible since T_1 is known *a priori*. We leave solving (2.80) to the reader. However, after few manipulations we found

$$\begin{bmatrix} T_1 \\ T_2 \\ T_3 \\ T_4 \end{bmatrix} = \begin{bmatrix} 1 \\ 1.7 \\ 2.01 \\ 2.11 \end{bmatrix}$$

There is an analytic solution for (2.63). The temperatures at nodes calculated using the analytic solution are $T_1=1$, $T_2=1.96$, $T_3=2.59$ and $T_4=2.89$.

Clearly, four element approximate solutions are not particularly accurate. The reader can readily implement this example in FEMLAB for arbitrary accuracy. The purpose of this four element worked example is to make concrete all the steps that are automatically done by FEMLAB upon specifying the problem (2.63) and using the default settings and options.

This example discussed the basics of FEM. However we left untouched many important issues. For an in depth study of FEM the reader is referred to [3] and [6]. The example is targeted to give an insight to what happens inside FEMLAB when you set the problem and ask it to solve. The availability of software packages like FEMLAB greatly reduces the need for understanding the fundamentals of the FEM. Instead of spending a considerable time on learning the method, one can concentrate on solving the problems and physics involved. However, it should be mentioned that an understanding of the core issues in FEM might help in describing the errors and interpreting solutions in some cases.

Exercise: Steady state heat transfer in 3-D

In section 2.1.1 we considered the steady state heat transfer equation with a distributed source, the Poisson equation. Here, we demonstrate the 3-D solution without the source – Laplace's equation. There is nothing particularly new in this example except the demonstration of 3-D modeling. Since all of the models in this book are run on a relatively low performance PC, complicated 3-D modeling would tax its resources. Consequently, this is the only 3-D example in the book. In 3-D modeling it is especially important to conserve memory by taking full advantage of symmetries in your geometry. In this problem, we will model the steady heat transfer within a hexagonal prism with differentially heated (or cooled) basal and side planes. The basal planes are held at the hot temperature (T=1) and the side faces are held at the cold temperature (T=0). Since the steady state solution is sought, the thermal diffusivity is immaterial – as long as the medium is conductive, it achieves the same steady state. Figure 2.14 shows the 3-D geometry and mesh for our model.

Figure 2.14 does not resemble the hexagonal prism. But since the differential equation (2.6) (with f(x)=0) and the geometry admit six-fold periodic symmetry, solutions to (2.6) on Figure 2.14 are periodically extendible to the full hexagonal prism. And all solutions to (2.6) on a hexagonal prism are periodically reducible to a solution on Figure 2.14. It is important when attempting to exploit geometrical symmetry to make sure that the equation and boundary conditions, in general the entire model, shares the symmetry property. Furthermore, one should be careful about solutions that break symmetries inherent in the model description and domain. Nonlinear problems can admit solutions that break the symmetry of the equations and the boundary and initial conditions – by bifurcations typically. All the different solutions to a nonlinear

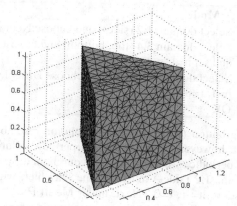

Figure 2.14 One-sixth segment of a hexagonal prism and standard mesh.

system must satisfy the symmetry conditions collectively, but may violate them individually. Here we are safe – Laplace's equation is linear and it is not an eigenvalue problem.

Start up FEMLAB and enter the Model Navigator:

Model Navigator
• Select 3-D dimension
• Select Classical PDEs → Laplace's Equation
• OK

This application mode gives us one dependent variable u, in a 3-D space with coordinates x,y,z. Now we are in a position to set up our domain. Pull down the Draw menu and select Add/Edit/Delete work plane. Accept the x-y plane and defaults. Enter a triangle with vertices (0,0), (1,0), (0.5,0.8) by adding line segments. Edit the last vertex (double clicking) to amend the point to (0.5, 0.866025). This should give a fair representation of an equilateral triangle for the basal plane of our hexagonal prism. Use the palette to "coerce to solid" CO1. Now for the fun part. Select from the Draw menu: extrude. Accept the defaults, in particular the distance 1 in the z-direction. The product of this drawing activity should give the equilateral triangular prism of Figure 2.14. This is the second easiest drawing technique (extrusion) among those available; the easiest is to select primitive 3-D objects.

All of the interest in Laplace's equation lies in the boundary conditions. First, let's set up a mixed set of Neumann boundary conditions for the symmetry edges and Dirichlet conditions for the fixed temperature faces. Pull down the Boundary menu and select Boundary Settings.

Boundary Mode

- Select domain 1,2 (sides) and choose Neumann
- Select domain 3,4 (top, bottom) and choose Dirichlet and set h=1, r=1
- Select domain 5 (back) and choose Dirichlet and set h=1, r=0
- OK

Click on the Solve button (=) to arrive at a solution resembling Figure 2.15. This solution is not, however, the most general periodic solution possible. To make the solution periodic, we need to alter our boundary conditions. The conditions on domains 1,2 (sides) must become Dirichlet, with one set to h=1 r=0 and the other set to h=-1, r=0. Next, under **Mesh Parameters**, symmetry boundaries 1 2 must be set. This is the standard recipe for periodic boundary conditions, but 3-D adds a new twist. If you try the above, FEMLAB should issue an error "NaNs or Infs encountered during mesh generation." I am grateful to Shu-Ren of COMSOL who realized that the geometry of each periodic face must be identical to machine precision so that each face meshes exactly the same for the faces to coincide. Although $\frac{\sqrt{3}}{2} \approx 0.866025$, this is not exact to machine precision. The solution is to edit the model m-file and insert the MATLAB command for the above number, so that internal precision is used. The model m-file should be edited near the top under the geometry specification making the obvious replacement. A similar solution to Figure 2.15 is found. Note that the periodic solution requires only 6723 elements, by comparison to the no flux BC model which used 7352 elements. This is a modest saving, but worthwhile nonetheless.

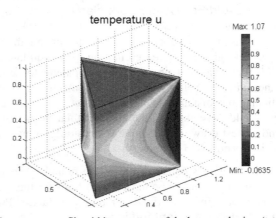

Figure 2.15 Temperature profile within a segment of the hexagonal prism (tetrahedron plot).

```
% Geometry
p=[0 0.5 1;0 sqrt(0.75) 0];
rb={1:3,[1 1 2;2 3 3],zeros(3,0),zeros(4,0)};
wt={zeros(1,0),ones(2,3),zeros(3,0),zeros(4,0)};
lr={[NaN NaN NaN],[0 1 0;1 0 1],zeros(2,0),zeros(2,0)};
tmp=solid2(p,rb,wt,lr);
g1=extrude(tmp,'Distance',1,'Scale',[1;1],'Displ',[0;0],'Wrkpln',[0
1 0;0 0  ...
1;0 0 0]);
```

An exercise for the reader. Compute the flux across all boundaries. What would you expect the sum to be theoretically? Why is the discrepancy appreciable?

2.2 Summary

The flexibility of FEMLAB and FEM analysis in treating higher dimensional problems and canonical PDEs was explored. The ease with which point sources, quasi-linear terms, and periodic boundary conditions are treated was demonstrated. An overview of how the stiffness matrix, load vector, and general (boundary) constraints are dealt with by the FEM approach was presented. Worked examples of the FEM approach illustrated the principles.

References

1. Constantinides A. and Mostoufi N., Numerical Methods for Chemical Engineers with MATLAB Applications, Pretice Hall, Upper Saddle River NJ, 1999.
2. Freiden R.B., Physics from Fisher Information: A Unification, Cambridge University Press, 1998.
3. Reddy J.N., An Introduction to the Finite Element Method, McGraw-Hill Inc, New York, 1993.
4. Strang, G. Introduction to Applied Mathematics, Wellesley-College Press, Massachusetts, 1986, p. 37.
5. Mitchell, A.R. and Wait, R., The Finite Element Method in Partial Differential Equations, Wiley-Interscience, New York, 1977.
6. Chung T.J., Computational Fluid Dynamics, Cambridge University Press, Cambridge, 2002.

Chapter 3

MULTIPHYSICS

W.B.J. ZIMMERMAN

Department of Chemical and Process Engineering, University of Sheffield, Newcastle Street, Sheffield S1 3JD United Kingdom

E-mail: w.zimmerman@shef.ac.uk

Multiphysics is a recent conceptualization to categorize modeling where different physicochemical mechanisms are prevalent in a given application, where these mechanisms are modelled by wholly different field equations. But to be multiphysics, the field equations must couple. In this chapter, we treat models of multiphysics for thermoconvection and non-isothermal chemical reactors as examples of the genre of multiphysics. Applications in later chapters show largely multiphysics modeling since "single physics" models are likely to be well studied in their core disciplines. We also take the opportunity to introduce the concept of parametric continuation, which is an essential mechanism for arriving at the solution to highly nonlinear problems by inching there by starting from nearby solutions in function space or even linear systems. The model is then altered to treat the differential side wall heating of water between walls held at the freezing and boiling points (without boiling) and the full dependency of buoyant force on temperature. Simulations in large cavities show the beginnings of stratification in temperature. Next we treat a non-isothermal tubular reactor that couples mass and energy transport. Finally, we treat chemical reaction in the pores of a solid pellet with diffusion from a bulk flow.

3.1 Introduction

FEMLAB makes a big selling point of *multiphysics* modelling as a key advantage of its software package. Not long ago I described one of the important features of the burgeoning research area of microfluidics as requiring skills in multiphysics modelling. A respected colleague asked pointedly, "What's that? Physics that happens on multiple scales?" So multiphysics is jargon that may not be uniformly recognized in the sciences and engineering. Not wanting to use the term wildly, we shall define multiphysics modelling here to mean any complete, coupled system of differential equations that has more than one independent variable of different physical dimensions (vector equations count as one equation). The FEMLAB definition is actually an operational definition – "Does FEMLAB have a single application mode for it or can you only describe it by coupling more than one application mode?" In FEMLAB's Model Navigator, you can create a multiphysics model by coupling two or more application modes (under the multiphysics tab).

Of course either definition is a Byzantine notion, so let's make it concrete by examples. Are fluid dynamics multiphysics? Yes, but only on the technicality that pressure is an independent variable which has different units to velocity. Is it multiphysics in FEMLAB? No, because there is a single Navier-Stokes application mode. Is heat transfer multiphysics? No, there is only one independent variable – temperature, and only one FEMLAB application mode. What about thermofluids? Yes, as velocity, temperature, and combustion conversion are three independent variables with different units, and there are three transport equations coupled. Many typical research areas in chemical engineering are multiphysics: physicochemical hydrodynamics, magneto-hydrodynamics, electrokinetic flow, multiphase flow, double diffusion, and separations.

FEMLAB deals with specific common multiphysics applications by creating application modes that are a full description of single field models, but can be readily coupled to other application modes. The user provides the coupling by specifying PDE terms and boundary and initial conditions symbolically. FEMLAB does the "bookkeeping" to make sure that application modes have different specifications for the field variables and derived quantities that are commonly computed for a typical application. For two coupled application modes, FEMLAB assembles the FEM description (through the sparse matrices K,N,L,M) for each mode, including the user specified coupling terms. If the pre-made application modes do not cover the user's coupled system, then the user can adapt as many coefficient form, general form, or weak form systems as necessary to describe their dynamics.

A few examples will illustrate multiphysics modelling to much better effect. Chapter 8 of the book of Ramirez [1] has a wealth of multiphysics PDE models with chemical engineering applications. They are computed on simple domains with finite difference methods coded with full detail in MATLAB. We shall adapt several such examples to FEMLAB models here. But first we will attempt some simple buoyant convection problems.

3.2 Buoyant Convection

Coupling momentum transport and heat transport is a well studied area of transport phenomena. The governing equations are

$$\frac{\partial u}{\partial t} + u \cdot \nabla u = -\frac{1}{\rho} \nabla p + \nu \nabla^2 u + \frac{\alpha g}{\rho} T$$

$$\nabla \cdot u = 0 \tag{3.1}$$

$$\frac{\partial T}{\partial t} + u \cdot \nabla T = \kappa \nabla^2 T$$

Here, the dependent variables are described as follows: u is the velocity vector, p is the pressure, and T is the temperature. The independent variables are spatial coordinates (implied in the differential operators) and time t. Everything else is a parameter (v, ρ, α, κ, g) with fixed value once the fluid and venue are selected. If there is no imposed moving boundary or pressure gradient, then the whole of the motion is created by temperature gradients and is termed buoyant (or free) convection. If there are imposed velocities or pressure gradients, then the application is termed forced convection. Either case can be studied by the same multiphysics mode created in FEMLAB, but are historically considered different physical modes.

In buoyant convection, there are two dimensionless parameters that govern the dynamical similarity of the problem, the Prandtl number that is a function of the fluid, and the Rayleigh number that gives the relative importance of temperature driving forces to dissipative mechanisms:

$$\Pr = \frac{v}{\kappa}$$

$$\mathrm{Ra} = \frac{\alpha g\left(\delta T\right)h^{3}}{\rho v \kappa} \tag{3.2}$$

where h is the depth of the fluid, δT is the applied temperature difference, α is the coefficient of thermal expansion, g is the gravitational acceleration vector (g is its magnitude), ρ is the density, v the kinematic viscosity, and κ is the thermal diffusivity.

Batchelor [2] showed that differentially heating any sidewall automatically induces buoyant motion, so the canonical buoyant convection problem is the hot wall/cold wall cavity flow. This problem is always taken as a test case for the development of new numerical methods for transport phenomena. We will develop a FEMLAB model for it in this section. This problem is treated in [3], but the variations on the theme treated here are original.

Launch FEMLAB and in the Model Navigator, select the **Multiphysics** tab.

Model Navigator
• Select 2-D dimension
• Select Physics modes⇒Incompressible Navier-Stokes >>
• Select ChE ⇒Convection and conduction >>
• Select PDE modes ⇒ Coefficient form >>
• OK

The last mode, the coefficient form, will be used to solve directly for the streamfunction from the streamfunction vorticity Poisson equation:

$$\nabla^2 \psi = -\omega \tag{3.3}$$

You may have noticed that the Incompressible Navier-Stokes application mode will print "flowlines." But to my eye, they are streaklines of randomly positioned particles, rather than the streamlines (contours of streamfunction) that are traditionally interpreted in two-dimensional flow. Adding equation (3.3) is straightforward, and not particularly expensive to compute.

Pull down the options menu and select Add/Edit constants. The Add/Edit constants dialog box appears.

Add/Edit Constants

- Name of constant: T0
- Expression: 0
- Name of constant: T1
- Expression: 1
- Apply / OK

At this stage we will leave out the constants *Ra* and *Pr*. For simplicity throughout, we will keep *Pr=1*, which is a good approximation for many gases. By enforcing the range of the temperature to lie between 0 and 1, i.e. a dimensionless temperature, all of the dynamics are controlled through the Rayleigh number.

Pull down the Options menu and set the grid to $(0,1) \times (0,1)$ and the grid spacing to 0.1,0.1. Pull down the Draw menu and select Rectangle/Square and place it with unit vertices $[0,1] \times [0,1]$.

Now for the boundary conditions. Pull down the Boundary menu and select Boundary Settings.

Boundary Mode

- Select domain 1
- Use the multiphysics pull down menu to select the IC NS mode
- Set boundaries 1-4 with No-Slip
- Use the multiphysics pull down menu to select the CC mode
- Set bnd 1 with T=T0; bnd 4 with T=T1; keep 2 and 4 no flux
- Use the multiphysics pull down menu to select the coeff mode
- Keep Dirichlet conditions on bnd 1-4: h=1; r=0
- Apply / OK

Now pull down the Subdomain menu and select Subdomain settings.

Subdomain Mode

- Select domain 1
- Use the multiphysics pull down menu to select the IC NS mode
- Set $\rho=1$; $\eta=1$; $F_x=0$; $F_y=-1$
- Use the multiphysics pull down menu to select the CC mode
- Set $\rho=1$; $\kappa=1$; $c=1$; $u=u$; $v=v$
- Select the init tab; set $T(t0)=T0+(T1-T0)*x$
- Use the multiphysics pull down menu to select the coeff mode
- Set $c=1$; $da=0$; $f=vx-uy$
- Apply
- OK

Now pull down the Mesh menu and select the Parameters option. We will need to pack elements into the corners for best resolution.

Mesh Parameters

- Select more>>
- Max element size near vertices: 1 0.05 2 0.05 3 0.05 4 0.05
- Remesh
- OK

There should be 792 elements. Click on the = button on the toolbar to Solve. Now plot the temperature profile. Is it what you expected? How does it compare with the initial condition.

Now plot the streamfunction. Surprised by the complexity? Now look at the scale. Why so small? Recall that we set $F_y=-1$ (gravity is in the negative y direction). This has the effect of adding hydrostatic pressure only. So there is no back action on the momentum equation from the imposed differential sidewall temperatures. So what we have here is a plot of velocity noise generated by round-off error. It is always important to look at the scale of contour/surface plots to assess whether we are interpreting noise!

You may have had some difficulty getting FEMLAB to converge to a solution. When I originally wrote this example in FEMLAB 2.2, it converged fairly rapidly. Yet when done with FEMLAB 2.3, it took a long time. There are two contributions to the slow convergence – (1) the new scaling feature for the error estimate under the Solver Parameters, and (2) the lack of a pressure datum point. The first (scaling factor) was unexpected. Basically, FEMLAB hopes to aid convergence by scaling each contribution to the error automatically. But since our velocity field has the true solution of a zero velocity field, numerically we find the approximate solution as noise around zero. The automatic scaling feature is trying to "resolve" the noise, which is not particularly sensible. Solution – turn the feature off! Select under **Solver Parameters** the **Scaling** box, and in the pop-up dialogue box, select "none" rather than the default automatic.

The second point about the pressure datum is discussed in detail in Chapter 5. The solutions here are convergent, but greater accuracy and faster convergence result by specifying a pressure datum.

To implement a buoyant force that varies with temperature, edit the appropriate subdomain setting for the IC NS mode to be:

$$F_y = Ra\,T \qquad (3.4)$$

Actually, the proper dimensionless term is as below, but given our scaling for temperature set in the constants, they are equivalent. If you wish to use temperatures with units, then edit T0 and T1 appropriately, and use this substitute for (3.4). Later, we will use temperatures with units, so I recommend its use:

$$F_y = Ra \times \left(\frac{T - T_0}{T_1 - T_0} \right)$$

Add *Ra* to the constants list and set it to *Ra=1*. Now use the Restart toolbar button, which uses the noise velocity field as an initial condition. This is actually a useful technique for introducing noise as an initial condition. Figures 2.1 and 2.2 show the streamlines and isotherms at steady state.

A quantity of central interest in thermal convection studies is the heat flux. The natural dimensionless measure of heat flux is the ratio of the total time averaged rate of heat transport to the conductive rate of heat transport, termed the Nusselt number:

$$Nu = \frac{\kappa(\delta T)/L + \langle \overline{uT} \rangle}{\kappa(\delta T)/L} = 1 + \frac{L\langle \overline{uT} \rangle}{\kappa(\delta T)} \qquad (3.5)$$

The overbar represents spatial average and the brackets time average. FEMLAB permits the computation of the separate terms in the numerator as subdomain integrations under the Post menu. Compute the dfluxT_cc (conductive heat flux) and cvfluxT_cc (convective heat flux) integrated over the whole domain. What is the corresponding Nusselt number? Does this surprise you given the scale of the streamfunction in Figure 2.1? Now try simulating *Ra=50*. Do you get a converged solution?

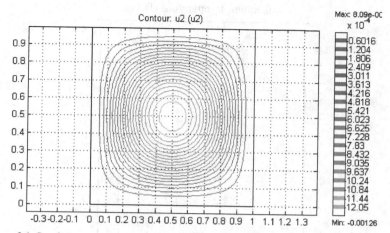

Figure 3.1 Steady state streamlines of hot wall/cold wall buoyancy driven convection for *Ra=1*.

Buoyancy Driven Cavity Flow: Parametric Continuation

Our solution strategy for the hot wall/cold wall problem to reach *Ra=50* was to build up elements of the solution piecemeal. Were we to try to start at Ra=50 directly, we would find that FEMLAB cannot find a solution. Why not? For a nonlinear problem, the initial condition may not be in the "basin of attraction" for the desired solution, so Newton's Method could career far off. For it to work well, Newton's Method must start near a solution. For instance, the initial solution for hydrostatic pressure and velocity noise for *Ra=0* was an essential step. As a fully linear problem, it was readily solvable. It serves the important purpose of introducing an asymmetric velocity profile (due to the numerical noise of truncation error). This permits the solution for Ra=1, which is qualitatively similar to the Ra=0 in that it has a circulation, though a massive change in scale. Even then, though qualitatively similar to the Ra=1 solution, Ra=50 was too far a leap from Ra=1 to converge. The notion of traversing the solution space to introduce various topological features consistent with the target solution so as to be in its basin of attraction is similar to the established concept of parametric continuation. In parametric continuation, one restarts the simulation with a parametric value close to that of the saved, converged solution. Because the solution at the new parameter values is not expected to be much different than at the old parameter values, the Newton solver should converge rapidly. This methodology only fails if the new parameter is close to a bifurcation point – in which case multiple solutions are possible. The Jacobian used by Newton's Method is then very close to singular, so convergence may not be achieved. Or if it is, which of the multiple solutions that is selected may not be *a priori* predictable.

Contour: temperature (T)

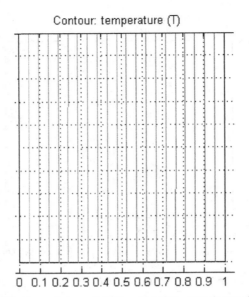

0 0.1 0.2 0.3 0.4 0.5 0.6 0.7 0.8 0.9 1

Figure 3.2 Isotherms between 0 (left) and 1 (right) at steady state for *Ra=1*.

Parametric continuation is typically used for one of two purposes. One is to map the response of some feature of the solution over a range of parameters. The second is to reach a target solution for which jumping to the solution from any arbitrary initial condition is non-convergent. So parametric continuation is metaphorically crawling along the limb of a tree, rather than expecting to jump and arrive safely. Parametric continuation can fail to converge as one ramps up a complexity parameter (like a Rayleigh or Reynolds number), and the complexity of the solution at smaller scales becomes unresolved. Thus, parametric continuation identifies at which parameter values refining the mesh is important. In this section, we will use parametric continuation to map the Nusselt versus Rayleigh numbers, using the power of MATLAB programming of FEMLAB subroutines.

FEMLAB 2.2 did not have a built in parametric continuation feature, but FEMLAB 2.3 introduced it. Yet building your own MATLAB m-file for parametric continuation is not especially difficult. We start by saving the model M-file for the current state of the FEMLAB simulation. We have solved for Ra=0, Ra=1, and attempted to solve for Ra=50. We have computed the subdomain integrations for conductive and convective fluxes. All the FEMLAB commands to do this are in the model M-file, and many more besides. SaveAs "convection.m" and then open this file with your favorite editor of MATLAB's m-file editor. You will want to delete all the PostPlot commands, and the entire Ra=50 attempt. Then you will need to add a looping structure, storage, and output.

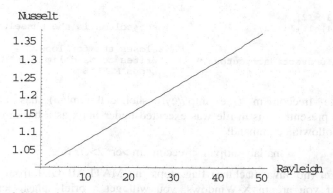

Figure 3.3 Nusselt number versus Rayleigh number found by parametric continuation.

Lines for storage (added at the beginning as the first executable statements):

```
%%%%%%%%%%%%%%%%%%%%%%%%WBJZ parameters and storage%%%%%%%%%%%%%%%
   Rayleigh=[1:1:50];                     %sets up a 50 long list
   output=zeros(length(Rayleigh),4); %storage for output of Nusselt
%%%%%%%%%%%%%%%%%%%%%%%%%%%%%%%%%%%%%%%%%%%%%%%%%%%%%%%%%%%%%%%%%%%
   Lines for looping (altering the Ra=1 computation):
%%%%%%%%%%%%%%%%%%%%%%%%%%%%%%%%%%%%%%%%%%%%LoopingStructure%%%%%%%%%
for j=1:length(Rayleigh)                  %loops until end statement
% Define variables
fem.variables={...
          'T0',      273,...
          'T1',      373,...
          'Ra',      Rayleigh(j)};        %replaces 1 with j-th Rayleigh
Lines for output (added at the end of the programme):
% Integrate on subdomains                 %was generated automatically
I1=postint(fem,'cvfluxT_cc',...
          'cont',        'internal',...
          'contorder',2,...
          'edim',        2,...
          'solnum',      1,...
          'phase',       0,...
          'geomnum',1,...
          'dl',          1,...
          'intorder',4,...
          'context','local');
% Integrate on subdomains
I2=postint(fem,'dfluxT_cc',...
          'cont',        'internal',...
          'contorder',2,...
          'edim',        2,...
          'solnum',      1,...
          'phase',       0,...
          'geomnum',1,...
          'dl',          1,...
          'intorder',4,...
          'context','local');
output(j,1)=Rayleigh(j);                  %First column is the Rayleigh
output(j,2)=I1;
```

```
output(j,3)=I2;
output(j,4)=1.+I1/I2;              %Last column is the Nusselt

end                                %closes the for-loop
dlmwrite('convect.dat',output,','); %writes comma-delimited ASCII
quit                               %stops MATLAB
```

The m-file freeconv.m (see http://eyrie.shef.ac.uk/femlab) has all these alterations present. This m-file was executed under linux as a background job using the following command:

matlab –nojvm <freeconv.m >err 2>err &

The –nojvm (no Java machine) flag stops the MATLAB GUI from loading, although if you are in X-Windows, you will get a brief splash screen for MATLAB. The command above generated the data in the file convect.dat used in Figure 3.3 to plot the Nusselt vs. Rayleigh number dependence. The file err contains the re-direction of the standard output (usually the screen) and the standard err (usually the screen) units. When called with input re-direction from an m-file, MATLAB does not launch the GUI, but evaluates the m-file programme directly. This is the most efficient way to conduct MATLAB computations, as the processor and memory are not tied up multitasking the GUI, so can pay better attention to your computation. The parametric continuation in Rayleigh number m-file generates a series of entries in the err file of the form:

Iter	ErrEst	Damping	Stepsize	nfun	njac	nfac	nbsu
0				1	0	0	0
1	7.644421253e-07	1.0000000	0.1057807343	2	1	1	2

These entries all show error estimates of $O(10^{-7})$, which implies that the first iteration on each new Rayleigh parameter value converges in one Newton iteration step. Since the convergence is so good, we could probably take bigger steps if the goal were only to reach the endpoint Rayleigh value. However, adapting the stepsize in the continuation parameter would need to be automated on the convergence performance, with substantial logic encoded for trapping non-convergent continuation steps. The simple linear continuation conducted here is easiest to code.

When using parametric continuation to arrive at the final parameter value, running the MATLAB m-file as a background job does not help. The FEMLAB GUI is not entered and will not be accessible with the solution results. FEMLAB has its own MATLAB workspace, however, so if you wish, you can run your edited m-files in the FEMLAB GUI. Just open the File Menu, Select Open (m-file) and FEMLAB will evaluate all your commands sequentially. Even those commands that execute non-FEMLAB functions in MATLAB. For instance, if you run freeconv.m in this way, you will eventually arrive at Ra=50, even though the for command for looping is not a FEMLAB command.

Although parametric continuation is therefore possible in the FEMLAB GUI, it is probably too time consuming to wait for the GUI to process 50 or 100 solutions at a time. So the user will probably want to invest some time in learning MATLAB programming tools and gaining a handle on the FEMLAB function library. Fortunately, FEMLAB's logging feature which records the FEMLAB commands issued by the GUI provides an excellent starting point for constructing your own FEMLAB programme. We have already given several applications of MATLAB programming with FEMLAB functions, but a full description of either MATLAB programming or FEMLAB functions is beyond the scope of this book.

Chapter one (matrix operations) and the Appendix (vector calculus) provide only a rudimentary working capacity in MATLAB programming. We will continue to use this user defined programming style in the book, with sufficient explanation to guide the informed reader, and at least to inform the MATLAB novice of what power they are missing out on!

Variations on a Theme: Non-Monotonic Density

The governing equations for buoyant convection (3.1) assume the conventional Boussinesq approximation [4], i.e. that the velocity field is divergence free, that kinematic viscosity is constant, and that the only effects of density variation are felt by the body force in the Navier-Stokes equations, which was taken to depend proportionally to the coefficient of thermal expansion and temperature. The latter constraint, is too severe. The Boussinesq approximation only requires that density is a slowly varying function of position so that locally the velocity is divergence free. So a less restrictive set of governing equations is

$$\frac{\partial u}{\partial t} + u \cdot \nabla u = -\frac{1}{\rho_0}\nabla p + \nu\nabla^2 u + \frac{\rho(T)g}{\rho_0}$$

$$\nabla \cdot u = 0 \tag{3.6}$$

$$\frac{\partial T}{\partial t} + u \cdot \nabla T = \kappa\nabla^2 T$$

where $\dfrac{\rho(T)}{\rho_0}$ is a general function of temperature, and $\rho_0 = \rho(T_0)$. The Rayleigh number is no longer a constant, but depends on this function:

$$\text{Ra}(T) = \frac{gh^3}{\nu\kappa}\left(\frac{\rho(T) - \rho_0}{\rho_0}\right) = Gr'\left(\frac{\rho(T) - \rho_0}{\rho_0}\right) \tag{3.7}$$

where the gravity group *Gr'* now appears as a dimensionless parameter. The density function plays the role of a nonlinear expansivity (and possibly non-monotonic).

T °C	ρ	T °C	ρ	T °C	ρ	T °C	ρ
0	0.99987	20	0.99823	45	0.99025	75	0.97489
3.98	1.	25	0.99707	50	0.98807	80	0.97183
5	0.99999	30	0.99567	55	0.98573	85	0.96865
10	0.99973	35	0.99406	60	0.98324	90	0.96534
15	0.99913	38	0.99299	65	0.98059	95	0.96192
18	0.99862	40	0.99224	70	0.97781	100	0.95838

Table 3.1 Specific gravity of liquid water.

So what does one of these expansivity functions look like? See Figure 3.4 below.

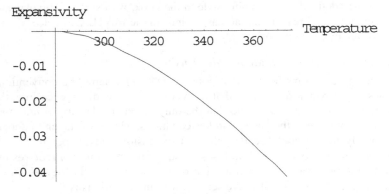

Figure 3.4 Expansivity $\dfrac{\rho(T)-\rho_0}{\rho_0}$ vs. Temperature (K) for water.

Next, how do we organize this data so as to use it in the FEMLAB GUI? You should find the m-file watrdens.m, which is a MATLAB function m-file that interpolates using cubic splines within Table 3.1 to find the dimensionless expansivity factor. This m-file is reproduced here:

```
function a=watrdens(ttemp)
%WATRDENS Interpolates the expansivity of water in 273 and 373
deg K
temp=[0 3.98 5 10 15 18 20 25 30 35 38 40 45 50 55 60 65 70 75 80
85 90 95 100];
dens=[0.99987 1. 0.99999 0.99973 0.99913 0.99862 0.99823 0.99707
0.99567 0.99406 0.99299 0.99224 0.99025 ...
0.98807 0.98573 0.98324 0.98059 0.97781 0.97489 0.97183 0.96865
0.96534 0.96192 0.95838];
temp=temp+273;
dens=(dens-dens(1))/dens(1);
a=interp1(temp,dens,ttemp,'spline');
```

Now for the FEMLAB implementation. Before you launch FEMLAB from MATLAB, make sure that you change the current directory to the one with your watrdens.m m-file. To make sure that it is available, try out some value between 273K and 373K:

```
>> watrdens(330)
ans =
   -0.0151
```

Now when you launch FEMLAB, it will inherit this current directory and have the waterdens m-file function at its disposal. Load the saved version of freeconv.mat from your distribution, and then Pull down the **options** menu and select **Add/Edit constants**. Replace the Rayleigh number entry with the gravity group Gr, and set it initially to Gr=0. Now edit the NS subdomain settings and set

$$Fx=-Gr*watrdens(T)$$

Now onto solving. Click on the solver button (=) on the toolbar. If FEMLAB hasn't already popped up the message "No differentiation rule for function watrdens", it will now. By default, FEMLAB computes symbolic derivatives of just about everything in sight in assembling the stiffness matrices, constraint matrices, and load vectors. So it naturally is annoyed at us for not telling it how to differentiate watrdens(T). FEMLAB has a place in its FEM structure for differentiation rules if you provide a function that can be differentiated analytically (fem.rules) which is used by the femdiff FEMLAB function. There is a handle on fem.rules (**Options** menu, **Differentiation Rules**) in the FEMLAB GUI, but we will use the MATLAB programming language.

I tried the following: Pull down the Solver menu, select Parameters, and uncheck the tick box F in the automatic differentiation section. Now click on the solve button. The solution progress window should now manifest as the nonlinear solver whirls away. Eventually it reports many NaNs and Infs in the solution, which should be interpreted as utter failure. So Plan B was necessary. I created a second MATLAB m-file function for the numerical derivative of watrdens(T):

```
function a=dwatrden(ttemp)
%DWATRDENS Interpolates the expansivity of water between 273 and
373 deg K
temp=[0 3.98 5 10 15 18 20 25 30 35 38 40 45 50 55 60 65 70 75 80
85 90 95 100];
dens=[0.99987 1. 0.99999 0.99973 0.99913 0.99862 0.99823 0.99707
0.99567 0.99406 0.99299 0.99224 0.99025 ...
0.98807 0.98573 0.98324 0.98059 0.97781 0.97489 0.97183 0.96865
0.96534 0.96192 0.95838];
temp=temp+273;
dens=(dens-dens(1))/dens(1);
pp=spline(temp,dens); % Piecewise polynomial form of a cubic spline
[br,co,npy,nco]=unmkpp(pp); % Breaks apart the pp form
```

```
sf=nco-1:-1:1;                    % Scale factors for differentiation
dco=sf(ones(npy,1),:).*co(:,1:nco-1);   % Derivative coefficients
ppd=mkpp(br,dco);                 % Build up pp form for derivative
a=ppval(ppd,ttemp);
```

This m-file (dwatrden.m) should also be loaded into your MATLAB current directory. It uses a MATLAB programming technique to compute the approximate first derivative of a cubic spline interpolation. Very clever. I wish I could take credit for it, but that goes to others [5]. It has been pointed out to me that I re-invented the wheel in that FEMLAB developers recognized the need for derivatives of interpolated functions, so they introduced their own interpolation function flinterp1, which works like the built-in interp1 but provides the derivative as well as the function. The help on flinterp1 shows how, abbreviated below.

```
>> help flinterp1

FLINTERP1 1D interpolation for use in FEMLAB.
    YI = FLINTERP1(X,Y,XI) interpolates to find YI, the values of
    the underlying function Y at the points in the vector XI.
    The vector X specifies the points at which the data Y is
    given. If Y is a matrix, then the interpolation is performed
    for each column of Y and YI will be length(XI)-by-size(Y,2).

    YI = FLINTERP1(X,Y,XI,METHOD) specifies alternate methods.
    The default is linear interpolation.  Available methods are:

        1) 'linear'    - linear interpolation
        2) 'nearest'   - nearest neighbor interpolation
        3) 'spline'    - piecewise cubic spline interpolation
        4) 'pchip'     - piecewise cubic Hermite interpolation

    METHOD is either a string or a scalar value.

    YI = FLINTERP1(X,Y,XI,METHOD,DER) differentiates the piecewise
    polynomial DER times.
```

Next I created a MATLAB m-file by hacking the model M-file generated by FEMLAB (waterdensity.m). They salient feature is that it specifies the "analytic" differentiation rules at the appropriate place in the FEMLAB subroutine:

```
% Differentiation rules
fem.rules={'watrdens(T)', 'dwatrden(T)'};

%With flinterp1, would use to specify tdata and densdata in the model
%m-file  functions  watrdens/dwatrden  to  be  read  in  so  that
%flinterp1(tdata,densdata,T) and linterp1(tdata,densdata,T,'spline',1)
%compute the function and first derivative.
```

This model m-file can now be read directly into the FEMLAB GUI from the file menu, the Open model m-file option. waterdensity.m computes the Gr=0 and Gr=1 solutions. You can now migrate to higher Gr by hand with parametric contiunuation. I wrote a second variation of the model m-file highgrav.m which migrates to higher Gr using automated parametric continuation and a doubly refined mesh. It arrives to $Gr=10^5$ after about a day on my fastest linux PC workstation as a background job. Expert criticism suggests that we should use a pressure datum introduced in point mode to improve the convergence rate. Chapter five discusses how to do this in detail, in the vicinity of equation (5.9).

The only new trick in this MATLAB program is saving the fem structure as a mat-file for later interrogation in MATLAB. See the line just before the end:

```
save highgrav.mat fem
```

The temperature isotherms tell an interesting story in Figure 3.5 that the fluid is starting to stratify, with cold fluid under hot:

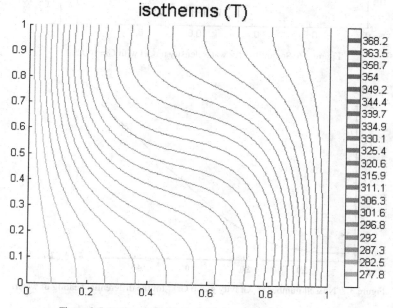

Figure 3.5 Isotherms for water density model with Gr=100000.

Similarly, the core recirculation warps as the flow in the cold region slows. Figure 3.6 gives the streamlines, which with the given scale, show a substantial strengthening.

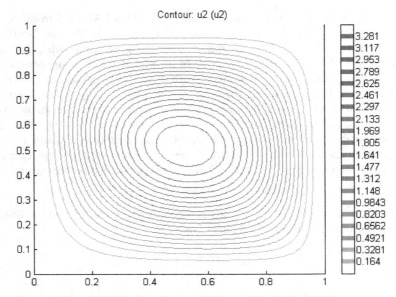

Figure 3.6 Sreamlines for water density model with Gr=100000.

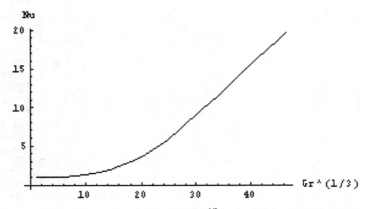

Figure 3.7 Nusselt number vs. Gravity group Gr$^{1/3}$ for the differential heating of water.

Figure 3.7 shows the high insensitivity of the Nusselt number to gravity group, i.e. large cavities are required before convective effects dominate. Note that the scaling of Gr$^{1/3}$ at high Gr was guessed on the grounds of heat flux scaling with the linear separation of the walls, L, which appears cubed in Gr. At large Gr, the near linear dependency suggests this asymptotic scaling, until the convective shearing becomes strong enough that the laminar flow breaks down.

3.3 Unsteady Response of a Nonlinear Tubular Reactor

Ramirez [1] [6] reports a simulation of the adiabatic tubular reactor where heat generation effects are appreciable. Generally, tubular reactor design estimates follow from steady-state 1-D ODE simulations. In the model of Ramirez, the reactor starts up cold or is subjected to perturbations of its steady operation which convect through the system before returning to steady operation. In this regard, such transient effects are important considerations for the safe, stable and controlled operation of tubular reactors.

Ramirez treats first order chemical reaction with heat generation. Thus only the mass transport equation for one species and energy transport equation, coupled through the temperature dependence of the reaction flux and the heat generation by reactive flux, need be considered. Interestingly, Ramirez solved the highly coupled, nonlinear equations by a technique of quasilinearization with finite difference techniques. The solution at the current time and the linearization of the equations about that solution are used to predict the profiles of concentration and temperature at the next time step. The procedure is iterated until convergence at the new time step is achieved. The prediction and correction steps involve solution of sparse linear systems. This is, of course, the same procedure as used by FEMLAB, except it is the finite element approximation and the associated sparse linear system that is solved iteratively by Newton's method.

Governing equations are given here in dimensionless form:

$$\frac{\partial \Theta}{\partial t} = \frac{\alpha}{\mathcal{D}} \frac{\partial^2 \Theta}{\partial x^2} - r_2 \frac{\partial \Theta}{\partial x} + B_1 \Gamma \exp\left(-QQ/\Theta\right)$$

$$\frac{\partial \Gamma}{\partial t} = \frac{\partial^2 \Gamma}{\partial x^2} - r_2 \frac{\partial \Gamma}{\partial x} - B_2 \Gamma \exp\left(-QQ/\Theta\right)$$

$$(3.7)$$

subject to boundary conditions on the reactor inlet and outlet:

$$-\left.\frac{\partial \Theta}{\partial x}\right|_{x=0} = r_1\left(1 - \Theta\big|_{x=0}\right) \qquad \left.\frac{\partial \Theta}{\partial x}\right|_{x=1} = 0$$

$$-\left.\frac{\partial \Gamma}{\partial x}\right|_{x=0} = r_2\left(1 - \Gamma\big|_{x=0}\right) \qquad \left.\frac{\partial \Gamma}{\partial x}\right|_{x=1} = 0$$

$$(3.8)$$

The former are called Danckwerts boundary conditions [7]. The initial conditions for temperature are uniform everywhere at $\Theta=1$. Ramirez [1] considers two different liquid phase reactions. The first is a reactor with an intermediate conversion at a single steady state. The second is a triple steady state. The Peclet numbers for heat and mass transfer are taken as identical for convenience, but this is not a realistic assumption which can be relaxed in FEMLAB without any difficulty. This problem has a long history in the chemical engineering literature, and the equal Peclet numbers are a legacy of an analytic approximation by Amundson [8] which provides validation for the solution.

Table 3.2 Parameters for Case 1.

$r_1 = \dfrac{vL}{\alpha}$	30	$B_1 = \dfrac{(-\Delta H)kC_0 L^2}{\rho c_p T_0 \mathcal{D}}$	7.84×10^7
$r_2 = \dfrac{vL}{\mathcal{D}} = Pe$	30	$B_2 = \dfrac{kL^2}{\mathcal{D}}$	1.2×10^8
$\dfrac{\alpha}{\mathcal{D}}$	1	$QQ = \dfrac{E}{RT_0}$	17.6

Table 3.3 Parameters for Case 2.

$r_1 = \dfrac{vL}{\alpha}$	30	$B_1 = \dfrac{(-\Delta H)kC_0 L^2}{\rho c_p T_0 \mathcal{D}}$	8.47×10^9
$r_2 = \dfrac{vL}{\mathcal{D}} = Pe$	30	$B_2 = \dfrac{kL^2}{\mathcal{D}}$	1.2×10^{10}
$\dfrac{\alpha}{\mathcal{D}}$	1	$QQ = \dfrac{E}{RT_0}$	23

Amundson's technique combines the mass and energy equations by linear transforms:

$$n = \Theta$$

$$\Gamma = \frac{B_2}{B_1}(n_{\lim} - n)$$

This, in turn, leads to equations (3.7) and (3.8), along with the BCs, being described identically, if and only if, the ratio of diffusivities is unity, $\dfrac{\alpha}{\mathcal{D}} = 1$:

$$\frac{\partial^2 n}{\partial x^2} - Pe\frac{\partial n}{\partial x} + B_2 (n_{\lim} - n)\exp(-QQ/n) = 0 \qquad (3.9)$$

subject to

$$-\frac{\partial n}{\partial x}\bigg|_{x=0} = Pe\left(1 - n\big|_{x=0}\right) \qquad \frac{\partial n}{\partial x}\bigg|_{x=1} = 0 \qquad (3.10)$$

Here, n_{lim} is the limiting dimensionless temperature that can be achieved upon exhaustion of the reactant, $\Gamma=0$. The Peclet number, Pe, is either the thermal or mass Peclet number (r_1 or r_2).

Amundson proposed a one dimensional search to the above boundary value problem, starting from a guess of $n(x=1)$ and shooting back to $x=0$. In both of the above cases, $n_{lim}=1.656$.

Let's first solve the single convection-diffusion-reaction equation (3.9) using FEMLAB. Because it is a boundary value problem, FEM has a natural advantage here.

Start up FEMLAB and enter the Model Navigator:

Model Navigator
- Select 1-D dimension
- Select Chemical Engineering Module $\rightarrow$ convection and diffusion
- Element: Lagrange - quadratic
- More >>
- OK

Pull down the Draw menu and select Specify Geometry.

Draw Mode
- Name: interval
- Start: 0
- Stop: 1
- Apply / OK

Now for the boundary conditions. Pull down the Boundary menu and select Boundary Settings.

Boundary Mode
- Select domain 1
- Check $-N \cdot n = Pe*(1-c)$
- Select domain 2
- Select $-N \cdot n = 0$
- Apply / OK

Pull down the options menu and select Add/Edit constants. The Add/Edit constants dialog box appears.

Add/Edit Constants

- Assign Pe, B_2, n_{lim}, and q with the values from Table 3.2
- Apply / OK

Now pull down the Subdomain menu and select Subdomain settings.

Subdomain Mode

- Select domain 1
- Set u=Pe; R= B2*(nlim-c)*exp(-QQ/c)
- Apply
- Select the init tab; set c(t0)=1.656
- Apply / OK

Now Pull down the Subdomain menu and select View as PDE Coefficients, then reselect Subdomain settings. We will tackle the steady state solution first. We could put d_a=0 as the mass coefficient now, but in fact the selection of the Solver (Stationary nonlinear or Time Dependent) will make the appropriate choice for us.

The mesh is extremely important here, as rapid variations near the boundary conditions are expected. Pull down the Mesh menu and select Parameters. By creating symmetry boundaries, the endpoints become equivalent.

Mesh Mode

- >>More
- Max size near vertices: 1 0.0001 2 0.0001
- Number of Elements in Subdomain: 1 1000
- Apply
- OK

This results in a 1312 element meshing. Note that "Max size near vertices" takes a vector entry with each pair of elements of the form: vertex number followed by maximum size. This constrains the elliptic mesh generator to give appropriately large or small elements as directed. Similarly, "Number of elements in subdomain" can be set by a MATLAB vector entry. Since there is only one subdomain here, only one pair (subdomain number, number of elements) can be specified.

Now for the Solver. Pull down the **Solver Menu** and select **Parameters**. Check the Stationary Nonlinear solver box, apply, and click on the Solve button. It takes FEMLAB 25 iterations to get there (this is a highly nonlinear problem), but it converges to 10^{-8} accuracy eventually. I played around with several meshes and parametric approaches before hitting on this one. Figure 3.8 holds the steady state solution for this case.

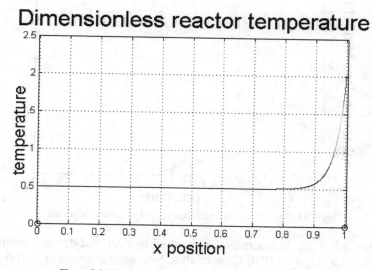

Figure 3.8 Steady state solution for parameters in Table 3.1.

At this point, we need to prepare for our stability study by creating an appropriate initial condition function m-file. The recipe is as follows:

From the File menu, export FEM structure to workspace as fem.

In MATLAB, save the data needed to build the steady solution:

```
>> [xs,idx]=sort(fem.xmesh.p{1});
>> u=fem.sol.u(idx);
>> u=u';
>> save amundson.mat xs u
```

Build a function m-file initcond.m

```
function a=initcond(x)
load amundson.mat
a=interp1(xs, u, x, 'spline')+0.05*sin(31.4159265*x);
```

Figure 3.9 shows the result of plot(xs,initcond(xs)):

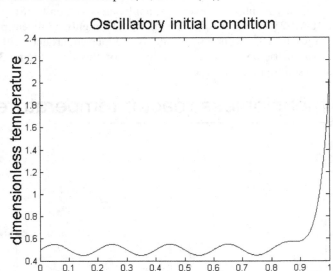

Figure 3.9 Steady solution with superimposed period one-fifth sine.

Now Pull down the **Subdomain** menu, then reselect **Subdomain settings**. Then un-check **View as PDE Coefficients**. Now enter initcond(x) as c(t0) and check the time-dependent solver in FEMLAB. Enter output times as 0:0.0001:0.003. Then under the **Post menu**, select **Plot Parameters** and **Animate**. Enjoy the experience of watching the perturbation dissipate and propagate out of the system.

Now for Table 3.3. Change the Add/Edit Constants under the Options menu to reflect the higher B_2 and QQ coefficients. n_{lim} stays the same. Use the stationary nonlinear solver from the oscillatory initial condition. What does the converged solution look like?

Converging from a solution that is not so similar to the steady-state is not so straightforward. Try the initial condition c(t0)=1. Does it converge with the stationary nonlinear solver? Now try the time dependent solver and set output times to 0:0.0001:0.01, then animate the solution. As you can see, the time dependent solver is attracted to the steady solution, but the stationary nonlinear solver wasn't "close enough" in solution space to fall onto the solution. Now perhaps it is. Check the stationary nonlinear solver, apply and cancel. Then click on the re-start button on the toolbar, which takes the initial solution as the last of the time dependent solutions. This should converge in about five iterations to the same solution found earlier.

Now try the initial condition $c(t0)=n_{lim}$. With the time dependent solver, set output times to 0:0.001:0.1, then animate the solution. You should be able to watch the initial condition pass completely out of the reactor and converge to the steady state solution found by the two previous methods. This should be a clear signal that the time dependent solver may be an essential tool in attacking non-convergence. Even in problems that have no inherent unsteady time scale, pseudo-time dependent solution may be essential to finding a converged stationary solution. If so, then we can be fairly certain that the steady state so found is stable, since it is attractive.

Exercise 3.1

This chapter is entitled "Multiphysics." The problem statement is definitely for two physics modes (heat and mass transport with reaction), yet due to Amundson's technique, the problem could be simplified to "single physics" for $\dfrac{\alpha}{\mathscr{D}} = 1$. Try implementing the calculation with the ChemEng Module modes convection and conduction (cc) and convection and diffusion (cd) with the same parameters, but as written in equations (3.7), (3.8). Take the initial condition to be uniform temperature $n=1$. Solve for the steady state after a long time, or use the steady solver. Compare you results with the Amundson technique solution given here.

3.4 Heterogeneous Reaction in a Porous Catalyst Pellet

It would be an injustice not to draw on the FEMLAB Model Library for an example of multiphysics. Although the chemical engineering curriculum does not contain many examples of multiphysics partial differential equations, the same cannot be said for the chemical engineering model library of FEMLAB. In keeping with the complementary of this text with the FEMLAB manual set, however, we must treat any problem that we extract differently. This section is inspired by the heterogeneous reaction modeling in a porous catalyst pellet, treated in [9]. The model uses the incompressible Navier-Stokes application mode and couples the results to the convection and diffusion mode through the multiphysics facility of FEMLAB in two-dimensions (c.f eqn (3.7)), adding the additional reaction term in the pellet subdomain, but without convection (reaction-diffusion model). This clearly counts as multiphysics by either definition, since there are two different PDEs with independent variables u,v,p, and c having different units. The twist that we add to the model is to decouple the multiphysics.

Because the reaction is taken to be isothermal and constant density, there is no back action coupling the concentration field into the Navier-Stokes equations. The mass transport requires knowledge of the velocity field to compute convection, but this does not change the momentum transport. So rather than computing both momentum transport and mass transport simultaneously, we can compute them sequentially. Why? Primarily because of the computing efficiency. If one requires several solutions over a range of mass transport/reaction parameters, but with the same flow field, then computing the flow field only once and importing the velocity field is the most computationally efficient method (or should be, if coded efficiently). Secondly, whatever platform you use to compute on is probably memory limited if you want to refine the mesh. For instance, because we computed the streamfunction explicitly in the buoyant convection example earlier, it was not possible to refine the mesh further without running out of system memory on a 1Gb RAM linux PC workstation. The final reason is that it illustrates further handles into the FEMLAB GUI through MATLAB programming, which is one of the reasons to read this text.

We visited the Incompressible Navier-Stokes (2-D) mode in §3.1, and in fact if we add a reaction source term to (3.1) and call concentration T rather than c, then those equations describe the model perfectly.

Launch FEMLAB and in the Model Navigator, select the Multiphysics tab.

Model Navigator
- Select 2-D dimension
- Select Physics modes⇒Incompressible Navier-Stokes >>
- OK

We will now follow the recipe on [9], p. 2-78ff to construct the configuration and Navier-Stokes solution around the pellet. Set up the axis and grid as follows Pull down the options menu and select Axis/Grid Settings.

Axis/Grid Settings

Axis		Grid	
Xmin	-0.001	X spacing	0.001
Xmax	0.003	Extra X	0.0009
Ymin	-0.001	Y spacing	0.001
Ymax	0.007	Extra Y	0.0021 0.0039

Next select the Add/Edit constants options and enter as below.

Add/Edit Constants

Name	Expression
ro	0.66
mu	2.6e-5
vo	0.1

We will now follow the recipe on [9], p. 2-79 to construct the geometry.

Draw Mode
• Deselect solid by double clicking on **SOLID** on the **Status Bar**
• Draw an arc C1 by clicking at the corners (0,0.0039), (0.0009,0.0039), (0.0009,0.003), (0.0009,0.0021) and (0,0.0021)
• Select solid by double clicking on **SOLID** on the **Status Bar**
• Draw a rectangle R1 with corners at (0,0) and (0.002, 0.006)

Now pull down the Subdomain menu and select Subdomain settings.

Subdomain Mode
• Select subdomain 1
• Enter PDE coefficients
• Set ρ=ro; η=mu
• Select subdomain 2
• Uncheck the "Active in this subdomain" tick box

Now for the boundary conditions. Pull down the Boundary menu and select Boundary Settings.

Boundary Mode
• Select domain 2
• Select inflow BCs with u=0 and v=vo.
• Set boundaries 1,4,6 with Slip/Symmetry
• Set boundary 5 with outflow BC p=0.
• Set boundary 7,8 with No-Slip
• Apply / OK

No pull down the Mesh menu and select the Parameters option. We will need to pack elements near the pellet for best resolution.

Mesh Parameters
• Select more>>
• Max element size near vertices: 6 1e-4 7 5e-5 8 5e-5
• Remesh and then Refine the mesh once more
• OK

There should be 6908 triangles. Click on the = button on the toolbar to Solve.

The flow field should look like Figure 3.10. It is convenient to save the geometry model using the "Export to file" feature on the File menu, since we will need it later for the reaction-diffusion model.

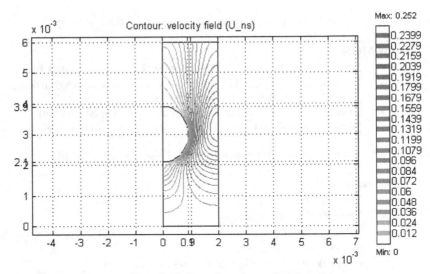

Figure 3.10 Streamlines surrounding the pellet.

In the subsequent steps of this chapter, I will encourage you to store this solution and then write a MATLAB interpolation file to read it in as a fixed velocity profile for the convection-diffusion equation. There is a better, GUI-based way to do this using the Multiphysics: Solve for Variables feature (see Solver Parameters, Multiphysics tab). Basically, you turn off the mass transport mode and solver for the velocity field. Then you turn off the Navier-Stokes mode, and solve for the mass transport using the restart button on the toolbar, which uses the velocity profile just computed as fixed throughout. This solves for the modes sequentially, rather than coupled. In some models, we have found that the sequential method converges to a solution, where the joint solution does not. Here, the exercise is worth doing to learn something about m-file "hooks" into FEMLAB.

Now we will export our solution to the MATLAB workspace (file menu, export to workspace, fem structure). Next we arrange the solution in a convenient format using postinterp:

```
[xx,yy]=meshgrid(0:0.00002:0.002,0:0.00002:0.006);
xxx=[xx(:)'; yy(:)'];
u=postinterp(fem,'u',xxx);
v=postinterp(fem,'v',xxx);
uu=reshape(u,size(xx));
vv=reshape(v,size(xx));
isn=find(isnan(uu));        %calls to postinterp in the pellet give
uu(isn)=zeros(size(isn)); %NaN.  This changes them to zero.
isn=find(isnan(vv))
vv(isn)=zeros(size(isn));
save pellet_flow.mat xx yy uu vv;
```

On to the MATLAB m-file function for the u-velocity

```
function u=pelletu(x,y)
%PELLETU Interpolates u from the FEM solution for the pellet
%    U = PELLETU(X,Y)
%    is interpolated on the rectangle [0,0.002] x [0.006]
load pellet_flow.mat xx yy uu vv
% Interpolate from rectangular grid to unstructured point.
u=interp2(xx,yy,uu,x,y);
```

Similarly for the v-velocity

```
function v=pelletv(x,y)
%PELLETV Interpolates u from the FEM solution for the pellet
%    V = PELLETV(X,Y)
%    is interpolated on the rectangle [0,0.002] x [0.006]
load pellet_flow.mat xx yy uu vv;
% Interpolate from rectangular grid to unstructured point.
v=interp2(xx,yy,vv,x,y);
```

These functions were used to produce the following pair of contour plots. Note that in this case, v, the vertical velocity, is the "flowwise" component, and u is the transverse velocity.

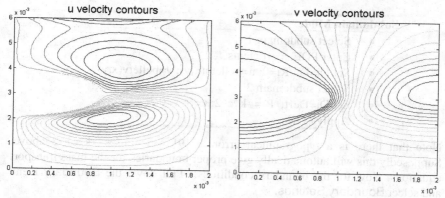

Figure 3.11 u-velocity around the pellet.
Figure 3.12 v-velocity around the pellet.

The function m-files are now ready for use in the reaction-diffusion mode.

Reaction/convection/diffusion model

If FEMLAB is already launched, select New from the File Menu, which will bring up the Model Navigator.

Model Navigator
- Select 2-D dimension
- Select ChemEng Modes⇒Cartesian Coordinates⇒Mass Balance⇒Convection-Diffusion⇒Time-Dependent>>
- OK

Now use the File menu to "Insert from File" and select your Geometry mat-file (pellet_geom.mat).

Next select the Add/Edit constants options and enter as below.

Add/Edit Constants

Name	Expression
D	1e-5
Deff	1e-6
k	100
clo	1.3

Now pull down the Subdomain menu and select Subdomain settings.

Subdomain Mode
- Select subdomain 1
- Enter PDE coefficients
- Set Di=D; Ri=0; u=pelletu(x,y); v=pelletv(x,y)
- Select subdomain 2
- Set Di=Deff; Ri= -k*c^2; u=0; v=0
- Apply / OK

Note that there is a typographical error in [9] which uses $-k*c1^2$ above. Supposedly this will automatically give proper boundaries for the mass transport equations. Now for the boundary conditions. Pull down the Boundary menu and select Boundary Settings.

Boundary Mode
- Select domain 2
- Select ci=clo.
- Set boundaries 1,3,4,6 with Insulation/Symmetry
- Set boundary 5 with convective flux BC
- Apply / OK

Select the standard coarse mesh (triangle button) from the Toolbar. Select the stationary nonlinear solver and click Solve.

Now pull down the Mesh menu and select the Parameters option. We will need to pack elements near the pellet for best resolution.

Mesh Parameters

- Select more>>
- Max element size near vertices: 6 1e-4 7 5e-5 8 5e-5
- Remesh
- OK

There should be 1727 triangles yet again. Click the Re-Solve toolbar button. Then refine the mesh to 6908 triangles. Again Re-Solve. The final solution should look like Figure 3.13 below (cf. [9], p. 2-83):

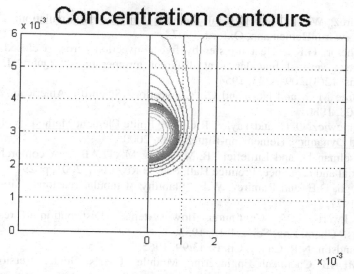

Figure 3.13 Contours of concentration (reduced by disappearance in pellet phase).

3.5 Discussion

This chapter introduced the concept of multiphysics and then ran away from it. One of the key messages from §3.3 and §3.4 should be to simplify the analysis of your PDE system. For instance, in §3.3, multiphysics was not necessary due to a change of variable to eliminate one PDE equation. Similarly, in §3.4, multiphysics could be dealt with by sequential treatment of single physics. This has the virtue of separating the work. It should be clear that the major difficulty in solving the pellet reaction/diffusion/convection problem was resolving the fluid dynamics of the hemispherical obstruction. The concentration profile was resolved on the most coarse mesh chosen.

Without too much aggrandizement, this chapter also shows the ease of solution of highly nonlinear problems by parametric continuation (typically automated by MATLAB programming). The chapter also shows how to include variable physical properties and complicated field behaviors by interpolation functions programmed in MATLAB m-file functions.

Multiphysics is a recurrent theme in this text, largely because "single physics" is well studied. Thus, inherently, multiphysics descriptions are required for state-of-the-art research models. So several more examples will follow.

References

1. Ramirez, W.F., Computational Methods for Process Simulation, 2nd ed., Butterworth Heinemann, Oxford, 1997.
2. Batchelor, G.K., "Heat transfer by free convection across a closed cavity between vertical boundaries at different termperatures." Quart. J. Appl. Maths. 12(3):209—233, 1954.
3. Velarde M.G. and Normand C. "Convection" Scientific American, 243(1): 92-108, 1980.
4. Zienkiewicz, O. C. and Taylor, R.L., The Finite Element Method. Volume 3: Fluid Dynamics, Heinemann-Butterworth, 2000.
5. Hanselman D. and Littlefield B.,Mastering MATLAB 6: A comprehensive tutorial and reference, Prentice Hall, Saddle River NJ, 2001, p. 294.
6. Clough, D.E. and Ramirez, W.F. "Stability of tubular reactors," Simulation 16, 1971.
7. Danckwerts, P.V. "Continuous flow systems. Distribution of residence times." Chem. Eng. Sci. 2:1-18, 1953.
8. Amundson, N.R. Can. J. Ch. E. 43:99, 1965.
9. FEMLAB Chemical Engineering Module, User's Guide, Version 2.2, p. 2-74.

Chapter 4

EXTENDED MULTIPHYSICS

W.B.J. ZIMMERMAN[1], P.O.MCHEDLOV-PETROSSYAN[2*]
G.A. KHOMENKO[2]

[1]*Department of Chemical and Process Engineering, University of Sheffield, Newcastle Street, Sheffield S1 3JD United Kingdom; [2]Labratoire d'Oceanografi,Cotiere du Littoral, ELICO, Universite du Littoral Coted'Opale, MREN, 32, Avenue Foch, 62930, Wimereux, France. *Permanent address: NSS Kharkov Institute of Physics and Technology, Ukraine*

E-mail: w.zimmerman@shef.ac.uk

Extended multiphysics is a feature that is conceptually complicated and original with FEMLAB. The concept is the linkage of two or more logical computational domains through coupling variables that can be used in either specifying the boundary conditions or subdomain PDE coefficients. The coupling variables can be found by subdomain or boundary integrations, internal or boundary values. These naturally arise in the multiple scale modeling of physical phenomena – the large scale model is coupled to subgrid cellular models, perhaps of a simpler parametric or lower spatial dimension. Extended multiphysics is ubiquitous in process engineering, however, because unit operations are conceptually separate domains, yet linked through at least inlet and exit conditions sequentially, but frequently linked more subtly through process integration. So the whole field of process simulation for optimization, design, retrofit, and control falls within the remit of extended multiphysics. Integration with Simulink gives the possibility of some unit operations being treated with distributed PDE models while others are treated with lumped parameters, yet with non-trivial levels of coupling requiring extended multiphysics modeling.

4.1 Introduction

If multiphysics, the subject of the last chapter, were a new concept to you, *extended* multiphysics must be a more alien concept indeed. So far, I have seen only one application of extended multiphysics – the Packed Bed Reactor model in the Chemical Engineering Module/Model library [1]. Initially I thought extended multiphysics was about coupling multiple scale models, as that is how it was done in [1]. This is a cutting edge area of research in multiphase flows/heterogeneous systems, because the dispersed phase can be treated as a point constraint in the domain of the bulk medium, but with information flowing in both directions. Usually the attempt is to treat such constraints parametrically, i.e. modeling the dependence of the small scale phenomena on bulk phase unknowns, and vice versa to complete the coupling of the scales. Usually, the small scale phenomena is too complex in its own right, for instance in the microhydrodynamics simulations of Grammatika and Zimmerman [2], to consider solving simultaneously with the bulk dynamics. So the coupling is through simple functional forms learned from simulations of the small scale dynamics, slaved to the large scale phenomena imposed on it. There are several drawbacks to the parametric slaving approach, but they are all summed up by "oversimplification". Fortunately, such models can be verified by experimentation that the physical systems can be well treated by the two scale approach. Traditional turbulence models are all heavily reliant on multiple scale modeling by parametrization. Since the multiple scale modeling techniques are specialized, perhaps extended multiphysics is not such a useful feature after all. To take advantage of it for complex modeling may require high performance computing.

Only belatedly did it occur to me that chemical engineering is awash with applications for extended multiphysics. First, let's give an operational definition for extended multiphysics in the FEMLAB sense: a model is categorized as extended multiphysics if it requires description of field variables in two or more *logically* disjoint domains. They are not likely to be physically disjoint domains since the physics must be coupled in some respect to warrant solving the problems in each domain jointly. FEMLAB allows the user to use several different geometries/application mode pairs in building up an extended multiphysics model.

So why is it that chemical engineering is awash with extended multiphysics? Look no further than your nearest flowsheet, say Figure 4.1.

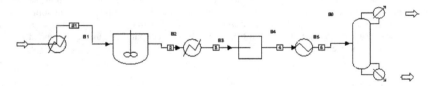

Figure 4.1 Flowsheet for a linear array of unit operations.

For the process in Figure 4.1,

"Cyclopropane at 5 bar and 30°C is fed at a rate of 1kmol/h. It is heated to a reaction temperature of 500°C by a heat exchanger before entering as CSTR (continuously stirred tank reactor). The reactor has a volume of 2 m^3 and maintains the reaction temperature of 500°C. The isomerization reaction:

$$C_3H_6 \rightarrow CH_3CH = CH_2 \qquad (4.1)$$

is first order with a rate constant of k=6.7×10^{-4} s^{-1} at 500°C. The products of the reaction are then cooled to the dew point by a second heat exchanger before entering a compressor. The compressor increases the pressure to 10 bar, the pressure at which separation of reacted propylene and unreacted cyclopropane will occur. The compressed gas is then condensed to bubble-point liquid and feed to a distillation column. The column has 31 ideal stages with the feed onto the stage 16. It operates with a total condenser and a molar reflux ratio of 8.4 producing a distillate flow rate of 0.292 kg-mol/h."

Sound familiar? Such scenarios populate modules on "Process Engineering Fundamentals." Why is it extended multiphysics? Each unit operation constitutes its own logical domain, connected to the others by entry and exit points. In the conceptual design stage of such a plant, the unit operations are treated by simplified models to permit facile exploration of the configuration space. Process integration by means of recycle and heat exchanger networks adds greater complexity to the flowsheet, and greater scope for economies in operating and capital costs. Eventually, however, the process engineer has to give detailed designs for such plant. These days that includes process simulation, typically including optimisation, parametric sensitivity studies and transient analysis. And even if the plant were designed a generation ago, process studies of this nature are common for retrofit and optimisation. In many cases, plant were over designed by 30-50% (since such flexibility is a common safeguard in design), so now that the plant is operational, efficiency savings of 30-50% should be achievable. Thus has grown the burgeoning field of process systems optimisation. And this is a regime for extended multiphysics. If any of the unit operations in Figure 4.1 are to be modelled in detail, that usually involves a spatial-temporal PDE where the simplified model used in design might have been a lumped parameter model. For instance, suppose the reactor in Figure 4.1 is CSTR reactor which is jacketed by a bath of its product liquid (at 500°C) before entering the heat exchanger proper. Temporal fluctuations in the reactor temperature propagate through the bath to the heat exchanger, requiring control action, which in turn leads to transients in the compressor operation. These feed into the distillation column. Presuming the separated unreacted cyclopropane is recycled back to the feed to the reactor, the temperature fluctuations into the distillation column will have translated into composition fluctuations in the recycle stream, which will then effect the reactor conversion, starting the whole cycle again. The plant should be designed to dampen fluctuations back to the set point, rather than reinforce them. Extended multiphysics is in play at every level of process coupling. In the linear flowsheet of Figure 4.1, it is possible to isolate the modelling of each unit operation, since the entry and exit points are the only overlaps. It is still extended multiphysics if you want to link them up in FEMLAB, but the linkages are simple. But if process integration enters in, then the linkages may be more thorough. For instance, in distillation columns, differential heating and cooling of stages can be done to influence separation efficiency (with multiple entry and exit points for various "fractions"). These streams can be crossed for heat integration and recycled for reactor integration. Thus, "lumped variables" of unit operations become distributed constraints for others. That FEMLAB can be called by Simulink for greater detailed modelling of some unit operations is a feature that allows better plant simulation. The commercial plant simulation packages, such as AspenPlus and HYSYS, have implemented links to computational fluid dynamics packages to improve detailed simulation of selected unit operations. This trend will be come a flood, as it is less expensive and safer to simulate "what if" scenarios than to implement them on real plant. Examples of extended multiphysics will make the concept clearer. We will start with a 1-D convection-diffusion-heterogeneous reaction model for a fixed bed supported catalyst system.

4.2 Heterogeneous Reaction in a Fixed Bed with Premixed Feed

Recently Mchedlov et al. [3] proposed a general lumped parameter model for heterogeneous reaction in a dispersed phase. The model focuses on situations where mass transfer is asymmetric, i.e. some species have greater mass transfer coefficients with the dispersed phase than others. Any number of physicochemical interactions could lead to this situation, but invariably it is in only slow flows, as through porous media, where kinetic asymmetry can survive. Turbulence usually leads to equal mass transfer coefficients for each species. Consider the reaction

$$u + v \rightleftharpoons w \tag{4.2}$$

which only occurs in the dispersed phase. The lumped parameter model gives three convection-diffusion-mass transfer equations in the bulk phase, which for steady operation read as:

$$U \frac{\partial u}{\partial z} = D_u \frac{\partial^2 u}{\partial z^2} - j_u (u, \tilde{u})$$

$$U \frac{\partial v}{\partial z} = D_v \frac{\partial^2 v}{\partial z^2} - j_v (v, \tilde{v}) \tag{4.3}$$

$$U \frac{\partial w}{\partial z} = D_w \frac{\partial^2 w}{\partial z^2} - j_w (w, \tilde{w})$$

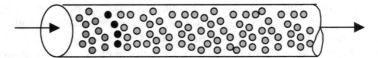

Figure 4.2 Schematic of a fixed bed with reaction largely localized.

The fluxes j take the traditional mass transfer coefficient form

$$j_u = \kappa_u\left(u - \tilde{u}\right)$$
$$j_v = \kappa_v\left(v - \tilde{v}\right) \qquad (4.4)$$
$$j_w = \kappa_w\left(w - \tilde{w}\right)$$

At steady state, these fluxes are all equal and thus give two constraints on the bulk variables u, v, w and on the disperse phase concentrations $\tilde{u}, \tilde{v}, \tilde{w}$. The sixth constraint is on the surface reaction, which is presumed to be in equilibrium (fast reaction kinetics and nearly irreversible):

$$\tilde{u}\tilde{v} - K\tilde{w} = 0 \qquad (4.5)$$

The boundary conditions will be taken as fixed concentrations of u and v at the inlet, no w, and outlet conditions with convection much greater than diffusion. For simplicity, since there are so many parameters, we will test just kinetic asymmetry of the mass transfer parameters and fix unit diffusivities $D_u=D_v=1$, mobile product $k_w=100$ and $D_w=0.001$, one of the reactants to have unit mass transfer coefficient $k_v=1$, and this leaves free parameters as the velocity U and mass transfer coefficient of the most resistive reactant, k_u, reactor length L, and equilibrium constant K. Since industrial interest lies in reactions that favor the products, we shall take $K=10^{-5}$ as a nearly irreversible reaction. Initially, let's consider a reactor of length L=5, velocity U=0.5, and mass transfer asymmetry with $k_u=0.2$. The inlet conditions will be u0=1 and v0=0.4.

Now to set up the FEMLAB Model:

Start up FEMLAB and enter the Model Navigator.

Model Navigator
• Select 1-D dimension
• Select PDE modes → general → time-dependent
• Element: Lagrange - quadratic
• Specify three independent variables
• More >> mode name: bulk ; name variables: U V W; ind var: z
• OK

Now pull down the Multiphysics and Select Add/Edit Modes.

Multiphysics Add/Edit Modes
• Select PDE modes → general
• Insert the mode name: surface
• Specify three independent variables: US VS WS
• Add across >>
• OK

There should have been two "Geom1: PDE general form" modes listed on the right hand column. Pull down the Draw menu and select Specify Geometry.

Draw Mode
• Name: interval
• Start: 0
• Stop: 5
• Apply
• OK

Pull down the options menu and select Add/Edit constants. The Add/Edit constants dialog box appears. Assign the constants as below.

Add/Edit Constants

u0	v0	Du	Dv	Dw
1	0.4	1	1	0.001
u	K	ku	Kv	Kw
0.5	1e-5	0.2	1	100

Now pull down the Subdomain menu and select Subdomain settings. Assign the model constants for the six equations as below:

Subdomain Mode

• Select bulk mode. Enter the PDE forms as below for each tab

	U	V	W
Γ	-Uz*Du	-Vz*Dv	-Wz*Dw
F	-u*Uz-ku*(U-US)	-u*Vz-kv*(V-VS)	-u*Wz-kw*(W-WS)
da	1	1	1
init	u0	v0	0

• Select surface mode. Enter the PDE forms as below for each tab

	US	VS	WS
Γ	0	0	0
F	ku*(U-US)-kv*(V-VS)	ku*(U-US)+kw*(W-WS)	US*VS-K*WS
da	0	0	0

• OK

Now for the boundary conditions. Pull down the Boundary menu and select Boundary Settings. Setup the boundary conditions as in the table below.

Boundary Mode

- Use the Mulitphysics mode to switch to mode bulk. Select domain 1 and enter Dirichlet BCs

U	V	W
R=u0-U	R=v0-V	R=-W

- Select domain 2
- Enter Neumann BCs (G=0) for all variables.
- Use the Mulitphysics mode to switch to mode surface. Select domain 1,2 and enter Neumann BCs (G=0) for all variables.
- Apply
- OK

The Neumann BCs for U,V,W require the normal component of Γ to vanish at the outflow boundary. Since $\hat{n} \cdot \Gamma_U = D_u \dfrac{\partial U}{\partial n}$ is the diffusive flux, this BC enforces a no diffusive flux boundary condition. So all the flux is convective, i.e. outflow. For the surface variables, however, $\Gamma = 0$ was specified, so entering zero Neumann conditions is a non-constraint (0=0).

Pull down the **Mesh** menu and select **Parameters**.

Mesh Mode

- >>More
- Max size near vertices: 1 0.0001 2 0.0001
- Number of Elements in Subdomain: 1 1000
- Apply
- OK

This results in a 1000 element meshing. Now for the Solver. Pull down the **Solver Menu** and select **Parameters**. Check the Stationary Nonlinear solver box, apply, then under Settings de-select the Automatic Scaling and select None. Now, click on the Solve button. It takes FEMLAB 9 iterations to get there (this is a highly nonlinear problem), but it converges to 10^{-8} accuracy eventually.

Figures 4.3 and 4.4 show the behaviour of the reactant concentrations. Figure 4.4 in particular requires interpretation. Because of kinetic asymmetry, $\tilde{u}$ and $\tilde{v}$ vanish in different sections of the reactor. Because v has greater mass transfer coefficient it populates the surface initially, $\tilde{u}$ reacts instantaneously as it arrives on the surface. As u is in bulk excess, however, eventually $\tilde{v}$ reacts away as well, until we reach the crossover point, where both surface reactants vanish. This is actually the point of greatest molecular efficiency, since any molecule of u or v that arrives on the surface reacts here. The theory of

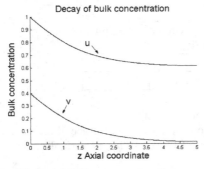

Figure 4.3 Bulk concentrations decay.   Figure 4.4 Surface concentrations exhibit crossover.

Mchedlov et al. [3] predicts the existence in parametric space of a crossover point, and gives a good approximation of its position X based on nearly irreversible reaction. Clearly, the actual profile requires solution of a two point boundary value problem with three conditions at either end. The system of equations (4.3)-(4.5) is a combined differential-algebraic system, which is inherently "stiff" due to the difficulty in satisfying the three nonlinear algebraic constraints simultaneously. Mchedlov et al. achieved it by shooting methods with stiff ODE integrators. The FEMLAB solution naturally permits the satisfaction of two point BVPs and analytically determines the Jacobian of the nonlinear system, automatically with its symbolic tools. Mchedlov et al. determined the general Jacobian for their system, but due to the simple stoichiometry, used elimination to reduce the problem to a third order reaction-convection-diffusion system with highly nonlinear constraints. In terms of programmer effort, the FEMLAB solution took an evening, the shooting method took several months.

Reactor-Separator-Recycle Extended Multiphysics

You would be forgiven for asking where in the above heterogeneous reactor model is the extended multiphysics. Although we saw rather clever use of FEMLAB to solve a differential-algebraic system, there is not yet any extended multiphysics coupling. So now let's consider our reactor as part of a very simple flowsheet with a separator and recycle.

The feed rates are taken as u_f, v_f. The reactor inlet rates are u_0, v_0. The reactor exit rates are u_e, v_e, w_e. The separator is taken as an ideal separator, but with a temporal response. For instance, a buffer tank where product w phase separates. The recycle rates are u_r, v_r. With steady operation, the separator outlet rates must equal the inlet rates. However, we are interested in the temporal response potentially, so we will model the separator as a buffer tank with an effective capacitance.

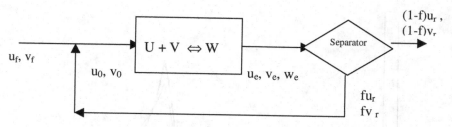

Figure 4.5 Simple flowsheet with heterogeneous reactor, separator and recycle.

Zimmerman [4] derived a model for imperfect mixing in buffer tanks due to stratification effects. A model flow configuration in a buffer tank with a two layer flow stratification was considered. The lower, denser stream is presumed to short-circuit to the outlet, driving a recirculating cavity flow in the upper layer. As the upper layer can be argued, due to strong convective dispersion, to be well mixed, mass transfer to the upper layer from the dense stream is the limiting step. In analogy with a plug flow reactor, a shell balance on the material fluxes in a slug of the lower stream leads to a lumped parameter mixing model with two limiting conditions: (1) no mixing at infinite superficial velocity of throughput; and (2) perfect mixing with infinite mass transfer coefficient. The time dependence of the model is readily described as

$$u_r = Eu_e + (1-E)u_c$$
$$\frac{du_c}{dt} = \frac{F}{V}(u_e - u_c)(1-E) \tag{4.6}$$

F and V are the volumetric throughput and the volume of the buffer tank, respectively. A similar set of equations holds for v. E is the lumped parameter that describes the capacitance of the buffer tank. The latter, equation (4.6), is the equation for the voltage response of the capacitor u_r in a driven RC-circuit with loading (1-E) u_e and RC time constant 1-E [5]. Perfect mixing, analogous to a stirred tank model, occurs when E=0, which then has the fastest possible response time constant. Figure 4.6 shows clearly that the concentration u_c in the upper layer is "charged" as the pulse passes and "discharges" after the pulse in the lower stream has passed. The outlet concentration Figure 4.7, however, for the imperfect mixing cases E>0 shows jumps up and down in concentration u_r due to the combination of the inlet stream short circuiting and the mass transfer to or from the upper reservoir, consistent with the first equation of (4.6).

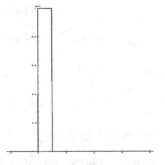

Figure 4.6a Pulse fed to buffer tank (u_e).

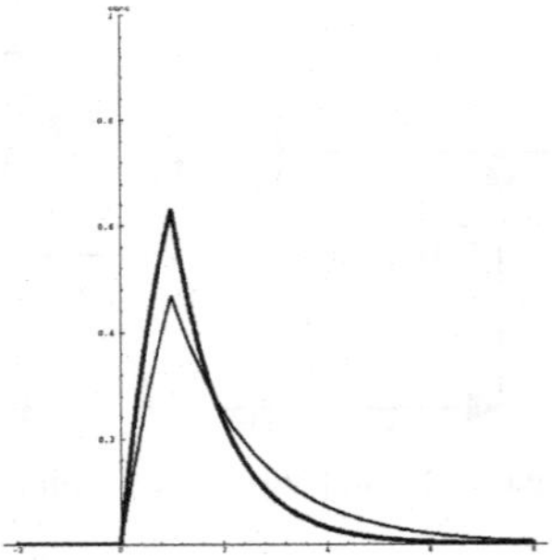

Figure 4.6b Pulse response of "charge" u_c in the capacitor for perfect mixing (thick line) and imperfect mixing E=0.5 (thin line).

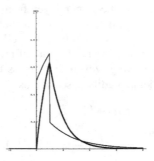

Figure 4.7 Outlet response of buffer tank to pulse in inlet to the pulse for perfect mixing (thick line) and imperfect mixing E=0.5 (thin line).

Simple mass balance can be used to compute the reactor inlet concentrations:

$$u_0 = u_f + fu_r$$
$$v_0 = v_f + fv_r \tag{4.7}$$

The non-trivial impact comes on reactor throughput. Taking the reactor to have unit area, the superficial velocity is given by:

$$U = U_{no \ recycle} \frac{u_0 + v_0}{u_f + v_f} \tag{4.8}$$

By the way, the fraction retained in the recycle, f, might not be unity in case the system needs a purge. It may be the case that the product take-off has the effect of stabilizing the system since the throughput U is greater than unity (the no recycle throughput), but a purge may be necessary to avoid an infinite recycle ratio due to build up of reactants or trace impurities. It is always good chemical engineering design practice to include a purge, and then minimize it in operation.

Implementing the Changes for Recycle

In principle, it should not be difficult to add a second weak subdomain mode for the buffer tank. In practice, it was rather frustrating, as I unearthed an apparent bug in the convection/diffusion application mode that took some ingenuity to create a workaround. Before finding the bug and workaround, however, I set up the FEMLAB model with several variations. This exercise turned out to be a *tour de force* in coupling variables.

The strategy is simply to create a 0-D domain (as we have done many times before) to implement the ODE in time (4.6) and appropriate couplings to import the reactor outlet concentrations and merge the recycle with the feed stream into the reactor inlet.

Pull down the Multiphysics menu and Select Add/Edit modes.

Model Navigator
• Select 1-D dimension
• Select weak subdomain mode, time dependent weak solver
• Element: Lagrange - quadratic
• Specify two independent variables: uc vc
• More >>
• OK

Pull down the Draw menu and select Specify Geometry.

Draw Mode
• Name: buffer
• Start: -2
• Stop: -1
• Apply
• OK

Now for the boundary conditions. Select mode w1. Pull down the Boundary menu and select Boundary Settings. Setup the boundary conditions as in the table below.

Boundary Mode
- Select domain 3,4 mode w1

	weak	dweak	constr
uc	0	0	0
vc	0	0	0

- Apply /OK
- Use the Multiphysics menu to select mode cd
- Set U=uinlet; V=vinlet on bnd 1
- Apply/OK

Zero-ing out the constraints returns the weak mode to natural (Neumann) boundary conditions.

Pull down the options menu and select Add/Edit constants. The Add/Edit constants dialog box appears. Assign the constants as below.

Add/Edit Constants
- Remove u, u0 and v0
- Set f=0; Ufeed=1.0; Vfeed=0.4; lump=0.5
- Apply
- OK

Now pull down the Subdomain menu and select Subdomain settings. Assign the model constants for the six equations as below:

Subdomain Mode
- Uncheck the "Active in this domain" box for domain 1
- Set domain 2

	weak	dweak	constr	init
uc	-ucx.*ucx_test+uc_test*(ue-uc)*(1-lump)	uc_time.*uc_test	0	ue
vc	-vcx.*vcx_test+vc_test*(ve-vc)*(1-lump)	vc_time.*vc_test	0	ve

- Apply
- OK

Now for the hard part. The coupling variables. Pull down the Options menu and select Add/Edit Coupling Variables. Enter the eight coupling variables as below:

Add/Edit Coupling Variables

Name	Type	Source	Integrand :Order		Destination
ue	scalar	bnd 2	U	: 1	sub 2
ve	scalar	bnd 2	V	: 1	sub 2
CU	scalar	sub 2	uc	: 1	sub 1:bnd 1
CV	scalar	sub 2	vc	: 1	sub 1:bnd 1
uue	scalar	bnd 2	U	: 1	sub 1:bnd 1
vve	scalar	bnd 2	V	: 1	sub 1:bnd 1
u0	scalar	bnd 1	uinlet	: 1	sub 1 & 2
v0	scalar	bnd 1	vinlet	: 1	sub 1 & 2

- Apply
- OK

Clearly, there are still undefined expression above, which are conveniently defined in the Add/Edit Expressions selection of the Options menu.

Add/Edit Expressions

Name	Type	Defined	Expression
uinlet	boundary	Geom1: bnd 1	Ufeed+f*CU*(1-lump)+f*lump*uue
vinlet	boundary	Geom1: bnd 1	Vfeed+f*CV*(1-lump)+f*lump*vve
u	geometry	Geom1	0.5*(u0+v0)/(Ufeed+Vfeed)

- Apply
- OK

So here is the faultline. The subdomain coefficients will only accept an expression defined on the *whole geometry* for the coefficient of x-velocity, called u, to which we assigned the expression as u above. Unfortunately, u0 and v0 cannot be defined as coupling variables in the *whole geometry* Geom1 in the GUI. Attempting to solve just elicits an error "Unknown variable or function u0." So here is the workaround. At this point, save a model m-file. Then edit the equivalent lines in the m-file

```
elemcpl{7}.src.bnd.var={'u0',{'uinlet'}};
elemcpl{7}.src.bnd.ind={3};
%elemcpl{7}.dst.g=1;
%elemcpl{7}.dst.equ.ind={[1 2]};
```

Comment out the dst fields. Do the same for v0. So what does this do? Page 5-73 of the Reference Manual tells us – "By default, the variables defined in src can be used everywhere in geometry 1. The optional field dst can be used to specify the domains of the definition for the variables in detail." So without the dst fields, the coupling variables u0 and v0 will be available at the geometry level, so that they can be used in defining u. Why the 'u' coefficient must be defined at the geometry level is certainly a bug, since other PDE coefficients need only be defined at the subdomain level. Read in this edited model m-file into the FEMLAB GUI to carry on. The definitions of u0 and v0 will now have the desired effects. Annoyingly, the Add/Edit Couplings dialogue window no longer launches with this alteration. But the computational model now solves.

Advice from COMSOL is that there is a simpler workaround for this apparent bug in the coupling variables, which allows you to solve this problem without playing around with m-files. You basically define a weak boundary mode (as opposed to a weak subdomain mode). The make sure to use non-conservative formulation for the convection-diffusion application mode. The reason for doing this is that the coupling variable or expression you define only needs to be accessed on a subdomain level, not on the boundary level.

Pull down the Mesh menu and select Parameters.

Mesh Mode
• >>More
• Number of Elements in Subdomain: 1 1000 2 10
• Apply
• OK

As is typical for FEMLAB, the pairs 1 1000 2 10 put 1000 elements in subdomain 1 and 10 in subdomain 2.

This results in a 1013 element meshing. Now for the Solver. Pull down the Solver Menu and select Parameters. Check the Stationary Nonlinear solver box, apply, and click on the Solve button. With no recycle and steady state, FEMLAB finds the same solution for the reactor as before. You can ramp up the recycle ratio gradually using the Re-Start button on the Toolbar. By f=10%, a modest variation in the position of the crossover point is noted. The effect of the recycle is to load even more U into the reactor, as well as to speed up the throughput somewhat.

The transient solution is a hard problem, but potentially more interesting as the "capacitor" takes a long time to charge in the buffer tank. Before undertaking the transient solution, however, we need to make some modest alterations to the PDE coefficients:

Pull down the Multiphysics menu, select surface mode, then pull down the Subdomain menu and select View as Coefficients. Finally, select Parameters.

Subdomain Mode (Surface Mode, View as Coefficients)
• Select the dweak tab, domain 2
• Replace US_time, VS_time, WS_time with zeros in the US,VS,WS coefficient boxes
• Apply
• OK

Simply, the mass transfer flux and surface reaction constraints are not time evolution equations. FEMLAB naturally gives these equations in the weak form a d/dt accumulation term in the time dependent solver. Without the time evolution terms, these algebraic constraints make the system rather stiff. The additional coupling variables also subtract from the "sparsity" of the system, thereby making the sparse matrix solvers strain harder to converge. So don't be surprised if the solver steps in extended multiphysics problems take longer.

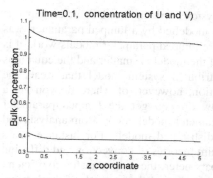

Figure 4.8 Bulk concentrations decay slowly.

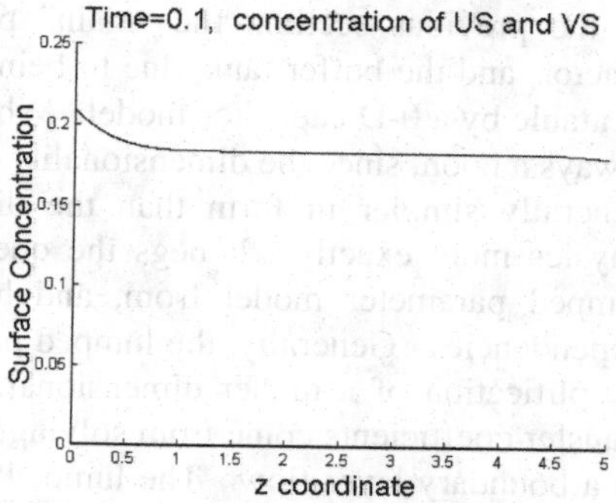

Figure 4.9 US is still depleted throughout the reactor.

The results are shown in Figures 4.8 and 4.9. One would typically expect that in a nonlinear reactive-diffusion problem at low Peclet number (about 0.5), the steady state would be well on its way at t=0.1. But the influence of the imperfect mixing in the buffer tank is apparent. The bulk and surface concentrations are slowly developing towards their steady state profiles, but the de-charging of the capacitor is a slow process, which will hamper the development of the steady state. If further system time lags were modelled, say the piping and control systems, then it is likely that this buffering resistance to developing the steady state from start up would be exacerbated.

Exercise 4.1

In §3.3, we added an oscillatory disturbance to the steady state solution of a tubular reactor and observed the transient response. Store the steady state solution to this heterogeneous reactor with buffer tank process, then create m-file functions for the initial conditions, adding an oscillatory disturbance to U only. How many m-file functions do you need to specify the steady state condition with oscillation? Comment about the transient response of the system.

Exercise 4.2

The "sluggishness of the buffer" tank model depends to a large extent on the ratio F/V in (4.6), which is an inverse time scale. In the FEMLAB model, implicitly, F/V was taken as unity. Explicitly add FoverV as a parameter, and explore the transient response when varying FoverV.

4.3 Primacy of the Buffer Tank

In the previous section, the "main" physics were in the 1-D heterogeneous reactor, and the buffer tank, due to being modelled by a lumped parameter, was treatable by a 0-D capacitor model. Where lumped parameter models work, it is always a boon, since the dimensionality of the model is smaller and the equations generally simpler in form than the distributed system model that treats the physics more exactly. It begs the question, however, of where do you get a lumped parameter model from, and how do you get the lumped parameter dependencies. Generally, the lumped parameter model comes from analysis and simplification of a higher dimensional, distributed model. For instance, mass transfer coefficients come from solving film theories of convection and diffusion in a boundary layer flow. The lumped parameter, the mass transfer coefficient, can be predicted from the shape of the particle and the strength of the laminar flow. In turbulent flows, the functional form of the mass transfer coefficient is found from empirical correlations. The buffer tank lumped parameter model of §4.1 was developed for a specified industrial application for assessing concentration fluctuations, and the lumped parameter was fitted from samples of inlet and exit conditions.

Certainly to treat a specific industrial unit operation, semi-empiricism is a reliable approach. In the case of the buffer tank that inspired [4], fluid density varied significantly with solute concentration (salinity), and thus the "capacitance" effect of the buffer tank was expected to be influenced by the rate of forced convection (throughput) F/V, viscous and mass diffusivity, and by the strength of free convection causing stratification, characterized by Reynolds, Prandtl and solutal Rayleigh numbers, respectively. Since buffer tank lumped parameter model of §4.1 only includes the throughput effects explicitly, the dependence of the lumped parameter E on Reynolds, Prandtl and solutal Rayleigh numbers is unknown. It is just taken as a constant found from representative conditions. Whether or not a lumped parameter model is sufficient depends on the type and accuracy of the predictions required from the process model.

If the buffer tank is small, or shocks in solute concentration fluctuations are prevalent upstream, the lumped parameter model may be insufficient in predictive powers. Greater detail in the modeling would then be warranted.

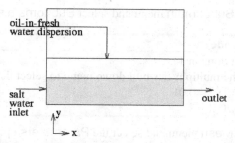

Figure 4.10 Schematic of a stratified buffer tank with potentially two types of inlets and one outlet.

In this subsection, we develop a two-component mass and momentum transport model for dense solute in a 2-D buffer tank. It is set up so that it can be augmented with a 1-D heterogeneous reactor with recycle model from the previous section. No lumped parameters are ever used in this model, as the detailed distributed effects of convection on mass transfer by coupling with diffusion and back action of density variations on convection are computed directly.

We will build up the model piecemeal, starting with the Incompressible Navier-Stokes model, then successively adding one mass transport mode, solutal Rayleigh effects, and then a second mass transport mode.

Component 1: Navier-Stokes flow field for cavity flow driven by free stream

Enter the Model Navigator.

Model Navigator
- Select 2-D dimension
- Select the Incompressible Navier-Stokes, time-dependent
- Element: Lagrange - quadratic
- Apply/OK

Pull down the Options menu and set the grid to $(0,5) \times (0,5)$ and the grid spacing to 0.5,0.5. Pull down the Draw menu and select Rectangle/Square (R1)and place it with vertices [0,0] and [0,5]. Now draw R2 with vertices [1,5] and [5,5].

Pull down the Boundary menu and select boundary settings

Boundary Mode
- Set up boundary 1 u=0; v=0, inflow
- Set up bnd 2,3,5,7 no slip
- Set up bnd 6 outflow p=0
- Apply/OK

Now pull down the Subdomain menu and select Subdomain settings.

Subdomain Mode
- Select domain 1
- Use the multiphysics pull down menu to select the IC NS mode
- Set $\rho=1$; $\eta=0.01$.
- Apply/OK

Now pull down the Mesh menu and select the Parameters option.

Mesh Parameters
- Select more>>
- Number of elements: 1000
- Remesh
- OK

Now to the Solver. Check that the Solver Parameters have set stationary nonlinear mode. Click solve. Now to save the results in a fashion suitable for restarting. Save the resulting FEMLAB workspace as tank_ns.mat.

Now let's follow our recipe for storing the solution as a MATLAB m-file function:

1. Export the fem structure to the MATLAB Workspace as fem.
2. Use postinterp to create a dataset for the u and v velocities and store to file.

```
[xx,yy]=meshgrid(0:0.05:5.0,0:0.05:5.0);
xxx=[xx(:)'; yy(:)'];
u=postinterp(fem,'u',xxx);
v=postinterp(fem,'v',xxx);
p= postinterp(fem,'p',xxx);
uu=reshape(u,size(xx));
vv=reshape(v,size(xx));
pp=reshape(p,size(xx));
save steadytank.mat xx yy uu vv pp;
```

3. Create m-file functions that interpolate in the dataset, e.g. tanku.m function u=tanku(x,y)

```
%TANKU Interpolates u from the FEM solution of the buffer tank
%    U = TANKU(X,Y)
%    is interpolated on the rectangle [0,0.002] x [0.006]
%
% Get the data
load steadytank.mat xx yy uu vv pp
% Interpolate from rectangular grid to unstructured point.
u=interp2(xx,yy,uu,x,y);
```

The tanku.m, tankv.m, and tankp.m m-file functions are now callable from FEMLAB's GUI as initial conditions for the tank. The velocity profile is a driven cavity flow, with free stream scaled so that the inlet velocity is unity. If you find this implementation of the initial conditions for the tank velocity field unwieldy, there is an alternative approach which uses the "Solve for variables" feature of the Solver Parameters to first select to solve only for the stationary velocity profile in the tank, without the mass transfer. Then with all variables turned back, solve with the restart button on the toolbar, taking the velocity field only as the initial condition from the previous solution. This mechanism was not available when this section was written with FEMLAB 2.2. To see it in action, chapter nine illustrates this methodology for electrokinetic flow.

Component 2: Passive scalar convection and diffusion equation

Purists will note that the Chemical Engineering Module comes with a convection and diffusion mode (cd). It is very good for the implementation of convective flux boundaries. However, if you wish to set the normal derivative of concentration to zero along an outflow boundary, it is clumsy to implement in cd mode. It is easier to implement in a standard coefficient mode, so we will tackle our solutal transport effects that way.

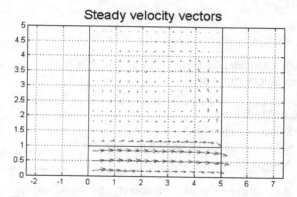

Figure 4.11 Cavity flow driven by a free current passing below.

Pull down the Multiphysics menu and select Add/Edit modes.

Multiphysics Add/Edit modes (c1)
- Select the coefficient mode, time-dependent
- Name the independent variable c1
- Element: Lagrange – quadratic
- Apply/OK

---

the **Boundary** menu and select boundary settings

Boundary Mode

- Set up boundary 1 Dirichlet h=1; r=1; (fixed c1=1)
- Set up bnd 2,3,5,6,7 Neumann q=0; g=0
- Apply/OK

Now pull down the **Subdomain** menu and select **Subdomain settings**.

Subdomain Mode

- Select domains 1&2
- Use the multiphysics pull down menu to select the IC NS mode
- Use the Init Tab to set up u(t0)=tanku(x,y); v(t0)=tankv(x,y); p(t0)=tankp(x,y)
- Apply/OK
- Select c1 mode
- Set coefficients c=1;a=0;f=0;da=1;α=(0,0);β=(u,v),γ=(0,0)

Now choose the stationary nonlinear solver and click on the solve button. Do not be surprised to find rapid convergence to a uniform concentration field and the same flow field as in Figure 4.11.

Component 3: Buoyancy effects of solutal mass transport

Save the FEMLAB model as tank_nsc1.mat. Now pull down the **Subdomain** menu and select **Subdomain settings**.

Subdomain Mode (ns)

- Select domains 1&2
- Set Fy=-0.25*c1 (Rayleigh number Ra=25)
- Apply/OK

Now choose the stationary nonlinear solver and click on the solve button. Again, do not be surprised that the steady state is a uniform profile, again with the driven cavity velocity field. Next try the time dependent solver with output times 0:0.1:1.0. The final concentration profile (Figure 4.13) just shows continual spreading of the concentration front, but no hint of a stratification forming (see Figure 4.12). The animation of the time series for the velocity vectors is suitably unenlightening – visually it never changes from the driven cavity vector field of Figure 4.11.

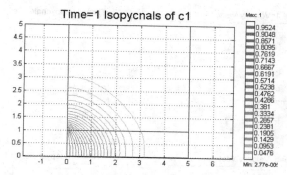

Figure 4.12 Isopycnals of c1 (t=1) from the initial state of cavity flow, solute-free buffer tank, and Ra=25.

So to achieve something more interesting, consider the no flow initial condition ($u(t0)=v(t0)=p(t0)=0$) with a no flow BC. Figure 4.13 shows the time history of with snapshots up to t=20 (diffusive time scale) of the free convection velocity and concentration profiles. Although the density stratification is weak, it is apparent that denser fluid stays below lighter fluid. Times 0-1, when animated, show the evolution of the gravity current as it spreads out along the bottom of the tank. The density front drives motion above and in front of it. Since c1=1 fluid entering is denser than the c1=0 fluid next to it, it literally falls over. Rottman and Simpson [6] have conducted laboratory experiments that beautifully illustrate the formation of gravity currents. Although at some time after t=1, the gravity current finds its way over to the constant pressure exit (whereupon it falls out), the gravity current continues to be the mechanism for driving the pseudo-steady flow. The fluid to the right is denser than the fluid to the left, so it just keeps on falling over. The initial push of fluid up and around that started the upper recirculation layer cycling does not maintain it. Rather, instead, it is the viscous drag from the gravity current layer that maintains the circulation above, much as how the free stream drives cavity flow.

The case of purely gravity current driven motion in a tank has not been studied before, so the two clear observations resulting from this model must be made. Firstly, the time to uniform concentration is extremely slow. The density variation with concentration not withstanding, one would expect nearly uniform concentration profiles after a few diffusion times, but in fact there were still substantial gradients after t=50. This is clearly due to the buoyant force opposing diffusive mixing, even in the presence of free convection which should, supposedly, enhance the mixing by dispersion. It is actually well known in the wave tank community that the ideal solution of fresh water/salt water can be used to set up any stable stratification density profile desired, simply because diffusion is such a weak mechanism that the profile is persistent. Turbulent mixing is another matter entirely. So the self-similar profile observed in Figures

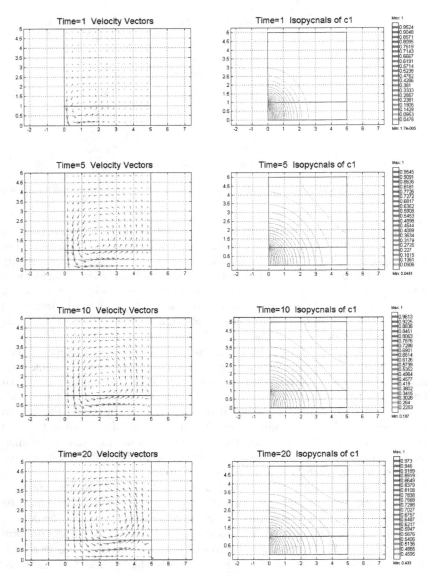

Figure 4.13 History of free convection in the buffer tank. Animation of times 0-1 shows the spreading out of the gravity current and the establishment of the upper recirculation region. All later times shown have apparently self-similar profiles. The strength of the velocity profile diminishes with time.

4.13 for both velocity vectors and concentration is indicative of the long-lived nature of the transient intermediate approach to uniform mixing. It makes a mockery of "steady-state" analysis, since it is not clear that steady state is ever achieved in finite time nor is it clear that the uniformly mixed state will result at all.

Fick's law, which models the non-equilibrium transport of species, would have us believe that the equilibrium endgame has concentration uniformly diffused everywhere from a steady source. In fact, there are two greater complications that preclude this. The first is that it is not concentration that is diffusing at all, but rather chemical potential, and in an external gravitational field. At equilibrium, these two potentials must be balanced. So a permanent concentration gradient is maintained against a gravitational field. This fact is responsible for the difference in composition between air at sea level and at Mile High Stadium. In a buffer tank, it is probably meaningless, as the gradient in concentration is minute. The second complication that is probably more important in most chemical plants is that few solutions are exactly ideal, and many show significant volume change on mixing. Zimmerman [7] has shown that non-ideal solutions can have the structure of their stratification selected on chemical equilibrium grounds, and that only ideal solutions can ever be expected to form uniform mixtures at equilibrium.

The second observation is of the form of the velocity profile established – recirculation layer over a current. This is exactly the form postulated by Zimmerman [4] for which the lumped parameter model of imperfect mixing in the buffer tank was derived, equations (4.6). Figure 4.14 shows the idealized flow configuration for a denser current driving an upper recirculating layer. The lumped parameter model presumes that the recirculation is strong enough that the upper layer becomes well mixed, according to a theory of Batchelor [8], and thus a single concentration characterizes it. In fact, it seems that the upper recirculation is weak, yet the concentration gradients are small in the upper layer.

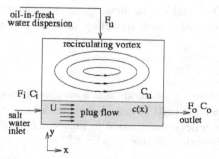

Figure 4.14 Plug flow across the tank bottom driving an upper recirculation layer.

As the assumptions are all qualitatively met by the finite element model, it would seem likely that the lumped parameter model would be an acceptable approximation in the case of purely free convection through the buffer tank. Were the point of this chapter to verify the applicability of the lumped parameter buffer tank model, then we could run a parametric study fitting E(Pr,Ra) for the free convection regime. The easiest route to fit E would be to compute outlet concentration by boundary integration over boundary 6 and fit the value of E which best fits the predictions of (4.6) to the simulated outlet times series. Since it is unlikely that a buffer tank would be operated under a purely free convection regime, it would not necessarily be useful information.

A second series of profiles are shown in Figure 4.15, which differs only from the model of Figure 4.11 in the boundary condition is taken as u=1, v=0 on the inlet (boundary 1) and the Rayleigh number is five times larger (Ra=25). It is actually the case that the recirculation layer above is much stronger in this model, since forced convection imparts more momentum to the upper layer than free convection. It should not come as striking, however, that the flow configuration for gravity driven and pressure driven flows are broadly similar. Only at early times, while the transient flow field is still establishing, does the forced convection flow differ from the gravity current driven flow qualitatively. Before diffusion has had much time to act, fluid in the upper layer is just dragged along by viscous forces, yet it is heavy enough to fall back into the lower layer and fall out the constant pressure outlet. Once the upper layer becomes significantly stratified, however, the fluid dragged by the current has enough momentum to "turn the corner" and establish the upper recirculation layer. Thereafter, the profiles look self similar for both concentration and velocity vectors, and in qualitative agreement with the basis of the lumped parameter model for imperfect mixing in buffer tanks. So one would expect (4.6) to hold on average across the outlet, with E(Re,Pr,Ra) as best fit "capacitance" constant found from time series analysis. Such analysis is beyond the scope of this chapter, but would be fruitful for modeling systems response in a complex flowsheet. In the next subsection, we link the 2-D model for the buffer tank to a 1-D model of the heterogeneous reactor, thereby justifying the description of the buffer tank modeling here in a chapter on "extended multiphysics."

Exercise 4.3

Compute the average outlet concentration at a number of times for a pulsed inlet concentration, i.e. U=1 for t∈[0,1] and U=0 thereafter. Compare qualitatively the collected data for outlet concentration to Figure 4.7. Is the behaviour closer to perfect mixing or imperfect mixing with E=0.5?

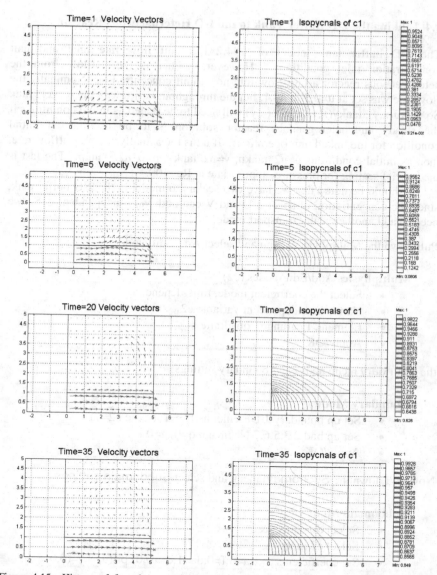

Figure 4.15 History of forced convection in the buffer tank. Early times do not show the establishment of an upper recirculation zone, which does not occur until about time t=5. Forced convection drags the lighter fluid with it at early times. It is only once stratification develops that recirculation follows. The later times are apparently self-similar in profiles.

4.4 Linking the 2-D Buffer Tank to the 1-D Heterogeneous Reactor

The 2-D buffer tank model of §4.2 has unit flow rate inlet and unit concentration inlet conditions.   However, neither is the outflow from the heterogeneous reactor a unit flow rate nor a unit concentration. Nominally the flow rate was set at u=0.5 in §4.1, but the throughput varies with the recycle rate. So tanku(x,y), tankv(x,y), tankp(x,y) will need to be scaled as initial conditions if the steady tank/heterogeneous reactor solution is to be used as an initial condition for the linked unit operations. If u2 is the actually reactor outflow rate, then as initial conditions u=u2*tanku, v=u2*tankv, p=ρ*u2^2*tankp. The last is the inertial scaling for pressure, appropriate at Re=100.

It should be noted that we have two scalars to model in the 2-D buffer tank, since the concentration of species U and V can both vary independently. So we need to add another scalar transport mode.

Pull down the Multiphysics menu and select Add/Edit modes.

Multiphysics Add/Edit modes (c2)
- Select the coefficient mode, time-dependent
- Name the independent variable c2
- Element: Lagrange - quadratic
- Apply/OK

---

the Boundary menu and select boundary settings

Boundary Mode
- Set up boundary 1 Dirichlet h=1; r=1; (fixed c2=1)
- Set up bnd 2,3,5,6,7 Neumann q=0; g=0
- Apply/OK

Now pull down the Subdomain menu and select Subdomain settings.

Subdomain Mode
- Select domains 1&2
- Select c2 mode
- Set coefficients c=1;a=0;f=0;da=1;α=(0,0);β=(u,v),γ=(0,0)

Before we can move on to the linked model, we also need to follow our recipe for building initial conditions for the reactor. Open flowsheet.mat from the FEMLAB GUI, and export fem structure to the workspace. Then in MATLAB, execute the following commands to store the steady solution:

```
xx=[0:0.05:5];
[is,pe]=postinterp(fem,xx);
[u,v,w,us,vs,ws]=postinterp(fem,'U','V','W','US','VS','WS',is);
save sheetdata.mat xx u v w us vs ws;
```

Next build six m-file functions along the lines of sheetu.m below:

```
function U=sheetu(x)
%SHEETU Interpolates u from the FEM solution for flowsheet.mat
%   U = SHEETU(X)
%   is interpolated on the interval [0,5].

% Get the data
load sheetdata.mat xx u v w us vs ws
% Interpolate from rectangular grid to unstructured point.
U=interp1(xx,u,x,'spline');
```

Everything is in place, so now let's start building the linked model, using the tank model as a template.

Pull down the Multiphysics menu and Select Add/Edit modes.

Model Navigator

- Select add geometry name: geom2
- Check 1-D dimension. Give the independent variable as z.
- Select PDE modes → general → time-dependent
- Element: Lagrange - quadratic
- Name mode bulk. Specify 3 dependent variables: U V W
- Add across >>
- Select PDE modes → general → time-dependent
- Name mode surface. Specify 3 dependent variables: US VS WS
- OK

Pull down the Draw menu and select Specify Geometry.

Draw Mode

- Name: reactor
- Start: 0
- Stop: 5

Now for the boundary conditions. Pull down the Boundary menu and select Boundary Settings. Setup the boundary conditions as in the table below.

Boundary Mode
- Use the Mulitphysics mode to switch to mode bulk. Select domain 1 and enter Dirichlet BCs

U	V	W
R=u0-U	R=v0-V	R=-W

- Select domain 2
- Enter Neumann BCs (G=0) for all variables.
- Use the Mulitphysics mode to switch to mode surface. Select domain 1,2 and enter Neumann BCs (G=0) for all variables.
- c1: Select domain 1: r=uout
- c2: Select domain 1: r=vout
- Apply/OK
- ns: Select domain 1: u=u1
- Apply/OK

conv refers to the convective flux >> diffusive flux boundary condition.

Pull down the options menu and select Add/Edit constants. The Add/Edit constants dialog box appears. Assign the constants as below.

Add/Edit Constants

Ufeed	Vfeed	Du	Dv	Dw
1	0.4	1	1	0.001
f	K	ku	kv	kw
0.02	1e-5	0.2	1	100

- InitU= 0.6147; InitV= 0.0147 (tank scalings)
- Apply
- OK

Note that the recycle rate is set to 2% -- this is the only variation from the steady operation and reflects the change that at time t=0, the recycle is turned on.

Now pull down the Subdomain menu and select Subdomain settings. Assign the model constants for the six equations as below:

Subdomain Mode (bulk and surface)
- Select bulk mode. Enter the PDE forms as below for each tab

	U	V	W
Γ	-Uz*Du	-Vz*Dv	-Wz*Dw
F	-u*Uz-ku*(U-US)	-u*Vz-kv*(V-VS)	-u*Wz-kw*(W-WS)
da	1	1	1
init	sheetu(x)	sheetu(x)	sheetw(x)

- Select surface mode. Enter the PDE forms as below for each tab

	US	VS	WS
Γ	0	0	0
F	ku*(U-US)-kv*(V-VS)	ku*(U-US)+kw*(W-WS)	US*VS-K*WS
da	0	0	0
init	sheetus(x)	sheetus(x)	sheetws(x)

- OK

Before we move on to linking the models, we need to scale our buffer tank initial concentrations.

Now pull down the Subdomain menu and select Subdomain settings.

Subdomain Mode

- c1: Select domain 1 & 2; select the init tab; set c1(t0)=InitU
- c2: Select domain 1 & 2; select the init tab; set c2(t0)=InitV
- ns: Select domain 1 & 2; set Fy= -0.25*(c1+c2)
- Select the init tab; set u(t0)=u3*tanku(x,y); v(t0)=u3*tankv(x,y); p(t0)= u3^2*tankp(x,y)
- Apply
- OK

Now for the hard part again. Extended multiphysics means coupling variables. Pull down the Options menu and select Add/Edit Coupling Variables. Enter the seven coupling variables as below:

Add/Edit Coupling Variables

Name	Type	Source	Integrand :Order	Destination
ue	scalar	geom1: bnd 6	c1 : 1	geom2: bnd 1
ve	scalar	geom1: bnd 6	c2 : 1	geom2: bnd 1
uout	scalar	geom2: bnd 2	U : 1	geom1: bnd 1
vout	scalar	geom2: bnd 2	V : 1	geom1: bnd 1
u0	scalar	geom2: bnd 1	supvel : 1	geom2: sub 1
u1	scalar	geom2: bnd 1	supvel : 1	geom1: bnd 1
u2	scalar	geom2: bnd 1	supvel : 1	geom1: sub 1&2

- Apply
- OK

Clearly, there are still undefined expressions above, which are conveniently defined in the Add/Edit Expressions selection of the Options menu.

Add/Edit Expressions

Name	Type	Defined	Expression
uinlet	boundary	geom2: bnd 1	Ufeed+f*ue
vinlet	boundary	geom2: bnd 1	Vfeed+f*ve
supvel	geometry	geom2: bnd 1	0.5*(uinlet+vinlet)/(Ufeed+Vfeed)

- Apply
- OK

Pull down the **Mesh** menu and select **Parameters**.

Mesh Mode (cd2, geom2)

- \>\>More
- Number of Elements in Subdomain: 1 1000
- Apply
- OK

This results in a 1000 element meshing. Now change the output times on the **Solver Parameters** time stepping tab to 0:0.1:1. Now doubt you are shortly visited with the error message "Unknown variable or function u" – the same fault line as before. We found that the convection velocity in a 1-D convection-diffusion model must be a geometry-wide expression to pass this hurdle. The workaround is to set u=0.5 (say), run the time integration long enough to get a solution. Save a model m-file, flowsheet2.m, and edit the coupling variable lines as follows:

```
elemcpl{5}.elem=elcplscalar;
elemcpl{5}.src.g=2;
elemcpl{5}.src.bnd.gporder={1};
elemcpl{5}.src.bnd.var={'u0',{'supvel'}};
elemcpl{5}.src.bnd.ind={1};
elemcpl{5}.dst.g=2;
%elemcpl{5}.dst.equ.ind={1};
```

Commenting out the destination makes u0 available throughout geometry 2. For good measure, I also edited the convection velocity variable line to read

$$equ.u=\{\{\{'u0'\},\{'u0'\},\{'u0'\},\{'0'\},\{'0'\},\{'0'\}\}\};$$

which replaced the hardwired u0=0.5 with the coupling variable. Now we read in the model m-file (**File Menu**, **Open** command), which computes the solution. Not surprisingly, the time dependent solution is visually invariant from the initial conditions, which were created as a solution to the no recycle problem. Two percent recycle does not make a major impact on the solution.

Exercise 4.4

Alter the initial condition so that U(t0)= sheetu(x)+0.02*sin(31.4159265*x). Does this oscillation grow or decay? What effect does the buffer tank have on the oscillation?

4.5 Bioreactor Kinetics

Recall that in §3.4, we treated heterogeneous reaction in a porous catalyst pellet, with a variation on the treatment in the Model Library [9]. In this section, we will try a different variation. In this section, a similar approach will be used to model reaction of a passive scalar occurring in a single cell. The reaction kinetics will be taken as typical of bioreactors – Langmuir-Hinshelwood:

$$r = \frac{kc_s}{\left(1+\sigma c_s\right)^2} \tag{4.9}$$

where r is the rate of disappearance by reaction, which only occurs within the cell. σ represents the finite capacity of the cell to hold the substrate concentration, which saturates at a value controlled by this parameter. The usual rate controlling step, however, is the transfer of the nutrient from the medium across the cell membrane. The overall mass transfer process is usually modelled with a first order resistance, with the flux j given by

$$j = K_{eff}\left(c_i - c_s\right) \tag{4.10}$$

At steady state, the rate of disappearance by reaction is equal to the flux of nutrient across the cell membrance, i.e.

$$\int_{\partial\Omega} K_{eff}\left(c_i - c_s\right) = r \tag{4.11}$$

Thus, the boundary condition on mass transport on the cell wall involves the concentration c_i on the boundary and the concentration within the cell itself, which is taken to be uniform. So the extended multiphysics here is to treat c_s in an additional 0-D space with reaction occurring only there, and coupling between the two spaces through the flux into the cell and through the boundary condition (4.10). Equation (4.11) can be seen as modeling the cell as a continuously stirred tank reactor (CSTR) with effective influx given by the integral, and irreversible reaction. The boundary condition (4.10) is ubiquitous in the chemical engineering literature, nevertheless, to the authors' knowledge, this is the first higher dimensional model that incorporates it as a boundary condition in a non-trivial way. If c_i is constant, (4.10) represents a simple mass transfer coefficient boundary condition of the Biot type that is easily included in any pde engine. But with c_i integrably coupled to the dynamics in the bounding domain, the implementation here, made possible by extended multiphysics in FEMLAB, is unique. Attempts by one of us to implement this boundary condition in other finite element solvers were once abandoned due to the complexity of the coding.

Reaction/Convection/Diffusion Model

If FEMLAB is already launched, select New from the File Menu, which will bring up the Model Navigator.

Model Navigator

- Select 2-D dimension
- Select ChemEng Modes⇒Cartesian Coordinates⇒Mass Balance⇒Convection-Diffusion⇒Time-Dependent>>
- OK

Now use the File menu to "Insert from File" and select your Geometry mat-file (pellet_geom.mat).

Next select the Add/Edit constants options and enter as below.

Add/Edit Constants

Name	Expression
D	1
Keff	1
sigma	40
c1o	1.

Now pull down the Subdomain menu and select Subdomain settings.

Subdomain Mode

- Select subdomain 1
- Enter PDE coefficients
- Set Di=D; Ri=0; u=pelletu(x,y); v=pelletv(x,y)
- Select subdomain 2, check inactive in this domain
- Apply / OK

Pull down the Boundary menu and select Boundary Settings.

Boundary Mode
- Select domain 2
- Select ci=c1o.
- Set boundaries 1,3,4,6 with Insulation/Symmetry
- Set boundaries 7,8 with $-n \cdot N = -K_{eff}(c_i - c_s)$
- Set boundary 5 with convective flux BC
- Apply / OK

Pull down the Multiphysics menu and Select Add/Edit modes.

Model Navigator
- Select add geometry name: geom2
- Check 1-D dimension
- Select PDE coefficient form (c1)
- Name the independent variable cs
- Element: Lagrange - quadratic
- More >>
- OK

Pull down the Draw menu and select Specify Geometry.

Draw Mode
- Name: cell
- Start: 0
- Stop: 1
- Apply
- OK

Pull down the Boundary menu and select Boundary Settings.

Boundary Mode
- Select domain 1&2
- Specify Neumann BCs with q=g=0
- OK

Pull down the Now pull down the Subdomain menu and select Subdomain settings.

Subdomain Mode

- Select subdomain 1
- Enter PDE coefficients
- Set c=1; da=1; f= flux-cs/(1+sigma*cs)^2
- Apply
- OK

Now for the hard part again. Pull down the Options menu and select Add/Edit Coupling Variables. Enter the two coupling variables as below:

Add/Edit Coupling Variables

Name	Type	Source	Integrand :Order	Destination
CS	scalar	geom2: sub 1	Cs : 1	geom1: bnd 7&8
flux	scalar	geom1: bnd 7&8	N_flux_c : 1	geom2: sub 1

- Apply
- OK

Select the standard coarse mesh (triangle button) from the Toolbar. Select the stationary nonlinear solver and click Solve.

Figure 4.16 shows the short time increase in intracellular substrate concentration, computed from the 0-D domain.

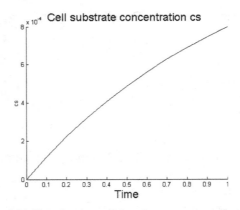

Figure 4.16 Short time intracellular substrate concentration profile.

4.6 Discussion

Undoubtedly the single most valuable new modeling feature of FEMLAB is its extended multiphysics modeling capabilities. As demonstrated in this chapter, this capability permits the modeling of logically linked but separated domains in different unit operations, or of different physical mechanisms within one process unit. The possibilities for extended multiphysics modeling have been modestly explored with regard to heterogeneous chemical reactors and bioreactors here. Nevertheless, it is clear that in the coupling available between logical domains, the extended multiphysics capabilities of FEMLAB can be instrumental in carrying out even simple models, such as the proper boundary conditions for a single cell bioreactor or others of the Biot type, which have eluded other modeling methods.

References

1. FEMLAB Chemical Engineering Module, User's Guide, Version 2.2, p. 2-83.
2. M. Grammatika and W.B. Zimmerman, "Microhydrodynamics of flotation processes in the sea surface layer," *Dynamics of Oceans and Atmospheres*, 34:327-348 (2001).
3. P.O.Mchedlov-Petrossyan, G.A. Khomenko, and W.B. Zimmerman, "Nearly irreversible, fast heterogeneous reactions in premixed flow." *Chemical Engineering Science*, 58: 3005-3023, 2003
4. W.B Zimmerman, "The modelling of imperfect mixing in buffer tanks." in Mixing VI, *IChemE Symposium Series*, H. Benkreira, ed, 146:127-134, 1999.
5. M.E. Valkenburg and B.K. Kinariwala, "Linear Circuits." Prentice-Hall Computer Applications Series, F.F. Kuo, Ed., p.162ff, 1982.
6. J.W. Rottman and J. E. Simpson, "The formation of internal bores in the atmosphere: A laboratory model.", Q.J.R. Met. Soc. 115:941-963, 1989.
7. W.B. Zimmerman, "The effect of chemical equilibrium on the formation of stable stratification." *Appl. Sci. Res.*, 59:298, 1998.
8. G.K. Batchelor, "On steady laminar flow with closed streamlines at large Reynolds number" J. Fluid Mech., 3:177--190, 1956.
9. FEMLAB Chemical Engineering Module, User's Guide, Version 2.2, p. 2-74.

Chapter 5

SIMULATION AND NONLINEAR DYNAMICS

W.B.J. ZIMMERMAN

Department of Chemical and Process Engineering, University of Sheffield, Newcastle Street, Sheffield S1 3JD United Kingdom

E-mail: w.zimmerman@shef.ac.uk

Eigensystem analysis of the linearized operator derived by FEM analysis (the stiffness matrix) is a powerful tool for characterizing the local stability of transient evolution of nonlinear dynamical systems governed by pdes and for parametric stability of stationary, nonlinear problems. Here we discuss how to perform such an analysis in the context of two complex systems – Benard convection and viscous fingering instabilities. The later are simulated from "white noise" initial conditions added to a base flow. The linear stability theory in both cases assumes that the noisy initial conditions include all frequencies, and thus whichever eigenvalue has the largest real part corresponds to the eigenmode that grows most rapidly. FEM eigenanalysis is shown to reproduce the predictions of linear stability theory with good agreement, but is more general in regimes of applicability.

5.1 Introduction

Modelling versus Simulation

So far, we have been concerned with the use of FEM for computational modeling. The model could be expressed as a well posed mathematical system, typically PDEs with boundary and initial conditions, possibly algebraic constraints. Such systems are theoretically deterministic, i.e. the state of the system can be known up to any arbitrary accuracy at any given time. By simulation, something different is usually understood – the physics of the system includes some element of randomness in its temporal development. So we don't expect a simulation to be perfectly accurate in all details. Simulations are expected to mimic the microscopic behaviour of complex systems, typically by posing interaction rules for subsystems from which the global, coordinated behaviour of the whole system emerges. Where the low level interaction rules of the system are particularly poorly physically based, the simulation predictions about global emergent properties must be validated by experiment, perhaps even semi-empirically fitted.

Equivalence?

With the above classification scheme, computational modeling and simulations would appear to be wholly distinct – models are deterministic and physically based; simulations are stochastic and semi-empirically based. This dichotomy blurs, however, with modern understanding of complex systems. SA Billings and coworkers [1], for instance, have developed a data analysis technique for patterns in spatio-temporal systems that can identify the best PDE system within a candidate class that captures the nonlinear dynamics in experimental systems. The technique finds a rule based description for cellular automata that is consistent with the complex system pattern development. By limiting the types of PDE terms available to the model, an inverse mapping from interaction rules to PDE description can be elucidated. So, the common usage of muddling the terms 'modelling' and 'simulation' is justified by this functional equivalence. No doubt this is the "new kind of science" that Wolfram [2] is espousing; dynamics can be equated to simulation schemes (new science) which are equivalent to (nonlinear) pde systems derivable from physical laws (old science). Where the new science wins is that the applicable physical laws may be two complicated to describe in full *a priori*, but those that are being expressed in the complex system may be easily identifiable by finding the interaction rules that are consistent with the global emergent behaviour. Koza and coworkers at Stanford [3] have long been proponents of the view that the trick is to find the computer program that meets the physical requirements. Genetic programming is an approach to letting the program consistent with the observations to assemble itself.

Bridging the Gap: Nonlinear Dynamics

The gap between modeling deterministic systems and simulating stochastic ones is bridged by the nature of nonlinear dynamics and complex systems. The principle feature of a class of nonlinear dynamics – chaotic systems – is that of extreme sensitivity to initial conditions. States that are not particularly far initially in some sense become very far apart eventually. Before chaos theory became better understood in the late 1970s, it was conventional wisdom that in dissipative systems, equilibrium states or periodic oscillations would be the time asymptotic attractors for all initial states of the system. When a dynamical system is extremely sensitive to initial noise or uncertainty, such a system is termed complex. The paradigm in fluid dynamics is turbulence. Shear instability of flows "at high Reynolds number" lead to complexity of the motion at millimeter scales all the way up to thousands of kilometers in the atmosphere, for instance. Even though the temporal state of the system is theoretically deterministic, our uncertainty in the initial state is such that the system is indistinguishable from a stochastic one for practical purposes.

So is there any point in using PDE based models to describe complex systems for which the complexity is practically indescribable? Of course, we can derive or pose PDEs for the dynamics of the statistics (traditional turbulence modeling) or to collate the statistics of the dynamics. In meteorology, the latter is termed ensemble forecasting, and it is an attempt to quantify likely behavior of emergent properties, rather than to average out uncertainty.

Stability and Eigenanalysis: Time Asymptotic Behavior

Key to the evolution of nonlinear systems is the notion of stability. A state of a system, $u_0(t)$, is said to be stable if small perturbations, δ, do not displace the new state of the system, $u(t)$, very far from the original state. The concepts of a 'state', how you measure 'small' and 'how far' two states are separated need to be precisely defined for stability (and therefore instability) to be a useful concept.

The operational definition of a state $u(t)$ is simply to list all of the degrees of freedom necessary to uniquely define a recurring pattern in the system. For a FEM model, this means giving the time dependence of a solution which is typically either stationary or periodic. The exception is that a chaotic attractor is also a 'state' of a dynamical system, deterministically known as a solution trajectory $u(t)$ from an initial state, but not uniquely defined as the attractor is an 'asymptotic state' – many initial conditions are attracted after a long time to this state. In fact, the states of FEM models are easier to describe than for the underlying pde system, which is inherently infinite dimensional. Once the trial functions and finite elements are chosen, a FEM model is finite dimensional and the degrees of freedom necessary to define a state is just the space of all possible solution vectors.

In terms of FEM models, it is also straightforward to describe the stability of a solution trajectory $u(t)$. Consider the FEM operator that maps the solution at time t to the solution at time t+Δt:

$$N\{\mathbf{u}(t)\} = \mathbf{u}(t + \Delta t) \qquad (5.1)$$

Conventionally, for small time steps, this operator can be linearized, so that when applied to the perturbed system, we can compute

$$N\{\mathbf{u}(t) + \boldsymbol{\delta}\} = \mathbf{u}(t + \Delta t) + \mathbf{L}\boldsymbol{\delta} \qquad (5.2)$$

where L is the Jacobian of N

$$\mathbf{L}_{ij} = \frac{\partial N_i}{\partial u_j}(\mathbf{u}) \qquad (5.3)$$

If L is a Hermitian matrix (if real, then symmetric), then the principal axis theorem says that the evolution of the perturbations can be exactly described in terms of the eigenvalues λ_i and the normalized eigenvectors ϕ_i of L as follows:

$$\mathbf{u}'(0) = \boldsymbol{\delta}$$

$$\mathbf{u}'(\Delta t) = \sum_{i=1}^{N} \boldsymbol{\delta} \cdot \phi_i \exp(\lambda_i \Delta t) \phi_i \qquad (5.4)$$

where $\mathbf{u}'$ is the change in the system trajectory due to the occurrence of the initial disturbance $\boldsymbol{\delta}$. Due to the exponential growth rate, one would expect that from any initial condition, the mode associated with the eigenvalue λ with largest real part would eventually dominate the long term evolution of the disturbance to the state $\mathbf{u}$. It simply grows the fastest or decays the least. If the state $\mathbf{u}$ were either stationary or periodic, then if there is any eigenvalue with positive real part, then (5.4) will grow without bound. So the system is unstable. In fact, as we have defined the state $\mathbf{u}(t)$, even a chaotic attractor is unstable according to this criteria. The difficulty with a chaotic attractor is defining unequivocally what the asymptotic state $\mathbf{u}(t)$ is. Consequently, an instantaneous point in phase space $\mathbf{u}$ that is part of a chaotic state $\mathbf{u}(t)$ is found to always have at least one unstable direction ϕ_i, but since it is difficult to distinguish between the time evolution of the state $\mathbf{u}(t)$ and the perturbation, $\boldsymbol{\delta}$, the global stability of the attractor cannot be found by local, linear analysis. The eigenvalues λ_i from the local analysis of a chaotic attractor are called Lyapunov exponents. Since negative real parts for λ_i imply that a trajectory $\mathbf{u}(t)$ is decaying, at least one Lyapunov exponent must have a positive real part at each point of a chaotic attractor.

As an aside, equation (5.4) helps us understand what it means for a perturbation to remain small and the degree to which two trajectories are close. A straightforward measure of closeness of two trajectories, $\mathbf{u}_1(t)$ and $\mathbf{u}_2(t)$, is the distance formula (or error):

$$E\{\mathbf{u}_1, \mathbf{u}_2\} = \|\mathbf{u}_1 - \mathbf{u}_2\| = \left(\frac{1}{N} \sum_{i=1}^{N} [u_{1i} - u_{2i}]^2 \right)^{1/2} \qquad (5.5)$$

where the sum is over the N degrees of freedom that defines a solution vector. For instance, the Newton solver attempts to converge successive solution approximations by sending E to zero. (5.4) implies that to be small $E\{\mathbf{u}',\mathbf{0}\}<\varepsilon$ for some tolerance ε for all time t. If all $\text{Re}\{\lambda_i\}<0$, this is achieved for $\varepsilon \geq \delta$. If any $\text{Re}\{\lambda_i\}>0$, this can never be achieved.

The error norm of (5.5) is just one of many weighted errors that can be defined, e.g.

$$E_W\{\mathbf{u}_1,\mathbf{u}_2\} = \|\mathbf{u}_1 - \mathbf{u}_2\| = \left(\frac{1}{N}\sum_{i=1}^{N} w_i\left[u_{1i} - u_{2i}\right]^2\right)^{1/2}$$

also defines a measure for any set of weights $w_i > 0$. The choice of all weights equal in (5.5) only makes sense for a convergence criteria if all degrees of freedom are expected to range over the same scale. One of the rationales for dimensional analysis of physical models is to condition all degrees of freedom to range over a unit scale. In any unconditioned model, the range of scales expected *a priori* for different degrees of freedom would not be expected to be identical. FEMLAB 2.3 introduced the "automatic scaling of variables" feature, that estimates the appropriate weights w_i automatically or permits user predefined scales. The release notes point out that in a structural mechanics application, displacements might be submillimeter, yet stresses could be megapascals. Without scaling of variables, numerically small quantities would have degrees of freedom contributing little to convergence criteria, and numerically large quantities would be unduly restricted by convergence criteria using formula (5.5).

In summary, excepting the case of chaotic states, linear theory can identify whether a stationary or periodic state is unstable. Regardless, it also identifies the mode(s) that are asymptotically attractive for the perturbation. For instance, if an eigenvalue is complex, then the frequency of oscillation of the perturbation can be predicted, along with the decay or growth rate. Furthermore, the eigenvector associated with the eigenvalue with greatest real part should be the pattern of degrees of freedom that a disturbance evolves into. Using FEM models, representing these eigenmodes is straightforward. They are elements of the space of all possible solutions, so any postprocessing that can be done on a solution can be done to an eigenmode as well. FEMLAB, for instance, can be "tricked" into displaying and analyzing eigenmodes as though they were solutions.

Chapter Organization

This chapter can only be a survey of the range of models that can be used in simulations. The theme of the chapter is to illustrate how features of the MATLAB/FEMLAB computational engines can be used for simulation. A strong undercurrent, however, is awareness of how nonlinear dynamics is important in computational modeling. Our first case study, Rayleigh-Benard convection, is simply a stationary nonlinear system for which convergence is difficult to achieve because of the inseparability of parasitic time dependent solutions excited due to numerical noise. Undoubtedly, users of FEMLAB have already found a straightforward application of nonlinear dynamics theory – conditioning the computational model on the basis of dimensionless parameters in the system. Our second case study illustrates the importance of resolving all scales of the complex system which naively range from the large scale of the geometric boundaries (dimensionlessly this is termed O(1) or order unity since the lengths are usually scaled by a geometric length) down to some small scale set by nonlinear processes coupled with dissipation. If the parameter that characterizes nonlinearity is called R, and complex behaviour increases with increasing R, then one expects creation of complexity down to lengths O(R^{-1}). Thus, in regions generating complexity, the mesh should be gridded with resolution O(R^{-1}). Novice modellers routinely fail to recognize that no satisfactory solution may emerge if all the physics generating complexity are not resolved. Some physical processes routinely generate large complexity parameters. Buoyant convection usually has large Rayleigh number *Ra*. Pipeline flows are almost always at large Reynolds number *Re*. Heat transfer almost always has large Peclet number *Pe*. Simply, given the small values naturally occurring for transport coefficients, human scale flows lead to large complexity parameters. Convergence to a solution does not guarantee that the dynamics of the model are resolved. Careful modeling requires mesh refinement studies until a claim that "refining the mesh does not change the result appreciably" is fully justified. Even experienced modelers can fall into the trap of unresolved computational models due to the large complexity parameter problem. For instance, if there are still unresolved motions, but little "sub-grid" energy transfer, it is convenient to think that laminar solutions to the buoyant convection problems in double diffusion are, for example, able to ignore small scale dynamics. Chascheskin et al. [4] argue cogently that there is never a stationary solution to the double diffusion problem with vertically heated walls. Internal boundary currents are automatically excited, leading to sharp fine structure layering the flow. This feature is not captured by high solutal/thermal Rayleigh number convection since it is not possible even with typical high performance computing resources. So the "large eddy" simulations with low subgrid fluxes may still be unresolved, even if there is little change on mesh refinement – fine structure may influence global dynamics.

5.2 Rayleigh-Benard Convection

Rayleigh-Benard convection is certainly the canonical problem for nonlinear dynamics and flow stability. You can visualize it by heating vegetable oil in your kitchen, sprinkling cocoa powder on the surface of a thin layer of oil heated from below in a frying pan. The hexagonal patterns are clearly visible unless your cocoa powder has congealed. Still an excellent reference on the history of the problem can be found in Drazin and Reid [5]. The gist of the problem is that a vertically decreasing temperature profile and no flow is identically a solution to the boundary value problem stated as

$$\frac{\partial u}{\partial t} + u \cdot \nabla u = -\frac{1}{\rho} \nabla p + \nu \nabla^2 u + \frac{\alpha g}{\rho} T$$

$$\nabla \cdot u = 0 \tag{5.6}$$

$$\frac{\partial T}{\partial t} + u \cdot \nabla T = \kappa \nabla^2 T$$

(c.f. equations 3.1) with boundary conditions of

$$T\big|_{z=0} = T_0$$

$$T\big|_{z=1} = T_1 \tag{5.7}$$

where the bottom temperature is usually greater than the top. The dimensionless groups that matter are still the Prandtl number and the Rayleigh number:

$$\mathrm{Pr} = \frac{\nu}{\kappa}$$

$$\mathrm{Ra} = \frac{\alpha g (\delta T) h^3}{\rho \nu \kappa} \tag{5.8}$$

where h is the depth of the fluid, δT is the applied temperature difference, α is the coefficient of thermal expansion, g is the gravitational acceleration vector, ρ is the density, ν the kinematic viscosity, and κ is the thermal diffusivity.

You can be forgiven for thinking that we have just turned the hot wall-cold wall problem on its side. In the case of *vertical* heated walls, motion is automatically induced even with an infinitesimal temperature difference. Fluid along a hot vertical wall must rise. For *horizontal* heated walls, however, a steady, linear temperature profile with no motion is an exact solution to the system. If the heating is from above, that makes perfect sense as hot light fluid will lie over colder dense fluid – gravity supplies a buoyant restoring force to any fluid element that might be displaced vertically. For heating from below, however, the stratification has dense fluid over light. Buoyant forces should overturn this top heavy profile. Yet, if viscosity and thermal conductivity are strong enough, they resist the motion. Stability theory identifies a critical Rayleigh number above which convection cells form. Below that $\mathrm{Ra_c}$, dissipation still damps out motion.

So let's explore these two situations by finite element analysis.

5.2.1 Heating from above

We could save time by just altering our buoyant convection example for the hot wall-cold wall problem by changing the boundary conditions. However, in order to visualize convective rolls in two dimensions (hexagonal cells are a 3-D phenomenon), we need to have an aspect ratio of about 3:1 for width:height of the layer or greater. So we will walk through the problem set up here.

Launch FEMLAB and in the Model Navigator, select the Multiphysics tab.

Model Navigator
• Select 2-D dimension
• Select Physics modes$\Rightarrow$Incompressible Navier-Stokes >>
• Select ChE $\Rightarrow$Convection and conduction >>
• OK

Since we anticipate heavy computing requirements ahead, we will forego the streamfunction post-processing luxury.

Pull down the options menu and select Add/Edit constants. The Add/Edit constants dialog box appears.

Add/Edit Constants
• Name of constant: Pr
• Expression: 1
• Name of constant: Ra
• Expression: -1
• Apply
• OK

Ra=-1 makes the heating from above, coupled with the setting below for Fy.

Pull down the Options menu and set the grid to $(0,1) \times (0,1)$ and the grid spacing to 0.1,0.1. Pull down the Draw menu and select Rectangle/Square and place it with unit vertices $[-1.5,1.5] \times [0,1]$.

Now for the boundary conditions. Pull down the Boundary menu and select Boundary Settings.

Boundary Mode

- Select domain 1
- Use the multiphysics pull down menu to select the IC NS mode
- Set boundaries 1& 4 with slip/symmetry and 2&3 as no-slip
- Use the multiphysics pull down menu to select the CC mode
- Set bnd 2 with T=1; bnd 3 with T=0; keep 2 and 4 no flux
- Apply
- OK

Now pull down the Subdomain menu and select Subdomain settings.

Subdomain Mode

- Select domain 1
- Use the multiphysics pull down menu to select the IC NS mode
- Set $\rho=1$; $\eta=$Pr; $F_x=0$; $F_y=$Ra*Pr*T
- Use the multiphysics pull down menu to select the CC mode
- Set $\rho=1$; $\kappa=1$; $c=1$; $u=u$; $v=v$
- Select the init tab; set $T(t0)= 1-y$
- Apply
- OK

Now pull down the Mesh menu and select the Parameters option.

Mesh Parameters

- Select more>>
- Max edge size general: 0.1
- Remesh
- OK

There should be 1028 elements. If you click on the = button on the toolbar to Solve you will find that the solution is poorly convergent. There are two contributions to this poor convergence. The first is the lack of a pressure datum. This problem has no imposed pressure on the boundary, so the solution is unique up to an arbitrary additive pressure constant. The pressure equation

$$\nabla^2 p = f\left(\frac{\partial u_i}{\partial x_j}\right) \qquad (5.9)$$

is therefore singular. A pressure datum is readily set by following a recipe discovered by Jerome Long of COMSOL UK Ltd.

Pull down the Point menu and select View as Coefficients. Then select Point Settings and the dialog box appears.

Point Settings / Coefficient View

- Select any vertex (say 4)
- Click on the Weak Tab
- Under constraint replace the first zero with -p
- Apply
- OK

The point datum is now set to p=0 on vertex 4. Note that any pressure value can be entered here as the constraint by replacing a zero by p_0-p. The order of the constraints does not matter. There are as many zeros as dependent variables in your model.

The second reason for the slow convergence is that the velocity field should be identically zero as the solution. However, noise around zero interacts strangely with the new feature of the solver that permits the scaling of the estimated error using the nonlinear and time-dependent solvers. So this feature must be disabled.

Pull down the Solver menu and select Solver Parameters. Click on the Settings button under "Scaling of variables." Check the None option. Now select the Stationary Nonlinear solver, and solve.

Internal Gravity Waves

The automatic scaling setting fails for a subtle reason. With Ra<0, any perturbation or numerical error excites a small amplitude internal gravity wave – an inherently time dependent phenomenon. So the stationary nonlinear solver cannot converge to the "internal waves" that are inherent in the Newton iterations. The automatic scaling setting senses that the proper velocity scale is that of the noise, and therefore tries to resolve and converge the internal gravity waves. Since these are small if the numerical error is small, they can be ignored, which is what happens if you disable the automatic scaling for the velocity field. That there are wave like solutions can be discerned from an eigenfunction analysis of the solution.

Export your solution to MATLAB using the export fem structure feature under the file menu. Although eigenfunction analysis is supported in FEMLAB, it is only for linear problems specified in the eigenfunction mode. You can, however, use the built-in analysis tools in MATLAB. Execute the following on the MATLAB command line:

```
>> sol2=femeig(fem, 'U', fem.sol.u,'Eigpar',20);
```

The arguments are described in the Reference Manual, however, it should be clear that fem is a fem structure, 'U' specifies that the next argument is a solution vector; fem.sol.u is that solution found for Ra=-1; 'Eigpar' is a flag that says the next argument describes the requested eigenvalue solver parameters (in this case the smallest 20 eigenvalues in magnitude).

This generates a structure sol2 with substructures sol2.u and sol2.lambda. You probably find that sol2.lambda has twenty elements, and that sol2.u is a matrix with twenty columns and a huge number of rows. Each column is an eigenvector associated with the same numbered eigenvalue. Femeig uses iterative sparse methods for generating eigenvalue/eigenvector pairs. By default, the smallest magnitude eigenvalues were selected. My list reads as

```
9.8695      17.0399 26.0309 -17.1321i   26.0309 +17.1321i
32.7180              29.1581 -25.5745i   29.1581 +25.5745i
39.4811      41.4966 34.8619 -30.5681i   34.8619 +30.5681i
47.8093              43.2471 -33.6591i   43.2471 +33.6591i
60.7142              54.2250 -35.6500i   54.2250 +35.6500i
74.1437
```

The first eigenvalue is clearly π^2. Since analytically, one can determine that $-\pi^2$ is actually an eigenvalue for *Ra=0*, we should note that this method of "tricking" FEMLAB into producing eigenvalue/eigenvector pairs produces the *negative* of the eigenvalue of the system. It follows that these eigenvalues are all causing perturbations to decay as they all have positive real part, $-\text{Re}\{\lambda_i\}$. Growing, or unstable eigenvectors, would have negative real part. Since the complex eigenvalues come in conjugates, the sign of the imaginary part is not material. However, the existence of imaginary components is equivalent to identifying oscillatory solutions. The interpretation of the eigenvalues here should be that the eigenvector would be expected to grow with a growth factor $\exp(-\lambda t)$ for small amplitudes of the eigenvector. So imaginary components are complex exponentials, equivalent to sines and cosines – oscillations. Nonlinear effects will dominate for large amplitude contributions of the eigenvector, c.f. (5.4). So what are these eigenvectors, really? In fact, they are best thought of as vectors of unknowns equivalent in some sense to solution vectors. So the vector associated with $\lambda = \pi^2$ is the slowest decaying component of the solution. Thus, if you wait long enough in a time dependent evolution, the only non-vanished component in the solution will be proportional to the eigenvector associated with this eigenvalue. Figure 5.1 shows this eigenvector (temperature and velocity fields). The salient feature of Figure 5.1 is that there are hot and cold regions with fluid falling in the cold region and rising in the hot region. Each region occupies a unit width approximately, with a halfwidth of transition zone. Thus to see the whole structure, the aspect ratio must be about 3:1. The parametric solver can be used to explore regions of *Ra<0*, but there are no situations where growing modes are excited. The best that happens is that for large negative *Ra*, the decay rate diminishes. In a perfect (inviscid) fluid, it would be identically

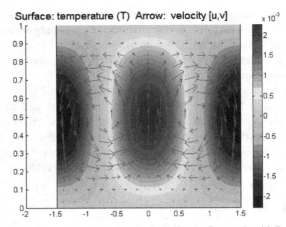

Figure 5.1 Temperature and velocity fields for the first mode with Ra=-1.

zero. As *Ra* increases in magnitude, dissipative effects become relatively weaker than the buoyancy, so the internal gravity waves become long-lived structures. Haarlemmer and Zimmerman [6] used wave tank studies to characterize the mixing properties of large amplitude internal gravity waves that are initially seeded in a concentration stratified fluid. They review the geophysical importance of this transport mechanism.

5.2.2 Heating from below

Heating from below changes the nature of the dynamical problem. As we found when heating from above, complex eigenvalues, equivalent to damped propagating waves were found. This is because vertical convection is opposed by the stable stratification of light fluid over heavy fluid, but gravity waves can propagate horizontally. If a patch of fluid is displaced vertically, it oscillates around its equilibrium position and can propagate right or left without loss of the original energy in an inviscid fluid. Lord Rayleigh [7] showed that when heating from below, the state of the fluid at rest is unconditionally stable. The same argument works in reverse to show that when heating from below, an inviscid fluid cannot remain at rest. But the state of rest can persist to high Rayleigh number in a viscous fluid with heat conduction – the dissipative mechanisms oppose the overturning motion until the heating differential is strong enough. The theory of Reid and Harris [8] describes the critical Rayleigh number for cells with upper and lower rigid boundaries occurs at $Ra_c=1708$ with a wavenumber of 3.117. The motion that is most unstable above Ra_c is supposed to be the onset of stationary cells in 3-D, and convection rolls in 2-D. Since the linear operator, and thus its FEM approximation as in (5.3), is self-adjoint, then all the eigenvalues are real. It follows that the unstable mode is not propagating, but stationary and growing in strength until it saturates.

To permit the possibility that waves might propagate, however, we need to change the horizontal boundary conditions from the earlier simulation which had no flux and slip boundary conditions on the horizontal bounding planes. Gravity waves cannot propagate through such planes, since they are transverse and require up-and-down motion. Furthermore, the model was stationary, so although complex eigenvalues are possible, transient motion was prohibited. To implement periodic boundary conditions, a minor change is necessary.

Recipe for Periodic Boundary Conditions

Pull down the Mesh menu and select the Parameters option.

Mesh Parameters

- Set symmetry boundaries: 1 4
- Apply
- OK

These boundaries are now equivalent, but not necessarily periodic. To make that constraint, we need to require that all variables are identical on the boundary nodes. Pull down the Boundary menu and select View as Coefficients, then select Boundary Settings. All four equations are displayed simultaneously in Coefficient View

---
Boundary Mode

- Select domain 1
- Click on the h-tab, and set the main diagonal to 1 and the off diagonal elements to zero. Make sure the r-tab has u,v,p, and T for the four equations.
- Select domain 4
- Click on the h-tab, and set the main diagonal to -1 and the off diagonal elements to zero. Make sure the r-tab has -u,-v,-p, and -T for the four equations.
- Apply
- OK

Since the boundaries 1 and 4 are equivalent, these two conditions are sequentially added to the boundary constraints. So the condition on u is that u on boundary 1 and −u on boundary 4 sum to zero. The same holds for v, p, and T, ensuring horizontal periodicity of all variables.

Because the domain is about as long as the most dangerous mode (wavenumber 3.117 implies that we have nearly a period of $2\pi/3.117 \approx 2$, i.e.

one wave per two units length), a domain of length 3 is sufficiently long to encompass one period of the unstable mode at supercriticality.

Our solution strategy is to compute the Ra=1 solution first using the linear solver, and then use the Parametric Solver to continue to high Rayleigh number, finding the unstable mode visually from plots of the velocity field. At first I thought that this does not yield a visually unstable flow, even up to Ra=10000 (see Figure 5.2). Why not? u=v=p=T=0 is a perfectly acceptable numerical solution, and the model finds solutions with small dimensionless convective flows, with velocity magnitudes of $O(10^{-2})$, for all values of the Rayleigh number attempted. Professor Bruce Finlayson and chemical engineering student Michael Johnson (private communication) pointed out that since the Nusselt number scales with the Rayleigh number, these are actually giving rise to appreciable convective heat flux. However, there is no specific threshold of Ra which is apparently an abrupt change in Nusselt number. To find the Rac, something else must be tried. The obvious strategy is to use transient integration to determine if, after a sufficiently long time, random small magnitude initial conditions have grown expontially large as in (5.4). The problem with this is that FEMLAB's Parametric Solver only applies to stationary models. The other solution is to compute the eigenanalysis for the system at each value of Ra in a parametric continuation of Ra to high Rayleigh numbers. We will do this two ways: one in the GUI, exporting solutions to the MATLAB workspace; the other in a MATLAB m-file with continuation implemented in a MATLAB loop structure. The results are edifying about the nature of the `femeig` command in the FEMLAB programming library.

GUI Methodology

Figure 5.2 was generated from solving the Benard problem using parametric continuation in the GUI. The linear solver for the Ra=1 problem was used, which is well conditioned. Parametric Solver was used to continue to high Rayleigh number. For eigenanalysis, we export our solution to MATLAB using the export fem structure feature under the file menu. The data structure for a parametric solution is different than for a single, stationary solution. For instance, for the case of a parametric solution [1801:100:10001], fem.sol is an array with three elements: u (the solution), plist (parameter list), and pname (the continuation parameter). Execute the following on the MATLAB command line:

```
>> fem.sol
        u: [6966x83 double]
    plist: [1x83 double]
    pname: 'Ra'
```

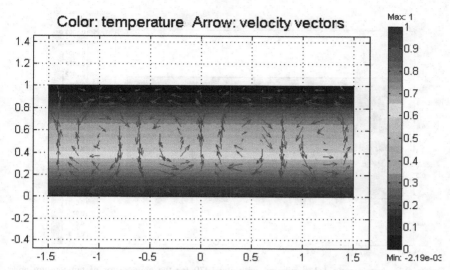

Figure 5.2 Aspect ratio 1:3 simulation with Ra=1970 for full solution of temperature and U_ns. Note that the maximum velocity field is still $O(10^{-2})$.

A transient solution has a similar structure, but with a tlist of output times. To access any of the solutions, the appropriate column is requested. For instance,

```
>> sol2=femeig(fem, 'U', fem.sol.u(:,83),'Eigpar',20);
```

yields the 20 smallest magnitude eigen pairs of the fem operator for the 17th solution, appropriate for the parameter Ra=10001. This feature would work very nicely if the `fem` structure were robust in substituting Ra=10001 in the stiffness matrix computed by `femeig`. Unfortunately, `femeig` takes the last specified value of the parameter Ra in the fem structure as a constant, which may have no relation to the final value in the parametric solver.

In our case, Ra=1 was specified as a constant, so the eigenfunction computed is for Ra=1 about the 83rd solution vector, which is still substantially close to zero everywhere. So, although the parametric solver is a good way to find solutions at high complexity parameter, it is not particularly good at interrogating them with eigenanalysis. Figures 5.2 and 5.3 were generated by using parametric continuation to solve up to Ra=1970, and then changing solver to the stationary nonlinear solver, exporting the single solution to MATLAB, and then performing eigenanalysis. The eigenvector was then substituted into fem.sol.u, and plotted with postplot. There is only one qualitative difference between Figure 5.1 and 5.3 – twice the number of rolls.

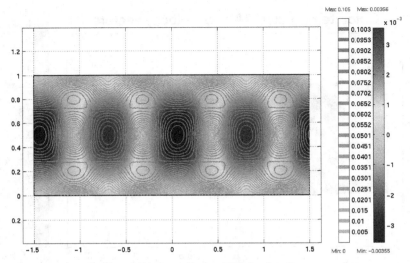

Figure 5.3   Aspect ratio 1:3 simulation with Ra=1970 for the eigenvector of temperature and streamlines associated with the largest eigenvalue.   Clearly the field variables have spatial periodicity 2.   The scale of either temperature or velocity U_ns is arbitrary, but the ratios are fixed.

Matlab m-File Methodology

But what do you do if you want to vary a parameter over a range of values and compile results for each individual parameter value? You still have to write your own looping structure in a MATLAB m-file. For instance, suppose we wish to find the critical Rayleigh number for a neutrally stable largest eigenvalue. We would need to compute `femeig` on each successive value of Rayleigh number, then substitute the old solution as the first guess for the new solution at higher Rayleigh number.

Ra	eigenvalue
1951	0.044447
1952	0.035951
1953	0.027477
1954	0.019062
1955	0.010725
1956	0.0040364
1957	-0.005567
1958	-0.014916
1959	-0.023515
1960	-0.032065

Table 5.1 Decay rates $-\lambda_i$ (largest eigenvalues) with Ra near critical for aspect ratio 1:3.

For example, the m-file bifurc3.m was generated by exporting the m-file of the benard problem, and then modifying the end to put a looping structure around the stationary nonlinear solver and eigensolver:

```
%%%%%%%%%%%%%%%%%%%%%%%%%WBJZ parameters and storage%%%%%%%%%%%%%%
Rayleigh=[100:100:4000];
output=zeros(length(Rayleigh),8);
%%%%%%%%%%%%%%%%%%%%%%%%%%%%Looping Structure%%%%%%%%%%%%%%%%%%%%%%%
for j=1:length(Rayleigh)

% Define constants
fem.const={...
        'Ra',       Rayleigh(j),...
        'Pr',       1};
if j>1
        init=fem0.sol;
end
% Solve nonlinear problem
fem.sol=femnlin(fem,...
        'out',      'sol',...
        'stop',     'on',...
        'init',     init,...
        'report',   'on',...
        'context',  'local',...
        'sd',       'off',...
        'nullfun',  'flnullorth',...
        'blocksize',5000,...
        'solcomp',{'p','u','v','T'},...
        'linsolver','matlab',...
        'bsteps',   0,...
        'ntol',     10.0e-007,...
        'hnlin',    'off',...
        'jacobian','equ',...
        'maxiter',25,...
        'method',   'eliminate',...
        'uscale',   'none');

% Save current fem structure for restart purposes
fem0=fem;

% Integrate on subdomains
I1=postint(fem,'cvfluxT_cc',...
        'cont',     'internal',...
        'contorder',2,...
        'edim',     2,...
        'solnum',   1,...
        'phase',    0,...
        'geomnum',1,...
        'dl',       1,...
        'intorder',4,...
        'context','local');

% Integrate on subdomains
I2=postint(fem,'dfluxT_cc',...
        'cont',     'internal',...
        'contorder',2,...
        'edim',     2,...
        'solnum',   1,...
```

```
            'phase',   0,...
            'geomnum',1,...
            'dl',      1,...
            'intorder',4,...
            'context','local');
output(j,1)=Rayleigh(j);
output(j,2)=I1;
output(j,3)=I2;
sol2=femeig(fem, 'U', fem.sol.u,'Eigpar',10);
output(j,4)=sol2.lambda(1);
output(j,5)=sol2.lambda(3);
output(j,6)=sol2.lambda(5);
output(j,7)=sol2.lambda(7);
output(j,8)=sol2.lambda(9);
end
save bifurc3.mat fem sol2;

dlmwrite('bifurc3.dat',output,',');
quit
```

The m-file script computes fluxes and the first ten eigenvalues for Ra in [100:100:4000], which shows a crossover between Ra values 1900 and 2000. Table 5.1 shows the eigenvalues homing in on the critical value of Ra=1956 for aspect ratio 3:1. Saving the solution and fem structure, as well as the eigenvalues in a mat-file permits the re-loading of the final solution in the FEMLAB GUI by importing from MATLAB into FEMLAB. bifurc3.m was computed as a UNIX background job from the command line:

matlab –nojvm <bifurc3.m >err 2>err &

since it takes a few hours to execute. The save command permitted subsequent perusal of the solution. The m-file script computes the total convective and conductive fluxes for each Rayleigh number solution. The critical Rayleigh number (circa 1956) corresponds to both the zero eigenvalue, but also an abrupt increase in convective heat transfer.

5.2.3 Agreement with thin layer theory

Recall, the theory of Reid and Harris [8] describes the critical Rayleigh number for cells with upper and lower rigid boundaries occurs at $Ra_c=1708$ with a wavenumber of 3.117 for an infinite layer. Since our layer has aspect ratio 1:3, we would not particularly expect agreement. Davis [9] computes the 3-D solution for finite aspect ratio boxes, and finds substantially higher critical Rayleigh numbers, approaching the theoretical predictions only at high aspect ratio. For this reason, we have reproduced our simulations here for aspect ratio 1:10. Figure 5.5 shows a periodicity of ten in a ten unit periodic layer for the critical mode, which is in agreement for the theoretical estimate of the wavenumber. Table 5.2 for the eigenvalues and Figure 5.6 for the Nusselt

Variation of convective flux

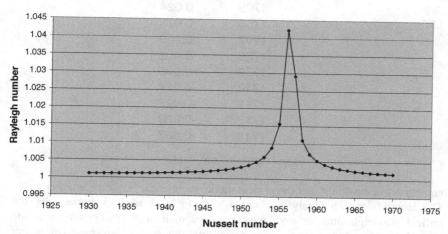

Figure 5.4 Convective flux variation for aspect ratio 1:3 over a range of Rayleigh numbers.

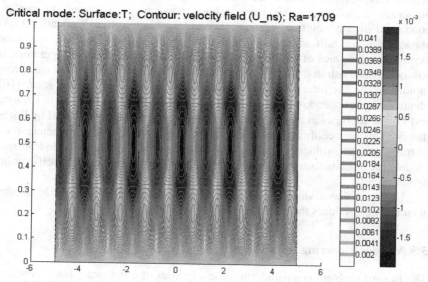

Figure 5.5 Eigenvector (temperature and U_ns) for aspect ratio 1:10 at critical Ra=1709 for near zero eigenvalue.

Ra	eigenvalue
1705	0.024
1706	0.016
1707	0.009
1708	0.002
1709	-0.006
1710	-0.014
1711	-0.022
1712	-0.029
1713	-0.037

Table 5.2 Decay rates $-\lambda_i$ (largest eigenvalue) with Ra near critical for aspect ratio 1:10.

number leave no doubt that the critical mode oocurs near Rayleigh number 1708. The triumph of eigenanalysis coupled with FEM is to find numerically the critical Rayleigh number and approximate length scale associated with the critical mode. Although the linear stability theory for this problem is not cumbersome, for many situations with non-trivial and three-dimensional base states, that cannot always be claimed. FEMLAB, through eigenanalysis, provides a consistency check on linear stability analysis of stationary states. The numerical technique is fair more robust, however. Eigenanalysis can be conducted on any solution, even of transient problems. Recall equation (5.4) shows that for self-adjoint operators, the eigenanalysis predicts the time asymptotic dynamics of the linearized system. For non-self adjoint operators, it has been demonstrated that pseudo-modes that are not eigenmodes can grow rapidly before the time asymptotic eigenmodes dominate. Trefethen *et al.* [10] identified spiral pseudo-modes as leading to transitions to turbulence in Couette and Poiseuille flows at much lower Reynolds Numbers than anticipated by linear theory. This was confirmed experimentally. The extent to which eigenanalysis of transient flow problems identifies the fastest growing pseudomode in transient models for instantaneous states is a largely unexplored area, for non-self-adjoint operators.

Here we have shown that for self-adjoint operators, the FEM model accurately reproduces the predictions of linear stability theory.

5.3 Viscous Fingering Instabilities

The Benard problem is a paradigm for instabilities of a stationary state. Viscous fingering is an instability of a non-homogeneous state in motion – a less viscous fluid displacing a more viscous fluid. Figure 5.7 shows the flow configuration for *miscible* viscous fingering, where diffusion tends to spread out viscous fingers and oppose their formation. Nevertheless, viscous fingering is a long wave instability – broad channels originally form along any displacement front, and then subsequently nonlinear interactions force fluid along these paths, leading to narrow channels.

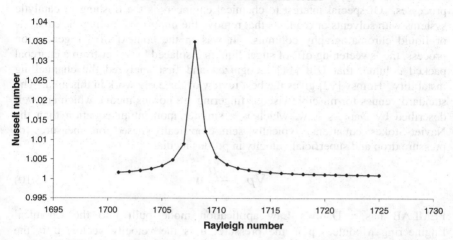

Figure 5.6 Convective flux variation for aspect ratio 1:10 over a range of Rayleigh numbers.

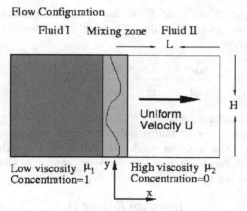

Figure 5.7 Fluid II (more viscous, concentration c=0) being displaced by Fluid I (less viscous, concentration c=1) in a porous medium with superficial velocity U. The mixing zone is the region of diffusive mixing and viscous finger formation.

The phenomenon is a recurring fundamental instability in many realms. Enhanced oil recovery, for instance the injection of dilute detergents into oil sands or flooding with CO_2 gas, as well as the remediation of contaminated acquifers are common geophysical applications. Miscible displacement and the concomitant pesky viscous fingering instability recur as well in regeneration processes. Of special interest to chemical engineers is the flushing of catalytic systems with solvents or oxidants that remove the impurities fouling the catalysts or liquid chromatography columns. It was in the context of a regeneration process, the 'sweetening off' of sugar liquors displaced by water from a charcoal packed column, that Hill [11] recognized and first analyzed the channelling instability. Homsy [12] gives the best review of the early work in this area. The standard venues for miscible viscous fingering are porous media, which are well described by Darcy's Law, which is a simpler momentum equation than the Navier-Stokes equations, typically semi-empirically based on measures of pressure drop and superficial velocity in porous media:

$$\nabla p = -\frac{\mu}{k}\mathbf{u} \tag{5.10}$$

FEMLAB has a Darcy's Law application mode built into the Chemical Engineering Module. p is the pressure; u is the velocity vector; μ is the viscosity, and k is the permeability of the medium. Along with (5.10), it imposes the conservation of mass for an incompressible fluid as

$$\nabla \cdot \mathbf{u} = 0 \tag{5.11}$$

The mixing as depicted in Figure 5.7 is due to convection and diffusion, also a built-in application mode in the Chemical Engineering Module, with concentration satisfying

$$\frac{\partial c}{\partial t} + u\frac{\partial c}{\partial x} + v\frac{\partial c}{\partial y} = D\nabla^2 c \tag{5.12}$$

where D is the molecular diffusivity. Additionally, in order to couple the mixing with the momentum transport realisitically, the fluid viscosity must depend on the concentration. The simplist model is monotonic dependence, which is a good model for glycerol-water, a common laboratory model system for the blending of viscous fluids:

$$\mu = \exp\left(R(1-c)\right) \tag{5.13}$$

Armed with these equations, we are now ready to simulate viscous fingering using the built-in application modes. Launch FEMLAB and bring up the Model Navigator. select the **Multiphysics** tab.

Model Navigator

- Select 2-D dimension
- Select ChE ⇒Cartesian Coords⇒momentum⇒Darcy's Law >> (dl)
- Select ChE ⇒Convection and diffusion>> (cd)
- Select solver **time dependent**
- OK

Pull down the Options menu and set the grid to $(-1,11) \times (-0.1,1.1)$ and the grid spacing to $0.5,0.1$. Pull down the Draw menu and select Rectangle/Square and place it with unit vertices $[0,10] \times [0,1]$.

Now for the boundary conditions. Pull down the Boundary menu and select Boundary Settings.

Boundary Mode

- Use the multiphysics pull down menu to select the dl mode
- Set bnd 2&3 as insulation/symmetry (no flux)
- Set bnd 1 with unit flux in and bnd 4 as unit flux out
- Use the multiphysics pull down menu to select the cd mode
- Set bnd 1 with c=1; bnd 4 with conv>>diff; keep 2 and 4 no flux
- Apply
- OK

Now pull down the Subdomain menu, and select Subdomain settings.

Subdomain Mode/ Coefficient View

- Select domain 1
- dl mode, set k=1, F=0, and $\eta = \exp(3*(1-c))$

- cd mode $D_i = 0.01$
- Now select View as PDE coefficients
- init tab: dl entry pinit(x),
  cd entry erfc(x)*(1.+0.05*sin(10*Pi*y))
- Apply
- OK

pinit.m is a m-file function placed in the MATLAB work directory which computes the initial pressure field consistent with the viscosity function, initial concentration profile and boundary conditions for velocity. It is a simple integration to find p(x):

```
function a=pinit(x)
presgrad=[
183.59
183.471
...
4.01706
2.00851
0.00000];
xlist=[0:0.1:10];
a=interp1(xlist, presgrad, x, 'spline');
```

$$pinit(X-x) = \int_0^x \exp\left(R\left(1 - Erfc(\xi)\right)\right)d\xi$$

where X is the domain length, taken here as X=10.

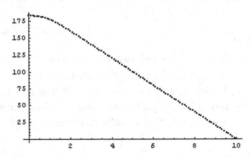

Figure 5.8 Initial pressure profile for concentration given by 1-erfc(x).

Many intermediate points are left out.

Now pull down the **Mesh** menu and select the **Initialize Mesh** option.

Select the time dependent solver and set output times to 0:5:200. Solve. The final concentration and pressure profile should look as Figure 5.9.

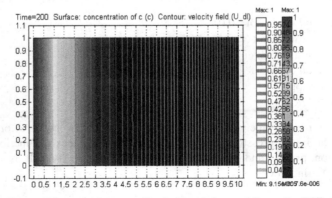

Figure 5.9 Concentration and pressure profiles for Darcy's Law and Convection/Diffusion model.

Apparently, the contours are all parallel, indicating stable evolution of the front. Figure 5.10 shows that the initial condition was wavy (seeded by the sine function initially), but animations show that this oscillation rapidly decays.

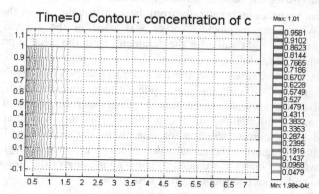

Figure 5.10 Initial concentration contours seeded with sine mode.

The concentration along the centerline smooths out regularly during the simulation. (see Figure 5.11).

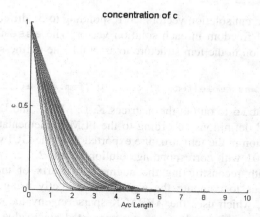

Figure 5.11 Centerline concentration profile history up to t=200.

From this simulation, one would conclude that R=3 and D=0.01 leads to stable spreading of the miscible displacement front. Yet this is not the result of Tan and Homsy [13], who found broad unstable fingers forming in the troughs of the sinusoidal initial condition. So why didn't we? To answer that question, let's conduct the eigensystem analysis of the FEM stiffness matrix.

Eigensystem Analysis

Export your fem structure to the MATLAB workspace (shortcut: CTRL-F). We have previously used the built-in `femeig` routine, which assembles the stiffness and constraint matrices automatically from the fem structure about the specified solution. Where the parametric solver was used in the Benard problem, however, getting the right value for the parameter from the p-list is problematic. Similarly, where the solution is transient, there is a t-list ordering the solutions recorded at successive times. Getting `femeig` to use the correct value of time, however, is not so difficult, as many pde systems that are evolved by the transient solver are *autonomous*, i.e. time does not enter the system of equations explicitly, but only through the initial conditions. Thus the solution at any time t can be used as the current solution for assembling the stiffness and constraint matrices, without regard to the actual value of t. To produce the eigenvalues in MATLAB without `femeig`, we need to know something of the structure of the FEM system and about manipulating fem structures.

The solution vectors are stored in fem.sol.u.

```
>>size(fem.sol.u)
ans =

       11130            51
```

There are 51 different solution vectors, corresponding to 51 different times, with 11130 degrees of freedom in each solution vector. The `assemble` command uses the information in the fem structure to assemble the stiffness and constraint matrices:

```
>> [K,L,M,N,D]=assemble(fem,'T',0.01,'U',fem.sol.u(:,2));
```

This tells `assemble` to output the matrices K, L, M, N , and D (see chapter 2 for their standard definitions according to the FEM implementation here) from the model definition in the fem structure exported from the GUI where it was set up, at time T=0.01 with corresponding solution vector 2, i.e. fem.sol.u(:,2). I experimented with reconstructing the augmented matrix of the stiffness and constraint matrices to compute the eigenvalues directly using the MATLAB `eigs` command, which uses the ARPACK sparse eigenvalue solver to find a requested number of eigenvalues about a requested eigenvalue, frequently the largest or smallest in magnitude. The FEMLAB reference information on `assemble` tells us that an eigenvalue λ satisfies:

$$-\lambda \begin{bmatrix} D & 0 \\ 0 & 0 \end{bmatrix} \begin{bmatrix} U \\ \Lambda \end{bmatrix} + \begin{bmatrix} K & N^{\dagger} \\ N & 0 \end{bmatrix} \begin{bmatrix} U \\ \Lambda \end{bmatrix} = 0 \qquad (5.14)$$

where D, K, and N are evaluated by linearization of the FEM equations around the solution $U=U_0$. Λ is the vector of Lagrange multiplers assuring that the constraints are satisfied in the eigenvector solution as well. Although FEMLAB provides an eigensolver only for eigensystems modes in the GUI, `femeig` accepts the appropriate arguments to solve the generalized eigensystem (5.14) for any solution U at any time, or for stationary nonlinear problems and the parametric solver, for any value of the parameter in the p-list.

For instance,

```
>> sol2=femeig('In',{'D',D,'K',K,'N',N},'Eigpar',20);
```

produces a list of the smallest 20 eigenvalues in magnitude and the associated eigenvectors (the eigen pairs) for the specified D, K, and N matrices. This is simpler than expressing a generalized eigenvalue problem as (5.14) in the appropriate format. Help on eigs gives the syntax as

```
[V,D]=EIGS(A,B) solves      the     generalized    eigenvalue    problem
A*V==B*Vector*D.    B   must   be   symmetric   (or   Hermitian)   positive
definite and the same size as A.
EIGS(A,K)    and    EIGS(A,B,K)    return    the    K   largest   magnitude
eigenvalues.
EIGS(A,K,SIGMA) and EIGS(A,B,K,SIGMA) return K eigenvalues based on
SIGMA:'LM' or 'SM' - Largest or Smallest Magnitude
If SIGMA is a real or complex scalar including 0, EIGS finds the
eigenvalues closest to SIGMA.
```

Here, D is a the diagonal matrix of eigenvalues, and V are the associated eigenvectors. Comparison with (5.14) gives the following assignments for appropriate input to eigs in terms of the block matrices K,N,D produced from assemble:

$$B = \begin{bmatrix} D & 0 \\ 0 & 0 \end{bmatrix}; \qquad A = \begin{bmatrix} K & N^{\dagger} \\ N & 0 \end{bmatrix} \qquad (5.15)$$

so you can produce the above block matrices using

```
>>A=[K N'; N zeros(size(N,1))];
>>B=[D zeros(size(N')); zeros(size(N)) zeros(size(N,1))];
>>[V,D]=eigs(A,B,'SM');
```

Either method (`eigs` or `femeig`) produces the list of smallest magnitude eigenvalues and associated eigenvectors.

0.0000	0.0002	0.0022	0.0057	0.0062	0.0121
0.0200	0.0260	0.0299	0.0417	0.0555	0.0613
0.0713	0.0891	0.0989	0.1009	0.1049	0.1088
0.1108	0.1111				

Since the eigenvalues reported by this method are the decay rates if positive, we can conclude that all the eigenmodes are decaying, although one is neutral. So the eigensystem stability analysis shows that the viscous fingering instability simulated here is unconditionally stable, even at parameter values that the analytic theory finds a long wave instability.

We can visualize the eigenmode solutions by tricking FEMLAB's built-in postplot facility to see them as solutions.

```
>>fem.sol.u=V(:,1)
```

corresponds to the eigenvector with the smallest magniture, $\lambda=0.0000$. A standard use of `postplot` to give concentration contours visualizes the eigenfunction.

```
postplot(fem,...
        'geomnum',1,...
        'context','local',...
        'contdata',{'c','cont','internal'},...
        'contlevels',20,...
        'contstyle','color',...
        'contlabel','off',...
        'contmaxmin','off',...
        'contbar','on',...
        'contmap','cool',...
        'geom',  'on',...
        'geomcol','bginv',...
        'refine', 3,...
        'contorder',2,...
        'phase', 0,...
'title', 'Time=200  lambda=0.0000 Contour: concentration of c  ',...
        'renderer','zbuffer',...
        'solnum', 15,...
        'axisvisible','on')
```

Figures 5.12 and 5.13 demonstrate that the eigenvectors found this way represent discernable patterns. That they all decay implies that there is no pattern formation due to instability. Note that the eigenmodes shown satisfy the pde and the appropriate homogeneous boundary conditions as well – no vertical flux (flat) and uniform outflow.

We are left in this subsection with the apparent disagreement between linear stability theory [13] and linearization of the full solution. Careful examination of the theory and the simulations, however, suggests that the simulations in FEMLAB are too restrictive in the imposition of uniform inlet and outlet boundary conditions. Logically, if the inlet and outlet boundary conditions are

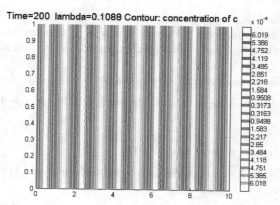

Figure 5.12 Vertical rolls in concentration (eigenvector) associated with eigenvalue λ=0.1088 at time t=200.

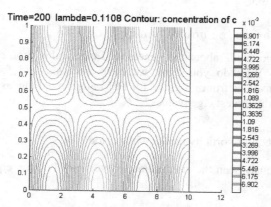

Figure 5.13 Cells in concentration (eigenvector) associated with eigenvalue λ=0.1108 at time. t=200.

uniform vertically, disturbances must decay. It follows that we must use a different modeling strategy for boundary conditions that is less restrictive to capture unstable growth dynamics in this situation. The easiest route to achieve this is through periodic boundary conditions along the vertical boundaries for the inlet and outlet. Vertical variation is then permitted, relieving the instability-killing uniformity constraint. But how can we achieve this boundary condition in the Darcy model? Pressure, as we see from Figure 5.8 is not periodic in this problem, which nixes directly imposing periodic boundary conditions. To make progress, we adopt the approach of Tan and Homsy [14] in transforming to a moving reference frame, where the streamfunction is nominally constant far enough away in both directions from the mixing zone for Figure 5.7. Since it is the same constant, this is legitimately a periodic quantity. The streamfunction-vorticity approach eliminates pressure, which is a non-periodic quantity, in favour of ψ and ω, which can be modelled as periodic in the flowwise direction.

Exercise 5.1: Computing generalized eigenvalues

Use `eig()` to compute the solution to the generalized matrix problem, see after (5.15), for

A=[1 0 0 0 0; -2 1 0 0 0; 1 -2 1 0 0; 0 1 -2 1 0; 0 0 1 -2 1];
B=[1 0 0 0 0; 0 1 0 0 0; 0 0 1 0 0; 0 0 0 0 0; 0 0 0 0 0];

Then compute `eig(A)` and compare. Why do you think you get these answers? Now use

B=[1 0 0 0 0; 0 1 0 0 0; 0 0 1 0 0; 1 1 1 1 1; 1 2 3 4 5];

What changes? Why?
Finally, try

B=[1 0 0 0 0; 0 1 0 0 0; 0 0 1 0 0; 0 0 0 1 0; 0 0 0 0 1];

What can you conclude about the generalized eigenvalue problem (5.14) from this exercise? Why do you think we have always asked for the smallest magnitude eigenvalues from `eigs()` for FEM augmented eigensystems? What if we asked for the largest eigenvalues?

5.3.1 Streamfunction-vorticity model with periodic BCs

We have previously seen the streamfunction-vorticity Poisson equation in (2.7) and (3.3):

$$\nabla^2 \psi = -\omega$$

where the streamfunction ψ is defined by its differential relationships:

$$\frac{\partial \psi}{\partial y} = u \qquad \frac{\partial \psi}{\partial x} = -v \qquad (5.16)$$

Additionally, by making a tranformation to the moving frame, x'=x-Ut, and then dropping the prime, we can write the momentum equation (5.10) and the convective-diffusion equation (5.12) in terms of the streamfunction ψ:

$$\nabla^2 \psi = -R\left(\frac{\partial \psi}{\partial x}\frac{\partial c}{\partial x} + \frac{\partial \psi}{\partial y}\frac{\partial c}{\partial y} + \frac{\partial c}{\partial y} \right) \qquad (5.17)$$

$$\frac{\partial c}{\partial t} + \frac{\partial \psi}{\partial y}\frac{\partial c}{\partial x} - \frac{\partial \psi}{\partial x}\frac{\partial c}{\partial y} = \frac{1}{Pe}\nabla^2 c \qquad (5.18)$$

Equation (5.17) can be viewed as the vorticity generation equation by direct comparison to (3.3). Any y-variation in concentration c or finite vertical velocity v creates vorticity on the RHS of (5.17), which then convects concentration in (5.18), potentially reinforcing the voriticity generation mechanisms on the RHS of (5.17), if diffusion is not strong enough to dissipate out the disturbance or if the nonlinear coupling parameter R is too strong for diffusion to overcome. The linear stability theory [13] quantifies for a given wavenumber of vertical disturbance, whether the relative opposing forces (R for instability by vorticity generation, 1/Pe for stability by diffusion) result in stabilization or destabilization in tandem. In general, there is a longwave instability that cuts-off at a given short length scale, smaller than which diffusion dominates and causes disturbances to decay. This longwave unstable packet is expected to manifest itself with the mode corresponding to fastest growth dominating.

Because of the change of variables and coordinate transform, we now expect that far enough away horizontally from the mixing zone, c becomes uniform and u=0, i.e. periodic boundary conditions can be used for c and y, if we apply a well known trick for c – domain doubling. If we use the mirror image of the initial condition for c, which was taken as a modification of the complementary error function on the positive x-axis, erfc(x)*(1.+0.05*sin(31.4159*y)), then c decays from unity at the origin in both directions, i.e. periodic at infinity, but effectively zero after a short distance, then both c and u can be approximated as periodic horizontally. The upper and lower bounding surfaces can be taken as either periodic (as in [14]) or no flux/no penetration. The latter pair of boundary conditions are adopted here. The FEMLAB model is specified as follows.

Launch FEMLAB and bring up the Model Navigator. Select the **Multiphysics** tab.

Model Navigator
- Select 2-D dimension
- Select PDE modes ⇒coefficient⇒mode name mom, variable si>>
- Select PDE modes ⇒coefficient⇒mode name condiff, variable c>>
- Select solver **time dependent**
- Apply/OK

Pull down the **Options** menu and set the grid to (-1.1,1.1) × (-0.1,1.1) and the grid spacing to 0.5,0.1. Pull down the **Draw** menu and select Rectangle/Square and place it with vertices [-1,1] × [0,1].

Pull down the options menu and select Add/Edit constants. The Add/Edit constants dialog box appears.

Add/Edit Constants

- Name of constant: R; Expression: 3
- Name of constant: Pe; Expression: 100

Pull down the Mesh menu and select the Parameters option.

Mesh Parameters

- Set symmetry boundaries: 1 4
- Total elements 2000
- Apply/OK

Now for the boundary conditions. Pull down the Boundary menu and select Boundary Settings.

Boundary Mode

- For both modes mom and condiff, specify:
- Set bnd 2&3 as Neumann (no flux, no penetration) q=g=0
- Set bnd 1 h=1, r=0
- Set bnd 4 h=-1, r=0
- Apply / OK

As we have seen before, symmetry boundaries with these h-values result in periodic boundary conditions.

Now pull down the Subdomain menu, and select Subdomain settings.

Subdomain Mode/ Coefficient View

- Select domain 1
- mom mode, set c=-1, α=0 0, γ= 0 0, β=R*cx R*cy, f=-R*cy
- condiff mode, set c=1/Pe, da=1, α=0 0, γ= 0 0, β=siy -six, f=0
  init tab: (condiff) erfcnois(x,y)
- Apply / OK
---

erfcnois.m is a m-file function

```
function a=erfcnois(x,y)
b=-0.5*erfc(40*(x + 0.1))*(1.+0.02*randn(1));
c=0.5*erfc(40*(x - 0.1))*(1.+0.02*randn(1));
a=b+c;
```

which creates an appropriate slug of concentration unity that rapidly smooths out in either horizontal direction, superposed with random (normally distributed) noise.

Now pull down the Solve menu and select the Parameters option. This pops up the Solver Parameters dialog window.

Solver Parameters
• Select time stepping tab, set output times 0:0.01:0.5
• Set time stepping algorithm ode15s
• Select general tab, set solution form to weak, solver type time dependent
• Solve
• Cancel
• OK

You should animate the solution. The final time should look similar to Figure 5.14. Compare with Figure 5.15 which shows the initial condition. The red band is the initial slug of unit concentration. Clearly, the nonuniform streamlines are due to the vorticity generated by the concentration noise. The effect of channeling of the less viscous fluid in the slug eventually becomes so pronounced that the upstream and downstream more viscous fluid connects through the slug, isolating islands of the less viscous fluid. In an immiscible fluid, a topological change forming droplets would have had to have occurred. Here, the interface remains smooth due to diffusive mixing, so channeling has occurred. The longer the slug, the less likely the complete channeling through the slug is to occur in a fixed time. The trailing front of the slug is stable since it has less viscous fluid displaced by more viscous fluid. As the animation shows, it is the instability of the leading front that becomes so pronounced, it eventually breaks through the slug. The success of this simulation in capturing unstable frontal dynamics , in comparison with the first model in the Darcy's Law mode, is predicted solely on the use of periodic horizontal boundary conditions. Otherwise, the model equations are dynamically equivalent to the built-in modes that were used before.

Figure 5.16 shows the formation of viscous fingers just on the leading edge with accompanying vorticity generation. Otherwise, the trailing edge simply has diffused out somewhat with still uniformity in the cross-stream direction.

Figure 5.17 shows the leading edge gouging out large, broad fingers into the slug. Viscous fingering is not a symmetric process – back-fingering of the more viscous fluid into the less viscous displacing fluid does not necessary mirror the dynamics of forward fingering. Here, the slug is poor in less viscous fluid. Figure 5.14 shows that we eventually use up the supply of finger-forming less viscous fluid if the slug length is too short in comparison to its breadth. Figure 5.17 has much shorter fingers than seen in [14] with a long slug.

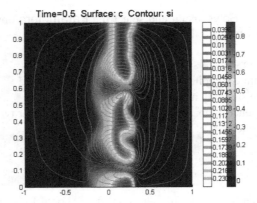

Figure 5.14 Concentration surfaces and streamlines at time t=0.5 for R=3 and Pe=100.

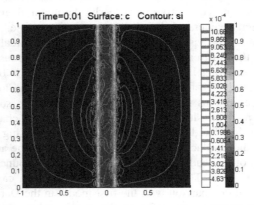

Figure 5.15 Concentration surfaces and streamlines at time t=0 for R=3 and Pe=100.

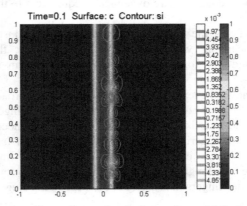

Figure 5.16 Concentration surfaces and streamlines at time t=0.1 for R=3 and Pe=100.

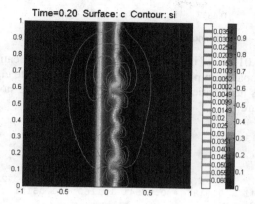

Figure 5.17 Concentration surfaces and streamlines at time t=0.2 for R=3 and Pe=100.

My claim that the leading front is unstable and the trailing front is stable can be validated by eigensystem analysis. The eigenmodes individually do not have fore-aft symmetry. For instance, Figure 5.18 shows one of the stable modes (λ=0.5099), yet the difference in fore-aft symmetry breaking is stark.

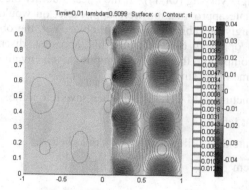

Figure 5.18 Eigenmode corresponding to the 20th eigenvalue for t=0.01. The positive x half plane is tiled with concentration hills and valleys within squeezed vortices. The left half plane is motionless and uniform in concentration.

Eigenvalues calculated for this case from `femeig` in MATLAB according to the previous recipe are, for example

```
sol2.lambda
0.0099      0.0395      0.0889      0.0892
0.0957      0.1579      0.1994      0.2140
0.2422 - 0.0056i
0.2422 + 0.0056i
0.2464      0.3389      0.3553      0.3715
```

```
0.3854      0.4109      0.4837
0.4869 - 0.0131i
0.4869 + 0.0131i         0.5099
```

Here, at time t=0.01, all of the eigenvalues are positive, indicating decay, some with a modest propagation phase velocity (complex conjugate eigenvalue pairs), yet the smallest is near critical. By time t=0.5, however, the story has changed. The eigenvalues are

```
sol50.lambda
0.0205      0.2113          0.2812 - 0.0607i   0.2812 + 0.0607i
0.4300                      0.3127 - 0.4465i   0.3127 + 0.4465i
0.5571                      0.5453 - 0.2961i   0.5453 + 0.2961i
0.8442                      0.8800 - 0.0650i   0.8800 + 0.0650i
-0.9224                 0.6947 - 0.7251i   0.6947 + 0.7251i
0.9183 - 0.5054i    0.9183 + 0.5054i 1.1160 - 0.0518i
1.1160 + 0.0518i
```

The presence of a negative eigenvalue represents a pure stationary growing mode. All other modes are decaying, yet possibly propagating (upstream and downstream with equal phase velocities).

During the evolution of the viscous fingers from the discrete slug (t=0) to the deeply channeled pattern (t=0.5), the decay rates change from fully stable ($Re(\lambda)>1$) to strongly varying ($O(-1)$). If the linear stability theory of [13] using the quasi-steady state approximation were applicable, on would expect gradual changes from strongly unstable to mildly unstable. Yet, the observed endpoint values show the opposite behaviour. This apparent discrepancy can be investigated by computing the smallest amplitude eigenvalues for the FEM operator at each time in the simulation.

Export the fem structure to MATLAB, and then save it to a file:

```
>>save vf_fem.mat fem
```

Now we will execute the m-file script vf_eigs.m as below:

```
load vf_fem.mat fem
times=[0:0.01:0.5];
output=zeros(length(times),21);
for j=1:length(times)
    [K,L,M,N,D]=assemble(fem,'T',times(j),'U',fem.sol.u(:,j));
    sol2=femeig('In',{'D',D,'K',K,'N',N},'Eigpar',20);
    output(j,1)=times(j);
    for k=1:19
        output(j,k+1)=sol2.lambda(k);
    end
end
dlmwrite('vf_eig.dat',output,',');
quit
```

This piece of code deserves several comments:

There is no direct way of calculating eigenpairs using `femeig` for transient models. Instead, we use assemble to create the three sparse matrices of equation (5.14) needed for the generalized eigenpair solution. `femeig` does accept these three matrices as inputs, and does not further reference the fem structure that created them. This recipe also works for parametric continuation, which has a p-list of parameters and an array of solutions for each parameter. Transient models have a fem.sol with a t-list of times and an associated array of the same length of solution vectors.

`femeig` creates a solution structure (see fem.sol) with subfields sol.lambda (list of eigenvalues) and sol.u (array of eigenvectors of same length as sol.lambda). eigenvectors have the same structure as solution vectors, but only satisfy the linearized, homogeneous boundary conditions and do not satisfy the pde itself. In particular, there is no requirement that the load vector (L) or the constraint load vector (M) be satisfied. Indeed, the code above does not use L or M in computing eigenpairs.

`femeig` sometimes has difficulty finding large decay rates. Even though I requested twenty 'SM' eigenpairs, after t=0.04, it can only find 19, and after t=0.42, only 18. femeig uses the sparse eigenanalysis routines of ARPACK, which is essentially iterative, to compute eigenvalues and eigenvectors. This package has difficulty in finding and distinguishing zero eigenvalues (associated with singular systems). Since [13] and [15] show theoretically and numerically that the linear stability theory has a neutral mode at zero wavenumber and at a finite cut-off wavenumber of the longwave unstable wave packet, k_c, the linear system is nearly singular and will have difficulty resolving these neutral (or numerically near-neutral) modes.

Figure 5.19 shows the eigenvalue with least real part at each instant in time, as computed from vf_eig.m. Due to the computational intensity of computing the eigenvalues of these large sparse matrices, it is recommended to execute this m-file script as a background job in MATLAB:

matlab –nojvm <vf_eig.m >err 2>err &
on the UNIX command shell.

From Figure 5.19, it can be seen that for a range of times shortly after t=0, up until t=0.15, the largest growth rate is roughly constant, but with substantial scatter. Our animations showed steady growth of the fingered instability during this interval. That there is substantial scatter is not surprising to me, as Zimmerman and Homsy [15] computed average growth rates versus wavenumber for 128^2 Fourier modes in a similar model but with anisotropic dispersion. They found growth of power in each mode was sporadic from time step to time step, but the average growth of power in all Fourier modes with the same wavenumber was exponential and agreed well with the linear stability theory of [13] while the disturbances were small. Figure 5.19 shows that individual unstable modes for short times have significant variation around the trend growth rate during the period in which macroscopically observable viscous fingers are developing. Thereafter, there is an alternation between periods in which diffusion dominates and those where the structure of the flow and concentration fields is varying rapidly and thus some eigenmode(s) are growing rapidly. That the slug of less viscous fluid is short, and therefore eventually becomes completely channeled, is a feature not investigated by [15]. Those authors treated miscible displacement of a single front as they stopped the simulations before the trailing front was fingered by the backfingering of the leading front with the more viscous material. Nevertheless, they found a wealth of nonlinear interaction mechanisms with varying scales and growth rates once the fingers become large enough to interact nonlinearly.

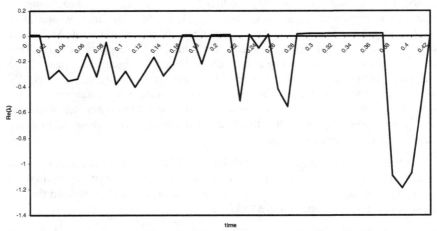

Figure 5.19 Decay rate of the eigenvalue with least real part, Re(λ) history. Recall eigs computes the negative of the growth rate as the eigenvalue, so nearly all the above are growing eigenmodes.

5.4 Summary

In this chapter, we explored how FEMLAB can be used to set up simulations and study nonlinear dynamics and stability. For stationary nonlinear problems, stability studies through eigensystem analysis give the growth rates and eigenmodes which are equivalent to the modes found in transient analysis of the same problem from initially noisy conditions. The most dangerous mode is expected to be observed asymptotically as long as it is smaller than nonlinear interactions. If the operator is non-self-adjoint, however, this is not necessarily the case (see [10]). So interestingly, eigensystem analysis informs about the results of *simulations*, even with stationary solutions. In the case of the Benard problem, the stationary solution returns the no motion base state, even in the situation that the eigensystem analysis identifies critical or growing modes. The second example, viscous fingering, is a paradigm for simulation of evolving instabilities – the base state is moving and changing with time, and the instabilities formed have complex nonlinear interactions. The eigensystem analysis in the uniform outflow Darcy's Law model did not show instability – neutral stability was enforced by the choice of outflow boundary conditions. This uniformity was relaxed in a model based on the streamfunction-vorticity generation equation and periodic boundary conditions that permitted unstable growth. In this example, noisy initial conditions were introduced directly in the simulation by a random number (normally distributed) modulating the concentration base state. As we claimed in the introduction, by simulation we normally expect some element of randomness is modelled. This case is the least controversial use of randomness in a simulation – noisy or uncertain initial conditions. Thereafter, the simulation is a completely deterministic model. In general, FEMLAB can be used for simulating more complicated stochastic processes by alternating random processes and deterministic ones. In this case, there is exactly one such cycle.

It should be noted that noisy initial conditions may not be necessary in such simulations simply due to the approximation error in FEM analysis and roundoff errors in truncation of fixed precision arithmetic. Since the user has control over error tolerances, stochasticity can be simulated by using unconverged or unresolved analysis, but this is a dangerous practice as the statistics of the noise so introduced may be unquantifiable, and the `simulation' may just be numerical instability. A more controlled simulation with quantifiable levels of noise is preferable.

As averse to classical linear stability theory, the application of FEM analysis and subsequent interrogation of the eigensystem analysis of the FEM operator is not limited to a specific type of basis functions – typically "normal modes." The advantage of normal modes is that the transform space that is dual to the physical space has useful measures as coordinates – wavenumber, for instance, specifies the lengthscale characterizing the associated eigenmode. With FEM eigensystem analysis, the growth rates are elucidated for whatever the natural growing mode(s) turns out to be, but the eigenmode does not have an unequivocal length scale, for instance. Where the normal modes are eigenmodes, the FEM methodology usually shows this qualitatively with regard to the patterns in the eigenmode. Figure 5.18, however, shows that normal modes do not necessarily get excited in systems that have FEM operators that are non-self-adjoint. I would speculate that this methodology for numerical computation of stability is far more likely to capture the pseudomodes of [10] for a non-self-adjoint problem than the linear stability theory.

This chapter introduces several new aspects of eigensystem analysis that can be done by using FEMLAB and MATLAB tools and a little user defined programming. The ease by which this can be done is a major advantage of the pde engine and programming language of FEMLAB. It is now common practice in stability theory, for instance of viscoelastic flows [16], across many disciplines [17], to compute via numerical methods the eigenvalues and eigenmodes of instabilities in transient conditions. Smith et al. [16] use the Arnoldi iterative method implemented in ARPACK [18] for their computation. The eigs() sparse eigensolver of MATLAB/FEMLAB does as well. This method, based on the Krylov subspace decomposition, becomes computationally cost effective with larger, sparse systems; the MATLAB/FEMLAB implementation of the ARPACK routines is robust and highly accurate.

References

1. D Coca and S A Billings, "A Direct Approach To Identification Of Differential Models From Discrete Data", *J Mechanical Systems And Signal Processing*, 13:739–755, 1999; D Coca, Y Zheng, J E M Mayhew and S A Billings, "Nonlinear system identification and analysis of complex dynamical behaviour in reflected light measurements of vasomotion", *Int J Bifurcation and Chaos*, 10:461–476, 2000.
2. S Wolfram, A New Kind of Science, Wolfram Media, Inc. ISBN 1-57955-008-8, 2002.
3. SC Roberts, D Howard, JR Koza, "Evolving modules in genetic programming by subtree encapsulation", *Genetic Programming, Proceedings Lecture Notes In Computer Science* 2038: 160-175, 2001.
4. C Sabbah, R Pasquetti, R Peyret, V Levitsky, YD Chashechkin, "Numerical and laboratory experiments of sidewall heating thermohaline convection", *International Journal Of Heat And Mass Transfer* 44 (14): 2681-2697, 2001; YD Chashechkin, VV Mitkin, "High gradient interfaces in a continuously stratified fluid in field of 2D adjoined internal waves" *Doklady Akademii Nauk* 362 (5): 625-629, 1998.
5. PG Drazin and WH Reid. Hydrodynamic Stability. Cambridge University Press, Cambridge, 1981.
6. GW Haarlemmer and WB Zimmerman, "Advection of pollutants by internal solitary waves in oceanic and atmospheric stratifications," *Nonlinear Processes in Geophysics*, 5:209–217, 1999.

7. Lord Rayleigh, "Investigation of the character of the equilibrium of an incompressible heavy fluid of variable density" *Proc. London Math. Soc.* 14:170–7, 1883.

8. WH Reid and DL Harris, "Some further results on the Benard problem", *Phys. Fluids* 1:102–110, 1958.

9. SH Davis, "Convection in a box: linear theory" *J. Fluid Mech.* 30:465–478, 1967.

10. LN Trefethen, AE Trefethen, SC Reddy, TA Driscoll (1993), "Hydrodynamic stability without eigenvalues," *Science* 261:578–583.

11. S Hill, "Channelling in packed columns", *Chem. Eng. Sci.* 1:247, 1952.

12. GM Homsy, "Viscous Fingering in Porous Media", *Ann. Rev. Fluid Mech.* 19:271, 1987.

13. CT Tan and GM Homsy, "Stability of miscible displacements in porous media: Rectilinear Flow", *Phys. Fluids*, 29:3549,1986.

14. CT Tan and GM Homsy, "Simulation of nonlinear viscous fingering in miscible displacement" *Phys. Fluids* 30:1239, 1987.

15. WB Zimmerman and GM Homsy, "Nonlinear viscous fingering in miscible displacement with anisotropic dispersion." *Physics of Fluids A* 3(8) 1859 (1991).

16. MD Smith, YL Joo, RC Armstrong, RA Brown, "Linear stability analysis of flow of an Oldroyd-B fluid through a linear array of cylinders," *J. Non-Newtonian Fluid Mech.* (to appear).

17. RT Goodwin and WR Schowalter, "Interactions of two jets in a channel: solution multiplicity and linear stability." *J. Fluid Mech.*, 313:55–82, 1996.

18. RB Lehoucq, DC Sorensen, C Yang, "ARPACK Users Guide: Solution of large scale eigenvalue problems by implicitly restarted Arnoldi methods, ftp://ftp.caam.rice.edu/pub/software/ARPACK

Chapter 6

GEOMETRIC CONTINUATION

W.B.J. ZIMMERMAN and A. F. ROUTH

Department of Chemical and Process Engineering, University of Sheffield,
Newcastle Street, Sheffield S1 3JD United Kingdom

E-mail: w.zimmerman@shef.ac.uk

Geometric continuation occurs if the mesh of the domain must change from one solution to the next due to variation of the geometry model. In this chapter, we take two examples as paradigmatic – the additional pressure loss in a channel due to various size orifice plates is an example of steady state geometric continuation. Conceptually, this problem is little different from the parametric continuation by Rayleigh number in the Benard problem of Chapter 5. The second example is a drying film with latex particles embedded in the fluid. Two variations on the theme are computed. The noncumulative model varies the front initial position and solves for the time evolution from a uniform surfactant concentration profile initially, with the front frozen at several different positions independently. The cumulative model takes the surfactant concentration profile from the previous front position as its initial value, and alternates solving the transport model with a point source and moving of the front position. This operator splitting technique is shown to be asymptotically convergent as the time increment for these two partial steps shrinks. On a minor note, the film drying model implements a weak term for the point source in a 1-D geometry model using the boundary conditions. The example is unique in that the FEMLAB manuals give only 2-D and 3-D point sources using point mode.

6.1 Introduction

We have already seen several examples of parametric continuation – the traversing in small steps of a range in a parameter, using the previous solution of a nearby parameter value as the initial guess for the solution at the new value of the parameter. As long as the parameter does not pass through a bifurcation point, the new solution should be smoothly connected to the old one if the step in parameter is small enough. Even if there is a bifurcation, however, the old branch may still be a solution, as we found with Benard convection in Chapter 5.

Geometric continuation is qualitatively different from parametric continuation in one important respect. In geometric continuation, the geometrical change of the domain leads to the requirement of re-meshing with each geometric parameter value. We should be careful to class as geometric continuation changes in a parameter that do not lead to a similar geometry. For instance, in pipe flow, it is well known that the flow is characterized by a Reynolds number:

$$Re = \frac{\rho U D}{\mu} \qquad (6.1)$$

This dimensionless parameter rolls the influences of the fluid density ρ, inlet velocity U, the diameter D, and viscosity μ into one parameter that describes the dynamic similarity of the flow. Thus changes in the pipe diameter for fully developed flow are not classed as geometric variation, but rather the more common parametric variation.

Just as in the last chapter where examples of simulations were given for stationary models and for transient models, in this chapter we will give examples of stationary geometric continuation and transient geometric continuation. In the former, distinct models are solved with slightly different domains and therefore different meshes. Therefore the solutions are incompatible (different degrees of freedom) from one geometric parameter value to the next. If the old solution is to be taken as the initial guess for new geometry, then mapping the old solution to the new domain in a consistent fashion must be done. In the examples given here, the system of PDEs for the stationary models are linear, so the solution can be determined directly in one FEM step. Thus mapping old solutions onto the new geometry has no additional value. The transient problem, however, involves a problem in a shrinking domain with a moving front. The domain changes after each time step, so the mapping of the solution at the old time step onto the new domain is essential to the model. Consequently, after each time step, re-meshing must be done as well. One class of problem where this is a crucial step is the free boundary problem. Film flows and jet flows, for instance, are cases where the position of the boundary is intimately related to the solution of the velocity field. The boundaries should be located wherever the stress balances are satisfied.

The 2-D incompressible, laminar Navier-Stokes equations can be solved by several standard means (finite difference, finite element, spectral element, lattice Boltzmann, and multigrid techniques) and have been implemented in standard simulation engines commercially with fixed boundary conditions and complex geometries. Standard computational fluid dynamics packages have two standard engines: (1) the grid generator to cater for complex geometry, and (2) the PDE engine, which can solve more general systems of transport equations that include the pressure as in the Navier-Stokes equations as a Lagrange multiplier for the continuity equation. These two steps are typically conducted separately. The grid is generated initially, and thereafter many simulations are conducted. FEMLAB is no different in this respect.

This paradigm for computational fluid dynamics does not deal particularly well with free boundary problems. An iterative scheme for coupling the flow solution to grid generation could be envisaged, but automation with standard packages is difficult to implement. Ruschak [1] described the now standard method of implementing boundary stress conditions with grid adaptation. Goodwin and Schowalter [2] have successfully implemented their simultaneous solution for the position of the mesh with the solution of the flow equations and boundary conditions using Newton iteration in the treatment of a capillary-viscous jet using finite element methods. In principle, FEMLAB could also do the latter, but in practice, the equations for the FEMLAB application modes would need to be augmented with the residuals for the movement of the grid positions. Standardizing the methodology for including these extra terms in all application modes whenever the grid is "active," i.e. there is a free boundary, would be a substantial re-write of FEMLAB. Given that the number of models that require free boundary computations, even in surface tension dominated flows, is rather few, such a general alteration to the package would not seem warranted. FIDAP, which does treat free boundary flows, uses the iterative flow solution/elliptic mesh regeneration methodology, rather than the simultaneous Newton iteration. In our transient model in a shrinking domain §6.3, we adopt the iterative approach to the variation of the geometric domain over time.

6.2 Stationary Geometric Continuation: Pressure Drop in a Channel with an Orifice Plate

In this section, we consider two related models that require geometric continuation. They are the orifice plate and the platelet in a duct filled with viscous fluid. They are related, as in fact there is only a slight change in the model from one case to the other.

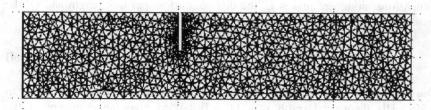

Figure 6.1 Mesh generated for the orifice plate in a duct filled with viscous fluid. The parameter representing the percentage of blockage is $\varepsilon=40\%$ (aspect ratio).

Although it is possible to consider the calculation of the flow around the orifice plate at arbitrary Reynolds number, the major effects in laminar flow are similar to those with artificially vanishing Reynolds number – the Stokes equations. The fundamental reason for this is that most of the dissipation occurs in the opening of the orifice plate, where flow is accelerated yet the small gap leads to strong viscous friction dominating the flow. So to a first approximation, we will model the momentum transport by the Stokes equations:

$$\nabla p = \mu \nabla^2 \mathbf{u} + \rho g$$
$$\nabla \cdot \mathbf{u} = 0 \tag{6.2}$$

where μ is the fluid viscosity and ρ its density, and all other symbols have their usual fluid flow interpretation. Equations (6.2) are dimensional, pseudo-stationary, and inertia-free. As they are also linear, they have been the subject of exhaustive analysis. Ockendon and Ockendon [3] is a good reference for the area. Homsy *et al.* [4] provides several excellent visualizations of the "pathologies" of viscous flow with vanishingly small Reynolds number. My attention to the problem of an orifice plate was drawn by Professor Dugdale [5], who arrived at the solution to (6.2) in the vicinity of a sharp-edged orifice by requiring the condition of optimum energy dissipation within the orifice itself, ignoring the dissipation on all other boundaries of the vessel. His argument is that since the orifice is so small, and all of the flow is forced through the orifice, nearly all of the energy must be dissipated through it, gives a dimensional argument that for a three-dimensional orifice with characteristic opening length a, the energy dissipation rate E must satisfy

$$E = c \frac{\mu Q^2}{a^3} = W = Q \Delta p \tag{6.3}$$

where Q is the volumetric flow rate and W is the rate of working. c is an unknown constant of proportionality that Dugdale calculates theoretically on the basis of the extremum of the energy absorption or can be found experimentally by measuring Q and pressure loss. In a two dimensional system, the analogous dimensional argument makes E' the dissipation loss per unit length and Q' the cross-sectional area flow rate, giving rise to the scaling argument

$$E' = c \frac{\mu Q'^2}{a^2} = W = Q' \Delta p \tag{6.4}$$

Dugdale reports experiments with molasses determining c in the range of 3.17 to 3.30. His theoretical result was 3.0. Bond [6] gives an argument of the similarity of orifice plates to Hagen-Poiseuille pipeflow in a pipe of length *2ka*, where *a* is the orifice radius, and his pressure drop equated to *k=0.631*, implying *c=3.21*.

One of us has been interested for some time in the drag on close fitting particles in tubes. For the same rationale leading to (6.3) or (6.4), close fitting particles in tubes have drag controlled by the gap width. Zimmerman for thin discs [7] (broadside motion) and for spheres [8] sedimenting in cylindrical tubes, reports on the rapid growth of drag as the particle is taken as having larger radius (smaller gap width a). By using perturbation methods in small particle radius *(1-a)* and summing the series expansion, it is possible to determine the nature of the singularity as the particle approaches scraping the duct wall. (6.4) would suggest a second order singularity, $O(a^{-2})$, on dimensional analysis alone for the thin disc in broadside motion by analogy with pressure loss and drag for 2-D or axisymmetric gaps. The sphere problem is not amenable to dimensional analysis, as the gap width changes with polar angle relative to the sphere's center. Bungay and Brenner [9] computed that the singularity for the drag on the sphere is $O(a^{-5/2})$. Using finite element methods, Harlen [10] found convergence difficulties with close-fitting spheres in a cylindrical duct, indicating the extreme difficulty in resolving large scale differences in numerical computations, even with linear models, when small length scales dominate the dynamics of the flow. It is my guess that much of the dynamics of close fitted particles with small gap width can be found by extrapolation of solutions for larger gap width.

In this section, we have proposed first solving for the additional pressure drop Δp due to the presence of the orifice plate with blockage factor ε obstructing the flow over the pressure drop for laminar flow in a channel without the orifice plate. The gap radius is related to the blockage factor, $a=1-\varepsilon$. The difference between this problem and the drag on a sedimenting particle is conceptually very small. For instance Shail and Norton [11] calculated both for the thin disc in broadside motion in a cylindrical duct, as well as the couple – the induced force that opposes rotation of a stationary disc. As these quantities are linearly related due to the linearity of (6.2), it is expected that the singular behavior of one mirrors that of the other as the gap width is squeezed.

Model of an Orifice Plate Inserted in a 2-D Channel

Launch FEMLAB and in the Model Navigator.

Model Navigator
• Select 2-D dimension
• Select Physics modes$\Rightarrow$Incompressible Navier-Stokes >>
• OK

Pull down the options menu and select Add/Edit constants. The Add/Edit constants dialog box appears.

Add/Edit Constants
• Name of constant: rho0   Expression: 0
• Name of constant: mu0   Expression: 1
• Name of constant: Umean Expression: 1
• Apply
• OK

The inlet boundary condition is fully developed Hagen-Poiseuille flow in a 2-D channel, with U_{mean} as the single parameter characterizing the inlet condition.

Pull down the Options menu and set the grid to $(0,1) \times (0,1)$ and the grid spacing to 0.1,0.1. Pull down the Draw menu.

Draw Mode
- Select Rectangle/Square R1 and place it with vertices $[0,1] \times [0,5]$.
- Next Draw a square R2 as a notch with vertices $\{(2, 0.95),(2,1),(205,1.),(2.05,0.95))\}$.
- Pull down the Draw menu. Select create composite object – channel with a notch -- Form the composite object C01 = R1 – R2.
- OK

Now for the boundary conditions. Pull down the Boundary menu and select Boundary Settings.

Boundary Mode
- Select domain 1 Set u=6*Umean*s*(1-s); v=0
- Set boundaries 2,3,4,5,6,7 as no-slip, u=v=0
- Set boundary 8 as "straight-out", i.e. no tangential velocity, p=0
- Apply
- OK

Now pull down the Subdomain menu and select Subdomain settings.

Subdomain Mode
- Select domain 1
- Set ρ=rho0; η=mu0; F_x=0; F_y=0
- Select the init tab and give u(t0)=Umean, v(t0)=0; p(t0)=0
- Apply
- OK

Accept the standard mesh parameters and hit the hesh button on the toolbar (triangle). Note that the output specification gives a pressure datum, so we would expect the pressure to be well conditioned.

Pull down the Solver menu and select Solver Parameters. Click on the Settings button under "Scaling of variables." Check the None option. I do this as a matter of course in problems where the mean flow is well conditioned. Furthermore, as our selection of density (rho0=0) forces this to be a linear problem, there is no point in complicating matters with scaling the variables to improve convergence. Linear problems are well-posed in terms of convergence – a single matrix inversion step. Now select the Stationary Nonlinear solver, and solve.

You caught that, right? The Stationary Nonlinear solver. Why? Because not only is this the default for the incompressible Navier-Stokes equations, the linear solver is not an option. It actually takes two iterations for FEMLAB to be convinced it has converged, although the initial error being 10^{-12} might have been a good clue! Figure 6.2 gives the arrow plot of velocity vectors for an $\varepsilon=0.05$ notch.

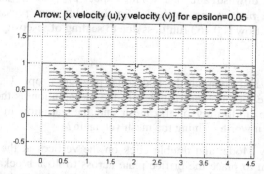

Figure 6.2 Velocity arrow plot for an $\varepsilon=0.05$ notch.

Figure 6.3 shows the pressure profile along the centerline of the channel generated

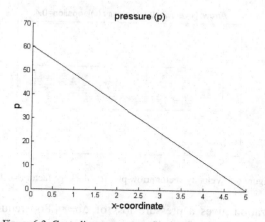

Figure 6.3 Centerline pressure profile for the $\varepsilon=0.05$ notch.

by crossplots on the Post Processing menu. The boundary integral on boundary 1 of the pressure gives the average pressure across the boundary, as it has unit length. This prints in the message dialogue box as 60.463. A quick calculation with fully developed laminar Hagen-Poiseuille flow gives Δp=60. The fully developed u-velocity profile is

$$u = 6U_{mean} y(1-y) \qquad (6.5)$$

Substitution into (6.2) yields the constant pressure gradient as $-12U_{mean}$. Over five unit lengths downstream, one would expect $p_{inlet}=60U_{mean}$ on the inflow plane to achieve $p=0$ at the outflow. So the additional pressure drop over Hagen-Poiseuille flow is 0.463 (unitless due to scaling of viscosity and velocity).

Exercise 6.1

Refine the mesh and compute the additional pressure drop. Use the standard refinement on the toolbar, and restart with the old solution as the initial guess. Comment on the uniformity of the mesh and the variation in the additional pressure drop. Is it worth refining the mesh yet again?

Now go to Draw Mode, and double click on the vertices at the bottom of the notch. Edit them to place the orifice plate across to 40% blockage of the gap, but with the same width (0.05). Solve. Figure 6.4 shows the arrow plot of velocity vectors. Clearly the velocity profile must "turn the corner", which causes substantially more disruption and by implication more dissipation of energy.

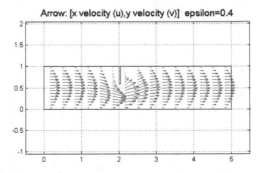

Figure 6.4 Velocity vector arrow plot for blockage factor ε=0.4.

Boundary integration gives a pressure loss of Δp=84.866 required to achieve uniform outflow with p=0. Note that boundary integration along the outflow boundary of the x-velocity gives 1, the value of U_{mean}. Figure 6.5 shows the isobars which clearly show rapid dissipation of pressure in the orifice. Also, just upstream of the plate, the maximum pressure occurs, due to the need to force flow "around the corner."

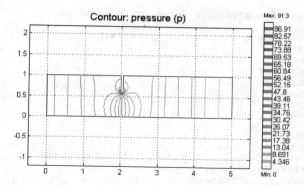

Figure 6.5 Isobars for ε=0.40 blockage factor. Note the high pressure gradients achieved in the orifice. Furthermore, the highest pressure 91.3, is greater than the pressure at the inlet, due to the blockage forcing fluid out of the positions upstream of the plate.

So how do we implement geometric continuation? In this case, all models, even without the use of nearby geometric parameters (blockage factor) converge in one iteration, since the problem is linear. However, the grid refinement studies are required to ensure resolution. First, export a model m-file. Then edit it to set up geometric parameters. The first part of my MATLAB m-file script reads as follows:

```
% FEMLAB Model M-file
% Generated 16-Apr-2002 20:25:17 by FEMLAB 2.2.0.181.
% WZ: Define a vector slot with a range of blockage factors
slot=[0.95:-0.05:0.25];
% WZ: Set up storage
output=zeros(length(slot),5);
% WZ: Now loop around the whole FEMLAB model m-file with j
for j=1:length(slot)

flclear fem
% FEMLAB Version
clear vrsn;
vrsn.name='FEMLAB 2.2';
vrsn.major=0;
vrsn.build=181;
fem.version=vrsn;

% Recorded command sequence
% New geometry 1
fem.sdim={'x','y'};
% WZ: Key section.  Note that I have edited occurrences of 0.95 for
% notch and inserted the variable slot(j)
% Geometry
clear s c p
p=[0 0 2 2 2.05 2.05 5 5;0 1 ...
slot(j) 1 slot(j) 1 0 1];
rb={1:8,[1 1 2 3 3 5 6 7;2 7 4 4 5 6 8 8],zeros(3,0),zeros(4,0)};
wt={zeros(1,0),ones(2,8),zeros(3,0),zeros(4,0)};
% The femlab recorded command sequence continues up to …
```

```
% Save current fem structure for restart purposes
fem0=fem;
% WZ: Now we compute our boundary integrals (I1) exit x-velocity
I1=postint(fem,'u',...
          'cont',     'internal',...
          'contorder',2,...
          'edim',     1,...
          'solnum',   1,...
          'phase',    0,...
          'geomnum',1,...
          'dl',       8,...
          'intorder',4,...
          'context','local');
% WZ: Boundary integral (I2) viscous stress on notch
I2=postint(fem,'Kx_ns',...
          'cont',     'internal',...
          'contorder',2,...
          'edim',     1,...
          'solnum',   1,...
          'phase',    0,...
          'geomnum',1,...
          'dl',       4:6,...
          'intorder',4,...
          'context','local');
% WZ: Boundary integral (I3) pressure at inlet
I3=postint(fem,'p',...
          'cont',     'internal',...
          'contorder',2,...
          'edim',     1,...
          'solnum',   1,...
          'phase',    0,...
          'geomnum',1,...
          'dl',       1,...
          'intorder',4,...
          'context','local');
% WZ: write out output to the array output
output(j,1)=slot(j);
output(j,2)=I1;
output(j,3)=I2;
output(j,4)=I3;
output(j,5)=length(fem.mesh.p);
% WZ: end our j-loop
end
% WZ: record results to file
dlmwrite('tubeori.dat',output,',');
quit
```

There is no difficulty executing this programme on the UNIX command line. It is also computationally "light" enough that you might want to run it in the GUI. You can do this by launching the m-file script (test.m in this case) from Open option on the File menu and selection Model m-file. FEMLAB has its own built-in MATLAB workspace, separate from that in the MATLAB command window that launched FEMLAB. So any m-file script that can run under MATLAB can run in FEMLAB as well. You get the added feature of watching the GUI execute your commands on the domain. You get the computational overhead of the GUI as well, which may not be a problem if your platform has sufficient

RAM memory. In my m-file script for this problem, I perform two mesh refinements, yielding over 4000 elements. This may be a significant inducement to run it in the background!

Eventually, Table 6.1 is generated by the results of this geometric continuation study.

Clearly, the additional pressure loss rises rapidly with increasing blockage factor. The viscous stress along the orifice plate shows a similar rapid rise (factor of 12, approximately) from $\varepsilon=0.05$ to $\varepsilon=0.75$.

$a=1-e$	exit velocity	viscous stress	Dp	dof
0.95	1	-0.54521	60.4025	4010
0.9	1	-0.73333	61.4221	4212
0.85	1	-0.88573	63.041	4362
0.8	1	-0.96572	65.3066	4640
0.75	1	-1.1072	68.3205	4710
0.7	1	-1.2737	72.2372	4992
0.65	1	-1.4233	77.3006	5202
0.6	1	-1.6601	83.8667	5400
0.55	1	-1.8649	92.4607	5594
0.5	1	-2.1666	103.9202	5804
0.45	1	-2.5329	119.5465	6054
0.4	1	-3.07	141.5609	6240
0.35	1	-3.7173	173.9259	6522
0.3	1	-4.7166	224.2228	6696
0.25	1	-6.178	308.6217	6906

Table 6.1 Average exit x-velocity, integrated viscous stress on the orifice plate, average pressure loss, and the number of degrees of freedom used in arriving at the FEMLAB solution as a function of the variation of blockage factor $a=1-\varepsilon$.

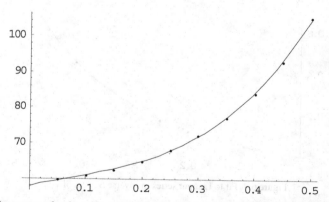

Figure 6.6 Δp versus ε for the orifice plate of thickness 0.05. The curve is the best cubic fit in the range $\varepsilon \in [0,0.5]$.

$$\frac{\Delta P}{\Delta P_0} = 1.0 + 0.6155 \ \varepsilon - 1.472 \ \varepsilon^2 + 6.955 \ \varepsilon^3 \qquad (6.6)$$

Although (6.6) fits the data well in the range shown, over the whole range, a Laurent series in inverse powers of (1-ε) gives a better fit to the ε=0.75 model than the cubic of (6.6), when the fit is only done on the range $\varepsilon \in [0,0.5]$.

$$\frac{\Delta p}{\Delta p_0} - 1 = \frac{0.395652 \ \varepsilon - 0.3832 \ \varepsilon^2}{(1-\varepsilon)^3} \qquad (6.7)$$

Figure 6.7 shows the fit of equation (6.7). The prediction by (6.7) is Δp=371, by (6.6) 214, and the model gives 309. The key feature of (6.7) is that, if you account for the coefficients in the numerator nearly being equal, it arrives at the predicted dependency by dimensional analysis only, equation (6.4).

Exercise 6.2: Sharpness effects

Dugdale's orifice plate was sharp. Ours has a thickness of 5% of the channel width. Try making the orifice plate sharper: 4%, 3%, 2%. What effect does this have on the additional pressure drop? According to [12], the detailed shape of the particle has a considerable effect on the drag force as the gap width becomes smaller. If the gap is flat, then Dugdale's dimensional analysis is correct, equation (6.3), but if the particle has finite curvature, then Bungay and Brenner's $O(a^{-5/2})$ result is recovered.

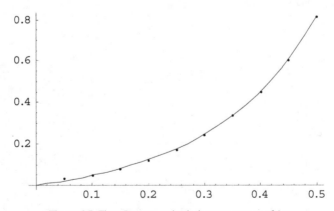

Figure 6.7 Fit to Laurent series in inverse powers of 1-ε.

Exercise 6.3: Platelet geometric continuation

(a) Change the top boundary condition of the orifice plate to be a symmetry boundary condition. This models a two-dimensional platelet with viscous flow past it. Try geometric continuation.

(b) Alter your m-file to use the solution to the last geometric configuration as the initial condition for the next. Does your m-file finish executing faster?

(c) This example isn't really multiphysics. Try adding the streamfunction-vorticity equation as in the buoyant convection example so as to compute streamlines.

The platelet problem was studied by Kim [13] with an analytically determined long perturbation series that was summed to yield the singular behavior of the drag force as the gap width becomes small.

6.3 Transient Geometric Continuation: Film Drying

In the previous section, geometric continuation did not require using the previous solution with a different geometry, varied slightly, as an initial condition for the new solution. Geometric continuation was carried out for the obvious reason of exploring the model for a range of geometric parameters that alter the domain. In this section, the solution of a transient problem is posed in the case that domain is changing over time, so the solution at the previous time is essential for the prediction of the solution at the current time. The application is to film drying. The model here is an idealization of experiments on film drying reported by Mallegol et al. [14].

Figure 6.8 gives the definition sketch of the film drying process. A thin film of liquid containing particles at an initial volume fraction of ϕ_0 is subject to evaporation from the top surface at a constant rate. If diffusion of particles throughout the film is small an accumulation at the top surface is observed, with particles packing at a volume fraction ϕ_m. Over time the thickness of the packed layer above the still fluid layer increases. The overall film thickness decreases linearly with time, and scaling time with the evaporation rate and initial film thickness allows the film surface to be described by $\overline{y}_{top} = 1 - \overline{t}$. A simple mass balance gives that the compaction front moves at a velocity α, given by

$$\alpha = \frac{\phi_m}{\phi_m - \phi_0}.$$

It follows that no further compaction can take place after time $\overline{t} = 1/\alpha$, in which case a steady film thickness is reached.

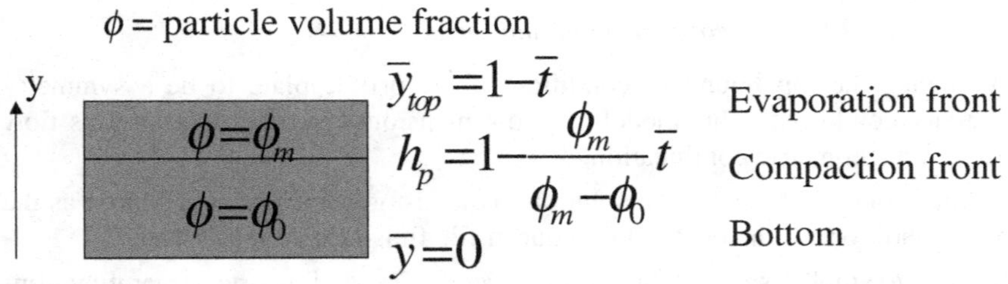

ϕ = particle volume fraction

$$\overline{y}_{top} = 1 - \overline{t} \qquad \text{Evaporation front}$$

$$h_p = 1 - \frac{\phi_m}{\phi_m - \phi_0}\overline{t} \qquad \text{Compaction front}$$

$$\overline{y} = 0 \qquad \text{Bottom}$$

Figure 6.8 Schematic of the two fronts in film drying: evaporation front at the top and compaction front in the interior.

There is also surfactant present in the film. This is taken as initially uniformly distributed at some concentration ϕ_{s0}. As the solvent evaporates from the film the non-volatile surfactant is trapped. This surfactant can either be in solution or stuck to the particles.

In the context of these packing dynamics, the surfactant concentration is also changing due to adsorption on to the packed particles. We note the following conditions on ϕ_s, the solvent concentration:

Initial condition: $\phi_s = \phi_{s0}$, initial surfactant concentration is known *a priori*.

Boundary condition: $\left.\dfrac{\partial \phi_s}{\partial y}\right|_{\overline{y}=0} = 0$, no surfactant flux across impermeable surface.

Boundary condition: $\left.\dfrac{\partial \phi_s}{\partial y}\right|_{\overline{y}=1-\overline{t}} = 0$, no surfactant across material surface – non-volatile surfactant is trapped.

Figure 6.9 shows an idealization of an adsorption isotherm for ϕ_s. Equation (6.8) is a rough representation of the adsorption isotherm giving the typical sigmoid shape. Langmuir isotherms are the most commonly fitted, but as long as the isotherm is differentiable, any will do.

$$\Gamma(\phi_s) = \frac{1}{2} + \frac{1}{2}\tanh\left(m(\phi_s - y_0)\right) \qquad (6.8)$$

The dynamic adsorption of surfactants in miscible displacement is a fundamental, recurring situation in the chemical and petrochemical industries. Enhanced oil recovery by detergent flooding has been practiced for more than twenty years. Liquid chromatography, where the adsorption-desorption isotherm is key to separation processes, is another common example. The desorption of the isotherm forced by the compaction front, however, is a unique feature of the

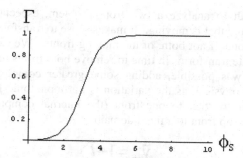

Figure 6.9 Absorption isotherm. Γ = amount of surfactant loading the particle surface. m=1 and y_0=3.

model presented here. Trogus *et al.* [15], in the context of enhanced oil recovery, proposed a kinetic model for adsorption/desorption rates, and Ramirez *et al.* [16] developed a two-equation (concentration and surfactant loading), 1-D spatio-temporal model for dynamic adsorption. Nevertheless, their transport model is still for a homogeneous porous media, where in ours, given below, the compaction front between the close packed and looser packed layers, serves as an impetus for desorption, and thus as a propagating point source of surfactant.

Posed for the first time here is a transport model for the surfactant:

$$\frac{\partial \phi_s}{\partial \overline{t}}\left(1+\phi\frac{d\Gamma}{d\phi_s}\right)+\Gamma(\phi_s)\phi_m\delta\left(\overline{y}-h_p(\overline{t})\right)=\frac{D_s}{H\dot{E}}\frac{\partial^2\phi_s}{\partial\overline{y}^2} \qquad (6.9)$$

where the first term on the LHS represents accumulation of surfactant, the second part of the factor being due to accumulation in the adsorbed phase; the second term represents a point source of surfactant being desorbed from the compaction front; the RHS represents a diffusion term. Since the equation is dimensionless, the coefficient of the diffusion term represents an inverse Peclet number:

$$\frac{D_s}{H\dot{E}}=\frac{1}{Pe} \qquad (6.10)$$

where D_s is the molecular diffusivity of the surfactant, H is the initial film depth, and E is the evaporation rate. The Peclet number is taken as unity for the purpose of example. In the simulations that follow, it will be varied systematically.

Representative values of packings are: $\phi_m=0.64$ $\phi_0=0.4$

It is rather difficult to analyze a two front problem, especially with one an effective point source that is moving. It makes sense to transform our coordinate system to remove one, if not both, of the moving fronts. We experimented with nonlinear coordinate transforms in time to remove both fronts to a fixed domain. Surprisingly, this was possible, adding some greater complexity to the PDE (6.9), but it is not physical, as the variation is not monotonic for the transform coordinate. Better to stick to one front (the internal compaction front) and transform away the top front to a fixed domain.

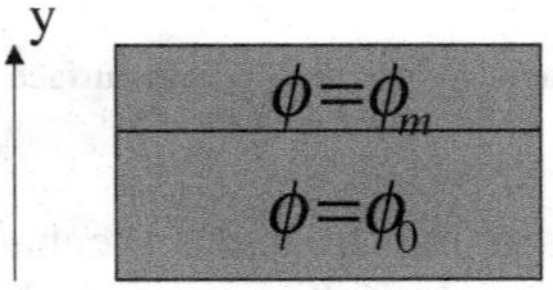

$$\bar{y}_{top} = 1 - \bar{t} \qquad \xi = 1$$

$$h_p = 1 - \frac{\phi_m}{\phi_m - \phi_0}\bar{t} \qquad \xi = h_\xi$$

$$\bar{y} = 0 \qquad \xi = 0$$

$\phi = \phi_m$

$\phi = \phi_0$

Figure 6.10 Coordinate transformation: one front.

The tranformation that achieves this is simple:

$$\xi = \frac{\bar{y}}{1-\bar{t}}; \quad \tau = \bar{t} \tag{6.11}$$

and results in a new specification for the compaction front:

$$h_\xi = \frac{1-\alpha\tau}{1-\tau} \tag{6.12}$$

Differentials are expressed in the new coordinates according to the chain rule:

$$\frac{\partial}{\partial \bar{y}} = \frac{1}{1-\bar{t}}\frac{\partial}{\partial \xi}; \quad \frac{\partial}{\partial \bar{t}} = \frac{1}{(1-\bar{t})}\frac{\partial}{\partial \xi} + \frac{\partial}{\partial \tau} \tag{6.13}$$

which results in a transformed PDE for surfactant transport:

$$\frac{\partial \phi_s}{\partial \tau} + \frac{\xi}{1-\tau}\frac{\partial \phi_s}{\partial \xi} - \frac{1}{Pe}\frac{\xi}{(1-\tau)^2}\frac{1}{\left(1-\phi\dfrac{d\Gamma}{d\phi_s}\right)}\frac{\partial^2 \phi_s}{\partial \xi^2} = -\frac{\Gamma(\phi_s)\phi_m}{\left(1-\phi\dfrac{d\Gamma}{d\phi_s}\right)}\delta\left(\xi - h_\xi\right) \tag{6.14}$$

The terms are now representative of, on the LHS: accumulation, pseudo-advection, and quasi-diffusion. On the RHS, the point source remanifests itself. The solution is "sensible" for $0 \leq \tau < \dfrac{1}{\alpha}$. The boundary and initial conditions are now expressed as an initial condition $\phi_s = \phi_{s0}$, BCs $\left.\dfrac{\partial \phi_s}{\partial \xi}\right|_{\xi=0} = 0$ and

$\left.\dfrac{\partial \phi_s}{\partial \xi}\right|_{\xi=1} = 0$.

Our FEMLAB modeling strategy is as follows:

- Solve PDE once for "frozen" compaction front at position just less than 1.
1. Move the compaction front by an infinitesimal amount
2. Keep the old solvent concentration profile and update by solving PDE with new front position
3. Go back to 1; iterate until compaction front hits bottom.

FEMLAB recipe for a single pass.

Launch FEMLAB and in the Model Navigator, select the **Multiphysics** tab.

Model Navigator
- Select 1-D dimension
- Select PDE modes $\Rightarrow$ General >> time dependent, Weak solution form
- OK

Options/Axis settings −0.1 to 1.1, grid 0.05

Draw/Specify Geometry: bottom 0 to 0.99, top 0.99 to 1, enter points name:front, start 0.99

Pull down the **options** menu and select **Add/Edit constants**. The Add/Edit constants dialog box appears.

Add/Edit Constants
- Name of constant: theta_m Expression: 0.64
- Name of constant: theta_0 Expression: 0.40
- Name of constant: alpha   Expression: theta_m/(theta_m−theta_0)
- Name of constant: Pe Expression: 1
- Name of constant: tau Expression: 0.01 (time step)
- Name of constant: slope Expression: 0.01 (isotherm parameter)
- Apply
- OK

Pull down the **options** menu and select **Add/Edit expressions**. The Add/Edit expressions dialog box appears.

Add/Edit Expressions

- isotherm subdomain 1&2, value: 0.5+0.5*tanh(slope*(u-offset))
- isobound bnds 1&2&3, value: 0.5+0.5*tanh(slope*(u-offset))
- dtherm subdomain 1&2, value: 0.5*slope*(sech(slope*(u-offset)))^2
- Apply
- OK

Now pull down the Subdomain menu and select Subdomain settings. Fill in the parameters as in the following table:

	Subdomain 1	Subdomain 2
Γ	-1/(Pe*(1-tau)^2)*ux	-1/(Pe*(1-tau)^2)*ux
F	-(x/(1-tau))*(1-theta_0*dtherm)*ux	-(x/(1-tau))*(1-theta_m*dtherm)*ux
da	1-theta_0*dtherm	1-theta_m*dtherm

Set the init tab with u(t0)=1.0 for both subdomains.

Now for the boundary conditions. Pull down the Boundary menu and select Boundary Settings.

Boundary Mode

- Select domain 1 & 3, check Neumann.
- Select domain 2, check enable borders
- Select weak tab. Weak term: isobound*theta_m*u_test
- Apply
- OK

As alluded to in the abstract, implementing a 1-D point source is unique in the FEMLAB literature in our experience. The examples in the Model Library are all in 2-D and 3-D, implemented through point mode. In 1-D, the only access to point residuals is through the boundary conditions, specifically the weak tab for point sources. In analogy with the Poisson model in Chapter 2, u_test evaluates as a Dirac delta function on the front (domain 2), with coefficient chosen to match (6.14). Although placed in the boundary condition, the residual adds the analytic equivalent to a point source to the augmented stiffness matrix (see Chapter 2).

Now pull down the Mesh menu and select the Parameters option.

Mesh Parameters

- Select more>> max element size 1 0.001 2 0.0001
- Remesh
- OK

There should be 1245 elements.

Now enter solver mode and select solver parameters. Select weak form. Set time stepping 0:0.001:0.01. Now solve. Then save a model m-file as the single pass solution. Figure 6.11 shows the history of the short time evolution of the surfactant concentration with the compaction front frozen at its initial position, $\xi=0.99$. In this single step, the compaction front has been translated in the first stage without diffusion, in the second stage computed here, it is permitted to diffuse without convection. This "operator splitting" technique, which divides the time step in to translation stages and convective-diffusion stages is not novel. Zimmerman and Homsy [17] give several references for its use. Figure 6.11 shows that during the convective-diffusive stage, the concentration grows at its peak due to the compaction front acting as a source, and spreads out underneath.

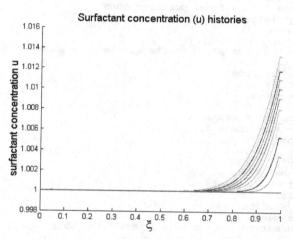

Figure 6.11 Surfactant concentration after first time interval t∈ [0.0:0.001:0.01] solving diffusion model in the transformed domain (ξ-coordinate) with frozen front.

Now for the complications. We will use our exported model m-file as a basis for controlling the movement of the front with an external loop around the subprogram for solving the diffusive transport equation with the front frozen. To do this, we need to restart the model each time step with the solution of the previous step with a different front position. We accomplish this below by interpolating the previous solution on a different mesh to the new mesh, which can be somewhat different owing to the changing position of the compaction front.

Now, the major features of the program, filmdry.m:

Initialization

```
% FEMLAB Model M-file
% Generated 08-May-2002 23:11:18 by FEMLAB 2.2.0.181.
xx=[0:0.001:1];
u=zeros(size(xx));
% Define parameters
Pe=1.;
slope=1.;
offset=2.;
theta_m=0.64;
theta_0=0.4;
alpha=(theta_m/(theta_m-theta_0));
endtime=1/alpha;
slottimes=[0.01:0.01:endtime];
% Set up storage for output
output=zeros(length(slottimes),length(xx));
% Roll out the frozen front position loop
for j=1:length(slottimes)
    tau=slottimes(j);
    frontpos=(1-alpha*tau)/(1-tau);
% Now just alter some parts of the single pass model m-file
...
front=point1(frontpos);

...
% Define constants
fem.const={...
        'theta_m',theta_m,...
        'theta_0',theta_0,...
        'alpha',   alpha,...
        'Pe',      Pe,...
        'tau',     tau,...
        'slope',   slope,...
        'offset', offset};
% Multiphysics
fem=multiphysics(fem);
% Extend the mesh
fem.xmesh=meshextend(fem,'context','local','cplbndeq','on','cplbndsh',
'on');
% Evaluate initial condition
if j==1
        init=asseminit(fem,...
        'context','local',...
        'init',    fem.xmesh.eleminit);
end
if j>1
        % Map initial solution to current xmesh
        fem0.sol.u(:,1:end-1)=[];
        umap={'u',1,1,1,0};
        init=asseminit(fem,'init', {fem0,umap});
end
...
% Save current fem structure for restart purposes
fem0=fem;
```

```
% Postinterp the solution
[is,pe]=postinterp(fem,xx);
[u]=postinterp(fem,'u',is);

output(j,:)=u-ones(size(u));
end
% Write the final output to file
dlmwrite('filmroll.dat',output',',');

quit;
```

Note that we write the transpose of the matrix output, purely for the convenience of the graphics package (GNUPLOT) which plots column vector data. The file filmroll.dat contains the current solution less one, which helps in maintaining accuracy, since by default, MATLAB writes five significant figures, but internally stores double precision floating point numbers (about twelve significant figures). The `postinterp` command generates an interpolation structure [is,pe], which can be applied to the solvent concentration 'u' or any other computed quantity, say 'ux' or 'ut'. This is the common way to extract detailed information about the solution from the fem structure. A frequently asked question in FEMLAB seminars is how to get the GUI to output a "data file" for import into the favorite graphics package. Figures 6.12 – 6.19 were generated using GNUPLOT on data output from MATLAB using `postinterp` on a fem structure.

The role of the two if structures, (if j==1 ...) and (if j>1 ...), are crucial to the model formulation. The default setting is to initialize with the initial fields u(t0)=1. This is done during the first time step using the (if j==1 ...) structure. For subsequent time steps, the (if j>1 ...) structure interpolates the solution from the last time step (fem0.sol) onto the new mesh and places the interpolated solution in the init field of the new fem structure. I wish I could take the credit for this fancy programming effort using little understood features of `asseminit()`, but in fact, all I did was to get the GUI to generate the commands automatically by altering a mesh and using the solve using previous solution toolbar button. The model m-file provided the appropriate code for `asseminit()`. Since it is easy to run filmdry.m with the default setting, we did this for Figure 6.12 and term it the "non-cumulative" model. Running with the interpolation scheme on for the cumulative effect of the point source at the compaction front as written above is termed the "cumulative" model. The non-cumulative model permits the understanding of the compaction front dynamics in the abstract.

Figure 6.12 demonstrates the model predictions for equal duration front translation "hops" and convective-diffusive relaxation steps of $\Delta t=0.01$. As we saw in Figure 6.11, the initial profile rises along the upper compaction layer, with the top surface having elevated surfactant concentration. The boundary condition (no flux) requires the flat profile. At subsequent times, the peak concentration associated with the compaction front is highest exactly at the front, and diminishes in height, as there is greater penetration downward of the surfactant flux released upon compaction (modelled by the point source term). As the compaction front approaches the bottom, the no flux boundary condition forces surfactant to accumulate along the bottom. It should be noted that the elevation in surfactant concentration does not reach 1% in this example. Perhaps the strength of diffusion keeps the compaction front broad and dilute in this example. The cumulative model, shown in Figure 6.13, shows a stronger aggregate effect, with maximum concentrations of up to 4%.

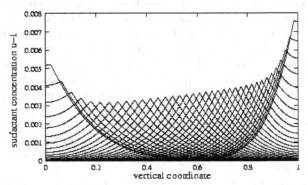

Figure 6.12   Non-cumulative model. Combined compaction front translation and convective-diffusive model for Pe=1, m=1, offset y_0=2. Shown are times t∈ [0.:0.01:0.375], the last time corresponds to the compaction front arriving at the bottom of the layer. Each time step is from a uniform surfactant concentration profile ϕ_s=1 but with translated front position.

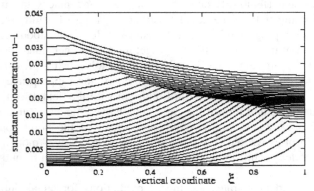

Figure 6.13   Cumulative model. Combined compaction front translation and convective-diffusive model for Pe=1, m=1, offset y_0=2. Shown are times t∈ [0.:0.01:0.375], the last time corresponds to the compaction front arriving at the bottom of the layer. This model builds on the profile of for the previous time step surfactant concentration profile ϕ_s. Although the governing equation is nonlinear, due to the small variation in surfactant concentration, Figure 6.13 is approximately the linear combination of the cumulative profiles up to time τ.

An Irish professor once remarked to me, "Anyone can do a calculation, the trick is figuring out if it's right." So how do we know that the cumulative model, Figure 6.13 is right? A checking point is whether it is convergent upon reduction of the time increment for moving the front, Δt. Clearly, computing Figure 6.13 at say three different values, successively cutting Δt, is going to be difficult to show on one figure, since Figure 6.13 is rather full already. It is probably sufficient to show a feature of the profile. The most prominent feature is the "ridge", of Figure 6.13, which corresponds to the maximum concentration of the profile at each time step, and therefore matches the front position seen in Figure 6.12. The maximum of a function is termed the L_0 norm. Because this problem is diffusive, getting the maximum right is a challenge. The L_2 norm is the most commonly used, which has the same connotation as a "root-mean-square" of the profile – an integral measure of size. It is the unscaled norm that FEMLAB uses in assessing the error of a model in its Newton solvers. Figure 6.14 demonstrates that the L_0 norm is time-asymptotically convergent, a necessary consistency check on the operator splitting scheme. Early times are divergent, since the front has had little time with small Δt to act as a source.

In Figures 6.15 and 6.17, we raise the Peclet number to Pe=100, to explore weaker diffusion.

The non-cumulative model in Figure 6.15(a) shows qualitatively the same behaviour as in Figure 6.12 – peak concentration associated with the compaction front, eventually accumulating along the bottom of the layer. The striking feature is that the peaks are much narrower in this example, resulting in 3—4 % elevation of surfactant concentration. In both of these cases, since the variation in surfactant concentration is so slight, the dynamics of the accumulation term is dominated by the slope of the isotherm at unity, and the dynamics of the point source are dominated by the value of the isotherm at unity.

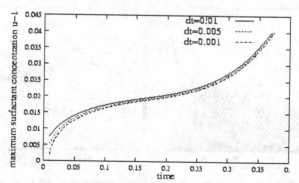

Figure 6.14 Cumulative model. Combined compaction front translation and convective-diffusive model for Pe=1, m=1, offset $y_0=2$. Shown are times $t \in [0.:0.01:0.375]$. Maximum surfactant concentration in the profile at a time for three different operator splitting increments $\Delta t=0.01$, 0.005, and 0.001. The time asymptotic convergence is a consistency check on the operator splitting scheme.

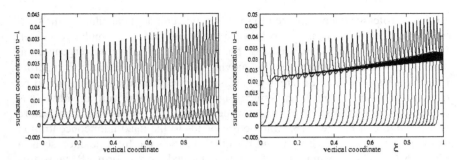

Figure 6.15 (a) Left: Non-cumulative model. (b) Right: Cumulative model. Combined compaction front translation and convective-diffusive model for Pe=100, m=1, offset y_0=2 for times t∈ [0.:0.01:0.375].

The cumulative model in Figure 6.15(b) shows qualitatively the same behaviour as in Figure 6.13 – peak concentration associated with the compaction front, eventually accumulating along the bottom of the layer. The difference is merely that the peak in Figure 6.15(b) is much more pronounced than in Figure 6.13. Strikingly, the range of peak heights is largely the same, 3—4.5 %, as in the more diffusive Pe=1 case and practically exactly the same in detail as in the non-cumulative model. This leads us to ask the question of whether the choice of Δt has a greater effect on high Peclet number models than on low. Figure 6.18 tests this for a Pe=100 case, with the conclusion that the default Δt=0.01 is insufficient for asymptotic convergence of the solution at high Peclet number. The splitting time increment Δt must be set tighter as diffusion becomes weaker, suggesting that the profiles should resemble Figure 6.16(b) with smoother ridges than Figure 6.17(b) with sharp peaks and trailing diffusive layers.

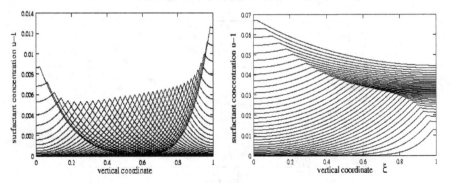

Figure 6.16 (a) Left: non-cumulative model. (b) Right: cumulative model. Combined compaction front translation and convective-diffusive model for Pe=1, m=0.7, offset y_0=2 for times t∈ [0.:0.01:0.375]. Note that the cumulative percentage variation runs from 1—7% in the cumulative model. Δt=0.01.

Since unit concentration is actually scaled by the initial uniform surfactant concentration in the layer prior to evaporation commencing, it is reasonable to vary the isotherm parameters m and y_0 to see the effect of changing the initial surfactant loading. Figure 6.16(b) tests this for Pe=1. This figure looks superimposable on Figure 6.12, with the exception that the vertical scale is stretched to accommodate the maximum surfactant concentration at 1.25% elevation, rather than the 0.75% with steeper isotherm (Figure 6.12).

Figure 6.17 shows that a 5—7 % surfactant rise is achievable with the flatter isotherm (than Figure 6.13). Figure 6.18 tests the asymptotic convergence as discussed before for high Peclet number situations. Clearly, $\Delta t=0.01$ is insufficient in Figure 6.17 for asymptotic convergence, casting doubt on the sharpness of the steepness of the peak in surfactant concentration actually accompanying the compaction front. A smoother ridge, as in Figure 6.16(b) is more consistent with the trace of the peak height for $\Delta t=0.001$ in Figure 6.18 below.

Curiously, Figure 6.19 shows that there is limited effect in flattening the isotherm further – hardly any dynamic change from Figure 6.17 at all. The peak heights are higher, 8—10 %, yet as the asymptotic convergence criteria of Figure 6.18 has not been met at $\Delta t=0.01$, it is likely that substantial smoothing will be achieved by shrinking the operating splitting time increment, as per Figure 6.18.

Because there is little variation in the surfactant concentration from the compaction front releasing adsorbed surfactant, the details of the isotherm, other than slope and value, do not enter the dynamics per se. They have the greatest effect in influencing the range of surfactant concentrations achievable, yet this is of limited influence given the small range. The key to this insensitivity is that

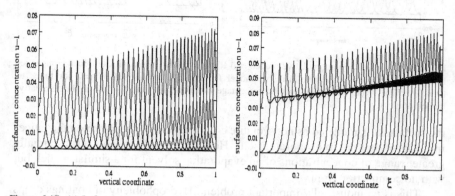

Figure 6.17 (a) Left: non-cumulative model. (b) Right: cumulative model. Combined compaction front translation and convective-diffusive model for Pe=100, m=0.7, offset y_0=2 for times $t\in$ [0.:0.01:0.375]. Note that the cumulative percentage variationruns from 6—8% in the cumulative model. $\Delta t=0.01$.

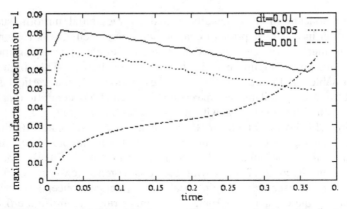

Figure 6.18 Cumulative model. Combined compaction front translation and convective-diffusive model for Pe=100, m=0.7, offset y_0=2. Shown are times t∈ [0.:0.01:0.375]. Maximum surfactant concentration in the profile at a time for three different operator splitting increments Δt=0.01, 0.005, and 0.001. The time asymptotic convergence is a consistency check on the operator splitting scheme. Comparison with Figure 6.14 leads to the conclusion that only Δt= 0.001 is asymptotically convergent.

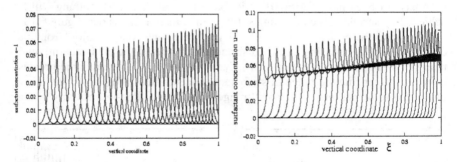

Figure 6.19 (a) Left: non-cumulative model. (b) Right: cumulative model. Combined compaction front translation and convective-diffusive model for Pe=100, m=0.5, offset y_0=2 for times t∈ [0.:0.01:0.375]. Note that the cumulative percentage variation runs from 8—10% in the cumulative model. Δt=0.01.

the surfactant is modeled as non-volatile constituent, able to adsorb on the latices in the layer, but not to evaporate itself. Howison et al. [18] show a substantially greater range of concentrations of an evaporating solvent for a similar model, but with no compaction front.

   This is an industrially important problem. Hydrophobicity of coatings drops considerably with large concentrations of surfactants. In addition channelling of water through pores of surfactant can seriously corrode material protected by an otherwise effective coating. For these reasons, models such as the one outlined above are central to the current research on coating efficiency.

Exercise 6.4: *Asymptotic surfactant concentration*

An interesting feature of the film drying application that is industrially relevant is the concentration of surfactant on the bottom surface when the front arrives, as a function of *Pe*. This has been commented on by many authors and they try to control it by varying the substrate chemistry, but our model suggests it is actually the particle/surfactant adsorption isotherm that controls this. This has many important ramifications for how to formulate industrial coatings. Write a MATLAB m-file script by altering filmdry.m to store only the final ($\bar{t} = 1/\alpha$) surfactant concentration value at ξ=0 with parametric variation from Pe in the range [1:5:100]. Try the isotherm parameters $m=0.7$, $y_0=2$ and $m=0.7$, $y_0=1$. Plot your data of u at ξ=0 versus Pe for both cases. Is the bottom surfactant concentration sensitive to the particle/surfactant adsorption isotherm?

6.4 Summary

In this chapter, we explored how FEMLAB can be used to set up simulations where the geometry model changes smoothly over either a parametric range or smoothly due to transient evolution of a front. The groundwork for these two situations was laid with previous discussion of parametric continuation. In particular, this chapter introduces an operator splitting technique to deal with transient geometric continuation, with geometry modification occurring during the first part of the time step, and a PDE being solved during the second part. The technique was shown to be self-consistent with asymptotic convergence tested for some parametric values and the simulation parameter – the increment over which the geometry is modified.

Although not novel, the transient model required re-starting the solution at one time step with the old solution at the last time increment. Yet, in order to do this, the old solution must be interpolated onto the new mesh, with potentially different numbers (and relevance) of the degrees of freedom. asseminit() was found to have sufficient power to do this, with the code supplied by the FEMLAB GUI programming interpreter, through a model m-file translation. By now, this should be a common technique for programming MATLAB routines calling FEMLAB functions – let the FEMLAB GUI provide the right commands.

The transient models for film drying were examples of two types of geometric continuation. The noncumulative model merely moved the initial position of the point source compaction front. Each front position was solved independently for surfactant concentration. The cumulative model read in the previous solvent concentration profile as its initial condition – the essence of parametric continuation and of transient integration. The operator splitting scheme developed here uses the best features of both types of continuation.

6.5   End Note: Solver Parameters for Problems with Pointwise Weak Terms

FEMLAB errors message(s) that inspired this note
•   Index into matrix is negative or zero.

Weak Point Terms

One of the most impressive features of FEMLAB is its ability to specify point source terms "semi-analytically" through the use of weak terms. Dirac delta functions, for instance, in the PDE are easily expressed through contributions to the FEM assembly through simple weak terms. The example in the User's Guide of a Poisson equation with a unit point source, or in Chapter 2 of the module workbook of a point vortex, both show how to implement a point singularity through weak terms. The only question posed here is what solver parameters are consistent with the weak source terms.

Solver Parameters

It is probably common sense that the weak solution form should be used on the Solver Parameters dialogue window if any weak terms or constraints are implemented. In FEMLAB 2.2, our film drying example worked fine with a numeric Jacobian and either coefficient, general, or weak solution form. In FEMLAB 2.3, however, this cocktail produces the error message "Index into matrix is negative or zero." The proper solution is found with the "exact Jacobian" option and weak solution form is recommended. A plausible explanation for the difference due to the change of versions has not been forthcoming, so this note should just serve as an advisory that the exact Jacobian and weak solution form are consistent with pointwise weak terms. COMSOL advise that the version inconsistency is probably due to the new "Automatic Scaling of Variables" feature.

The film drying example described here serves as a paradigm for treating pointwise weak terms. The FEMLAB model here is specified with only base FEMLAB application modes. It should be understood that by specifying a point source, the FEM implementation will smooth out the Dirac delta function on a sub-element length scale. Thus grid resolution may change the influence of the source at least locally. Thus the solution may never be "grid independent" in all details, but rather the model of the point source is intimately linked to the element meshing. Some other test than grid independence must then be passed to validate the model, typically experimental validation or matching behavior in limiting cases.

References

1. KJ Ruschak. "A method for incorporating free boundaries with surface tension in finite element fluid-flow simulators" *Int. J. Num. Methods Eng.* 15:639, 1980.
2. RT Goodwin and WR Schowalter "Arbitrarily oriented capillary-viscous planar jets in the presence of gravity." *Phys. Fluids*, 7(5): 954–963, 1995.
3. H. Ockendon and J.R. Ockendon <u>Viscous flow</u> Cambridge : Cambridge University Press, 1995.
4. GM Homsy et al., <u>Multimedia Fluid Mechanics (CD-ROM)</u>, Cambridge: Cambridge University Press, ISBN 0-521-78748-3, 2000.
5. DS Dugdale, "Viscous flow through a sharp-edged orifice," *Int. J. Engineering Science* 8:725–729, 1997.
6. WN Bond, *Proc. Roy. Soc.*, 34:139, 1922.
7. WB Zimmerman, "The drag on sedimenting discs in broadside motion in tubes." *International Journal of Engineering Science*, 40:7–22, 2002.
8. WB Zimmerman, "On the resistance of a spherical particle settling in a tube of viscous fluid" in preparation.
9. PM Bungay and H Brenner, "The motion of a closely-fitting sphere in a fluid-filled tube." *Int. J. Multiphase Flow*, 1:25, 1973.
10. OG Harlen, "High-Deborah-Number flow of a dilute polymer solution past a sphere falling along the axis of acylindrical tube" *J. Non-Newtonian Fluid Mech.* 37:157–173, 1990.
11. R Shail and DJ Norton, "On the slow broadside motion of a thin disc along the axis of a fluid-filled circular duct." *Proc. Camb. Phil. Soc.*, 65:793, 1969.
12. HA Stone, "On lubrication flows in geometries with zero local curvature," private communication.
13. MU Kim. "On the slow broadside motion of a flat plate along the centerline of a fluid-filled two-dimensional channel" *J. Phys. Soc. Jpn*, 53(1):139, 1984.
14. J Mallegol, J-P Gorce, O Dupont, C Jeynes, PJ McDonald and JL Keddie,"Origins and effects of a surfactant excess near the surface of waterborne acrylic pressure-sensitive adhesives." Preprint. 2002.
15. F Trogus, T Sophany, RS Schechter, and WH Wade, "Static and dynamic adsorption of anions and nonionic surfactants" SPE J. 17:337–344, 1977.
16. WF Ramirez, PJ Shuler, and F Friedman, "Convection, dispersion, and adsorption of surfactants in porous media," SPE J. 20(6):430–438, 1980.
17. WB Zimmerman and GM Homsy, "Nonlinear viscous fingering in miscible displacement with anisotropic dispersion." *Physics of Fluids A* 3(8) 1859 (1991).
18. SD Howison, JA Moriarty, JR Ockendon, EL Terrill, and SK Wilson, "A mathematical model for drying paint layers." *J. Engineering Math.* 32:377–394, 1997.

Chapter 7

COUPLING VARIABLES REVISITED: INVERSE PROBLEMS, LINE INTEGRALS, INTEGRAL EQUATIONS, AND INTEGRO-DIFFERENTIAL EQUATIONS

W.B.J. ZIMMERMAN

Department of Chemical and Process Engineering, University of Sheffield,
Newcastle Street, Sheffield S1 3JD United Kingdom

E-mail: w.zimmerman@shef.ac.uk

In this chapter, coupling variables are explored in great depth with regard to their role in solving inverse equations and integral equations of various types. Four important applications are taken as example studies – using lidar to detect position and spread of dense gas contaminant clouds, the inverse problem in electrical capacitance tomography, the computation of non-local heat transfer in a fiber composite medium, and the population balance equations in particle processing. En route, we encounter several features of FEMLAB not previously explored – coupling to optimization tools through MATLAB, extended meshes, using the time-dependent solver as an iterative tool for stationary nonlinear models, and the ability to selectively activate/deactivate multiphysics modes in coupled models. The latter is particularly useful if there is only one-way coupling (as in the hydrodynamics around the catalyst supported on the pellet in Chapter 3). In the case of the integral equations treated here, a fictitious dependent variable on an auxiliary domain is set up. The domain is used by coupling variables for various operations, but the dependent variable is never needed itself. So deactivating it results in better conditioning the FEM approximation to the integral equation.

7.1 Introduction

We are already familiar with boundary and subdomain integration – options available on FEMLAB's post processing menu. Boundary integration is useful for computing all manner of surface quantities: the charge on a body in electrostatics and the drag on a body in hydrodynamics, for example. Subdomain integration is typically used for averages and higher moments of combinations of the degrees of freedom defined in the domain. These features are reliable, and given the nature of the finite element method expressed through an integral property, the Galerkin method (see Chapter 2), FEMLAB naturally incorporates efficient and accurate integration schemes. Yet if the reader is interested in numerical integration of arbitrary integrands or ODEs, the built-in MATLAB schemes are generally sufficient (see Chapter 1) and do not warrant further discussion here. In this chapter, more complicated applications of integral equations and theory are explored with an eye to computation within

FEMLAB. Line integrals, integral equations, and integro-differential equations are the target applications in applied mathematics. These are all treatable with recourse to FEMLAB's coupling variable capability. Perhaps that is reason to have titled this chapter "Extended multiphysics II." The title selected, however, is probably more descriptive. As ever, we target illustrations in chemical engineering of the use of the FEMLAB features. The most important treated here are using lidar to detect position and spread of dense gas contaminant clouds, population balance equations which are exemplary of IDEs, the inverse problem in electrical capacitance tomography, and the computation of non-local heat transfer in a fiber composite medium.

Extended Multiphysics Revisited

When I read through the new features introduced with FEMLAB 2.2, I must admit to being skeptical of extended multiphysics as something that I was likely to use. Eventually, the utility of scalar coupling variables dawned on me, and provided the impetus for Chapter 4. It also spawned our interest in a new adventure for our research team, modeling microfluidics networks. Yet FEMLAB provides two other conceptual constructs for coupling variables – extruded and projected coupling variables. The examples of their use in the Model Library are nearly all about postprocessing, i.e. to express solutions in cross domain functionals to analyze particular features.

Rarely, however, coupling variables (extruded and projection) have been incorporated in the model and solved for simultaneously with the independent field variables. Multi-domain, multiple scale, and multiple process models are not common in engineering mathematics and mathematical physics. Typically, models are local in character – conceived of as a set of (partial) differential equations and boundary and initial conditions that are well posed. These are termed continuum models. Historically, this development has been predicated on the use of analysis techniques that have some scope for treating this class of models in closed form. Computational models, even in situations that are treatable by continuum methods, are approximated by discrete interaction rules that need not be local. Smooth particle hydrodynamics [1] and discrete element methods [2] are growing in popularity, but older methods like molecular dynamics simulations [3], Monte Carlo methods [4], microhydrodynamics [5], cellular automata [6] and exact numerical simulation in gas/plasma dynamics [7] bridge the continuum/discrete system gap in modelling distributed systems. Another set of techniques is based on optimization theory to satisfy pde constraints – penalizing the degree to which constraints are not satisfied. Mixed integer nonlinear programming [8], genetic algorithms [9] and genetic programming [10] are all suitable for treating models of mixed discrete/continuum systems. FEMLAB was formulated with a strong bias towards continuum systems with pde constraints. Yet, conceptually, extended multiphysics is not an afterthought for dealing with awkward situations. It permits treating discrete systems on an even footing with continuum systems characterized not only by pde constraints, but by integral constraints as well. Essentially, coupling variables permit nonlocal and discrete modelling.

In sections 7.3 (scalar), 7.4 (projection) and 7.5 (extrusion) we revisit coupling variables to explore FEMLAB treatment of inverse problems, line integrals, and integral/integro-differential equations, respectively.

Scalar Coupling Variables

Undoubtedly, scalar coupling variables are the conceptually easiest to grasp. In chapter 4, scalar coupling variables were used to link up a recycle stream in a flowsheet for a heterogeneous chemical reactor – the output of the reactor, suitably scaled, re-enters with the feed stream. An abstract 0-D element in a second geometry was created for the purpose of modeling a buffer tank that achieved the algebraic relationship between the recycle stream and the reactor outlet. Very simply, a scalar coupling is a single value passed to the destination domain, subdomain, boundary, or edge, where it is used anywhere in the description of the domain FEM residuals. The scalar coupling variable is created by an integration on the source domain. Since in our example, sources were 0-D (endpoints or the single element construct), the integrations were trivially the same as the integrand. Furthermore, that buffer tank model was artificial since the recycle relations could have been more readily incorporated in a weak boundary constraint without recourse to the second domain. So we have yet to see an example of scalar coupling variables where the source integration was non-trivial and the coupling itself essential. In the next subsection, we tackle an inverse problem where coupling is essential and intricate. An inverse problem has the connotation that there is an associated forward problem that is well-posed, but that the inverse problem is ill-posed. Our selected inverse problem is a tomographic inversion for electrical capacitance tomography.

Electrical Capacitance Tomography

Process tomography has matured as an engineering science in the past decade. One of the most common configurations is electrical capacitance tomography, frequently used for imaging processes with multiphase flows in cylindrical pipelines. Sensing of multiphase pipeline flows with information about the distributed flow of dispersed phases can be crucial to tight control of chemical and processing unit operations. Non-invasive and non-intrusive measurements of two-phase flow are notoriously difficult to obtain. The difficulty is often exacerbated by the highly time-varying flows some times encountered in gas-liquid flows in the oil and gas production industry. Accurate measurements of transients in the flow and instantaneous phase distributions cannot be achieved.

One possible way of obtaining such data is to measure the spatial electrical permittivity distribution of a flowing gas-liquid mixture using Electrical Capacitance Tomography (ECT). This will give information regarding the phase distribution about the pipe cross-section.

Tomographic instrumentation can provide images, non-invasively, of the distribution of components within a process vessel or pipeline. Electrical Capacitance Tomography (ECT) provides 2-D images of the *dielectric distribution* of the components within a process pipe. Non-invasive measurements of capacitance by electrodes - excited by a charge-discharge principle [11] - are used in a mathematical reconstruction algorithm to create images of materials having different permittivities. This procedure allows different phases to be determined. To date, process engineering studies involving ECT have been sparse, but some areas of application include fluidised beds and pneumatic conveying. McKee *et al.* [12] reported the use of capacitance tomography for imaging pneumatic conveying processes in two industrial pilot scale rigs. This work pioneered the application of ECT to *dense-phase* pneumatic conveying and demonstrated the potential of capacitance tomography as an aid to on-line process control. A good review of this area can be found in [13].

The tomographic imaging device involves three main sub-units: an array of sensors (typically 12 electrodes; 66 independent measurements), a data acquisition system and an image reconstruction system. Measurements of capacitance are obtained for all possible combinations of electrodes. For each electrode pair the following charge-discharge procedure is adopted: the active electrode is charged to a given voltage (15 volts) while the detecting electrode is earthed; the active electrode then discharges to earth while the detecting electrode connects to the input of a current detector. This detector then averages the resultant oscillating current from the detecting electrode, creating a voltage directly proportional to the unknown capacitance value.

The basic capacitance data acquisition system is based on the charge transfer principle. The discharging current flows out of the current detector producing a positive voltage output. The typical charge/discharge cycle repeats at a frequency of 1 MHZ, and the successive charging and discharging current pulses are averaged in the two current detectors, producing two DC output voltages.

Calibration of the instrument is performed before use of the electrode arrangement and involves the sensor device being filled with the material of lower permittivity. This procedure provides a reference value of permittivity. A change in the measurement sensitivity of the circuit then occurs when the pipe is filled with the material having the higher permittivity. A calibration procedure is needed for each type of material studied.

The image reconstruction process yields an image of the concentration distribution within the pipe by the use of a back-projection algorithm. Existing algorithm techniques for ECT are capable of producing images at a frame rate of 100 images per second and can, virtually, provide almost real-time information about the process. However, a limiting feature of the existing ECT system is the modest spatial resolution (about one tenth of the pipe radius). The major reason for this constraint is that the surface area of each electrode is large enough that, for all practical purposes, the electric field lines are parallel between the electrode pairs in the charge/discharge cycle. This convenience permits an easy image reconstruction by the back-projection algorithm. If more and smaller electrodes are used, there is the possibility of greater spatial resolution, but at the cost of a more complicated reconstruction algorithm. This algorithm would need to solve a Poisson equation with boundary data to find the internal permittivity field.

In this subsection, we give a flavor of the image reconstruction inverse problem with a toy model of a sparse system with large electrodes and distinct, rod-like inclusions in a 12-gon duct (see Figure 7.1 for the mesh).

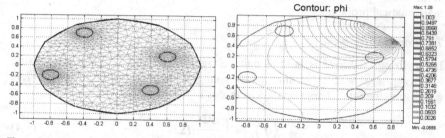

Figure 7.1 Left: mesh for four rod-like inclusions in a cylindrical duct, each with dielectric constant $\varepsilon_1=\varepsilon_2=\varepsilon_3=\varepsilon_4=0.05$ in a medium with unit dielectric constant ε_0. Right: Steady state contours of potential (voltage) when the boundary segment with unit normal (0.707,0.707) is held at unit voltage, and the segments with unit normal anti-parallel and perpendicular are held at ground, zero voltage, with all other boundary segments electrically insulated.

The electric charge density within the duct is related to the potential by the appropriate simplification to Maxwell's equations where there is no magnetic coupling [14]:

$$\nabla \cdot \varepsilon \nabla \Phi = -\rho^{(e)} \tag{7.1}$$

where $\rho^{(e)}$ is the total electric charge per unit volume, which is clearly zero within the bulk fluid and the inclusions, but non-zero on the electrode surfaces only, ε is the dielectric constant or permittivity of the medium, depending on the choice of scaling, and Φ is the electric potential (voltage). Using this (7.1) and applying over a thin control volume incorporating the interface between the electrode and the bulk fluid leads to this electric flux boundary condition:

$$-\varepsilon_s \frac{\partial \Phi}{\partial n} + \varepsilon_0 \frac{\partial \Phi}{\partial n} = q \qquad (7.2)$$

where ε_s is dielectric constant of the solid constituent of the electrodes and ε_0 is the dielectric constant of the bulk fluid medium. The LHS represents the electric flux out of the interface from the electrode side, the electric flux into the electrode from the bulk fluid, the difference balanced by the accumulated charge on the electrode at steady state. Rearranging (7.2) leads to the boundary status as

$$\frac{\partial \Phi}{\partial n} = \frac{q}{\varepsilon_0 - \varepsilon_s} = q' \qquad (7.3)$$

where we shall term q' as the charge on the electrode.

With these governing equations, we can define two related tomographic mathematical problems.

The Forward Problem

If the firing electrode is held at unit voltage (see Figure 7.1) and the sensing electrodes are held at ground (zero voltage), then the solution Φ to (7.1) computing the total charge on the electrodes i

$$\langle q' \rangle_i = \int_{\partial \Omega} \frac{\partial \Phi}{\partial n} d\Omega \qquad (7.4)$$

with known dielectric constants for the inclusions, is termed the forward problem. Figure 7.1 (right) shows the solution to the forward problem that we will shortly formulate in FEMLAB.

The Inverse Problem

Now suppose the same experiment is conducted, but that the dielectric field in the duct is not known *a priori*. The charges q_i' are measured on the electrodes and the permittivity field ε in the duct consistent (since Φ is a solution of (7.1)) with the measurements through (7.4) is sought. This is termed the inverse problem.

Modelling the Forward ECT Problem in FEMLAB

Launch FEMLAB and in the Model Navigator do the following:

Model Navigator

- Select 2-D dimension
- Select PDE Modes⇒General⇒Time-dependent >>
- Set the dependent variable as phi
- OK

Wait. Isn't the PDE system, with equations (7.1), BCs described in the caption of Figure 7.1, and outputs measured as boundary integrals (7.4), *stationary* and nonlinear? Shouldn't we be using the stationary nonlinear solver? Later, we will need the time-dependent solver. If we do not select it now, we will have to rebuild the model from scratch.

Pull down the options menu and select Add/Edit constants. The Add/Edit constants dialog box appears.

Add/Edit Constants

- Name of constant: e0   Expression: 1
- Name of constant: e1   Expression: 0.05
- Name of constant: e2   Expression: 0.05
- Name of constant: e3   Expression: 0.05
- Name of constant: e4   Expression: 0.05
- Apply
- OK

Pull down the Options menu and set the grid to (-1.1,1.1) × (-1.1,1.1) and the grid spacing to 0.1,0.1. Pull down the Draw menu.

Draw Mode

- Select Draw Arc. Now laboriously add arc points at the following positions:

(0,1),(0.2,1),(0.4,0.8),(0.6,0.8),(0.8,0.6),(1,0.4),(1,0),
(1,-0.4),(0.8,-0.6),(0.6,-0.8),(0.8,-0.6),(0.4,-0.8),(0.2,-1),
Now swap the signs
(0,-1),(-0.2,-1),(-0.4,-0.8),),(-0.6,-0.8),(-0.8,-0.6),(-1,-0.4),(-1,0),
(-1,0.4),(-0.8,0.6),(-0.6,0.8),(-0.8,0.6),(-0.4,0.8),(-0.2,1),

- Now double click on each vertex and edit it to the appropriate circular function value for angles $5\pi/12$ (0.258819,0.965926), $4\pi/12$ (0.5,0.866025), $3\pi/12$ (0.707107,0.707107), $2\pi/12$ (0.866025, 0.5), $\pi/12$ (0.965926, 0.258819). The trig identities for the second, third, and fourth quadrants are readily determined.
- Draw Ellipse (centered) at the following coordinates:

E1, edit center to (0.6,0.2) and set both semi axes to 0.1
E2, edit center to (-0.4,0.7) and set both semi axes to 0.1
E3, edit center to (0.4,-0.5) and set both semi axes to 0.1
E4, edit center to (-0.8,-0.2) and set both semi axes to 0.1

- Create a composite "Swiss cheese" object C01=C01-E1-E2-E3-E4
- Re-draw ellipses E1—E4 as above, to fill the "wholes" with four domains.
- Apply
- OK

The rough entry of the vertices is corrected by the ability to edit the analytical geometry features for graphical objects. Our closed curved is not quite a circle, but then neither are ECT systems once the electrodes are installed. The 12 segments of the boundary can be assigned individual boundary conditions and also are domains available for post processing. Now for the boundary conditions. Pull down the Boundary menu and select Boundary Settings.

Boundary Mode
- Set boundary 24 as Dirichlet, R=1-phi
- Set boundaries 5,6,21 as Dirichlet, R=phi
- Set boundaries 1,2,9,12,15,16,27,28 as Neumann, G=0
- Apply
- OK

The caption of Figure 7.1 explains that the system above has four Dirichlet boundary segments, one of which has fired, and the other three are held at ground. All other boundary segments are insulated. The charge on the four Dirichlet boundaries is computed according to (7.4).

Now pull down the Subdomain menu and select Subdomain settings.

Subdomain Mode	Γ		F	d_a
Subdomain 1	-e0*phix	-e0*phiy	0	1
Subdomain 2	-e2*phix	-e2*phiy	0	1
Subdomain 3	-e3*phix	-e3*phiy	0	1
Subdomain 4	-e4*phix	-e4*phiy	0	1
Subdomain 5	-e1*phix	-e1*phiy	0	1
• Apply				
• OK				

Accept the standard mesh parameters and hit the mesh button on the toolbar (triangle).

Pull down the Solver menu and select Solver Parameters. Click on the Settings button under "Scaling of variables." Check the None option. Now select the Stationary Nonlinear solver, and solve.

Figure 7.1 Gives the Post Plot of contours of constant phi between voltages of 0 and 1. Computation of (7.4) follows as below:

Post Mode			
Boundary integration: bnd 24	0.707107*(phix+phiy)	q1= 0.77067	
Boundary integration: bnd 21	0.707107*(phix-phiy)	q2=-0.30704	
Boundary integration: bnd 5	0.707107*(-phix-phiy)	q3=-0.16518	
Boundary integration: bnd 6	0.707107*(-phix+phiy)	q4= -0.30096	

The factor of 0.707107 and the signs are appropriate to form the normal derivative from the gradient according to the standard formula

$$\frac{\partial \Phi}{\partial n} = \hat{n} \cdot \nabla \Phi \qquad (7.5)$$

where the unit outward pointing normal is used.

Modelling the Inverse ECT Problem in FEMLAB

So far this example does not use coupling variables. To treat the Inverse ECT problem, we add a second conceptual domain.

Multiphysics Add/Edit Modes
• Select add geometry ⇒ g2 Select 1-D
• Select PDE Modes⇒General⇒Time-dependent >>
• Set the independent variables as u1, u2, u3
• Apply / OK

In Draw Mode, specify a geometry as the interval [0,1]. Then select Add/Edit Coupling Variables from the Options Menu.

Add/Edit Coupling Variables
Scalar add q1. Source Geom 1, bnd 24, Integrand: 0.707107*(phix+phiy); int ord 2 Destination Geom 2 subdomain 1 Check "Active in this domain" box.
Scalar add q2. Source Geom 1, bnd 21, Integrand: 0.707107*(phix-phiy); int ord 2 Destination Geom 2 subdomain 1 Check "Active in this domain" box.
Scalar add q3. Source Geom 1, bnd 5, Integrand: 0.707107*(-phix-phiy); int ord 2 Destination Geom 2 subdomain 1 Check "Active in this domain" box.
Scalar add q4. Source Geom 1, bnd 6, Integrand: 0.707107*(-phix+phiy); int ord 2 Destination Geom 2 subdomain 1 Check "Active in this domain" box.

Scalar add U1. Source Geom 2, subdomain 1, Integrand: u1; int ord 1
Destination Geom 1 subdomain 1-5 Check "Active in this domain" box.
Scalar add U2. Source Geom 2, subdomain 1, Integrand: u2; int ord 1
Destination Geom 1 subdomain 1-5 Check "Active in this domain" box.
Scalar add U3. Source Geom 2, subdomain 1, Integrand: u3; int ord 1
Destination Geom 1 subdomain 1-5 Check "Active in this domain" box.
   • Apply / OK

The coupling variables are of two types. The charges q1 through q4 that are explicitly to be found equal to the forward values by appropriate choice of the dielectric constants U1, U2, U3 which are coupled to the scalar value of u1, u2, u3 in Geom 2. The second multiphysics mode is to enforce the charge constraints. It should be noted that at steady state, charge (total electric flux out of the domain) must be net zero, which requires:

$$q_1 + q_2 + q_3 + q_4 = 0 \qquad (7.6)$$

It follows that there can only be three unknown dielectric constants. We arbitrary impose them as follows:

Subdomain Mode g1	Γ	F	d_a
Subdomain 1	-e0*phix -e0*phiy	0	1
Subdomain 2	-U2*phix -U2*phiy	0	1
Subdomain 3	-U3*phix -U3*phiy	0	1
Subdomain 4	-e4*phix -e4*phiy	0	1
Subdomain 5	-U1*phix -U1*phiy	0	1
• Apply			
• OK			

In the second multiphysics mode, we impose three constraints

Subdomain Mode g1	Γ	F	d_a
Init e1	-u1x	q1-0.77067	1
Init e2	-u2x	q2+0.30704	1
Init e3	-u3x	q3+0.16518	1
• Apply			
• OK			

The initial values are so that the search can start in the right region for the coupling variables. To round out the model, we need to specify the mesh in g2, which should be taken to be one element by giving the minimum element size as 1 in the mesh parameters menu. Finally, the boundary conditions for mode g2 should be Neumann (G=0) for u1, u2, and u3. This imposes that u1, u2, and u3 have spatially uniform, but unknown values, since the diffusive terms (Γ) do not change the imposed neutral boundary conditions. All of the dynamics come from the requirement that the charges are fixed at steady state, e.g. F1=q1-0.77067=0. Now to the solver parameters selection. Select Weak solution form. Depending on the mood of your platform, you should get a variety of error messages upon selecting the stationary nonlinear solver. The common error is "Inf or NaN repeatedly found in solution. Returned solution has not converged." The exact Jacobian does this; the numeric Jacobian takes longer to arrive at the same spot.

Error Message

A companion message to "Inf or NaN repeatedly found in solution. Returned solution has not converged." is the error message "Stepsize too small. Returned solution has not converged." The latter must be the most commonly encountered error message, as it is the symptom of many different ills: A short list includes:

1. Inconsistent model leading to a singular system. For instance, a badly posed boundary condition that can never be satisfied would never converge to a solution. The damping factor (i.e. step size) will be cut down until it reaches machine precision, but Newton's method will never provide a direction of decreasing error.
2. Unresolved physics. This pretty much means that you need more grid somewhere. Try the adaption option for the solver.
3. Your problem could simply be poorly posed or ill-conditioned. This is frequently due to large disparity in length scales or time scales at which *complexity* is generated in your problem. Try cutting down dimensionless *complexity* parameters like Reynolds, Rayleigh, or Peclet numbers to a size appropriate to your grid resolution or pack elements into supposed locations of boundary layers.

In the case of the ECT inversion problem, both explanation 1 and 3 fit the problem, as we explore further below.

The iterative solver should give a variation on the error theme – the preconditioning matrix has three rows that are all zero. If you try the linear solver, however, the story is different. It finds the solution for phi and quite readily determines values of the q_i near the imposed values. The dielectric constants u_i, however, are all extremely large magnitude, $O(10^{14})$. As we discussed in chapter 1, this behavior is consistent with a singular linearized operator, specifically with three zero rows.

How can this happen? Easily. The coupling variables q1, q2, and q3 are not differentiated correctly to form the contributions necessary for the Jacobian to be non-singular. They are treated as pseudo-constants that are not updated during the Newton solver operation. Consequently, the three equation model in g2 is singular. The linear solver only finds a solution because the matrix is non-singular due to numerical truncation at the double precision limit. But since it is ill-conditioned, the solution found is extremely large in magnitude.

The only solver which can bring out the nonlinear coupling through the coupling variables is the time-dependent solver. In this case, an extremely small time step is taken (note that the time-dependent equations are not singular as long as the F-constraints are not actually met).

Time Dependent Solver: Specifying Multiphysics Coupling on the Boundaries (and in Point Mode)

First a note about the time-dependent solver. One would think that the mixing of field variables in the boundary conditions (and point conditions) should be as straightforward as for the subdomain mode, with just a little care taken with the choice of Solver Parameters. Indeed, if you specify apparently linearly coupled boundary conditions in general or coefficient mode, selection of the "weak solution form" should permit accurate solution without difficulty for either stationary linear, stationary nonlinear, or time-dependent solvers. Even if you specify an apparently nonlinear coupling in the boundary conditions, the stationary nonlinear solver with weak solution form should handle it. But try it with the time-dependent solver and *general* solution form, and you should get the error message "Nonlinear constraints are not supported for time-dependent models." Switching to weak solution form gets mixed results. With one such set of boundary conditions, the time dependent solver simply ignored the condition and solved for the homogeneous Dirichlet condition instead in our electrokinetic flow model (see chapter 9). With a very similar condition, the time-dependent solver hangs without ever making the first time step. In the first case, the result was deceptive since a wrong solution is found. In the second case, the hung solver is disconcerting. Hence this note, to clarify how to treat nonlinear couplings in boundary conditions.

Weak constraint mode.   Specify your nasty nonlinear boundary condition as you like in the appropriate application mode. Then, follow this recipe:

1. Go to the multiphysics tab, and add a new application mode called weak,boundary constraint. Specify as many variables (lm1, lm2, lm3 ...), i.e. Lagrange multipliers, as you have nonlinear boundary constraint couplings.
2. Go to boundary mode and check Active in this domain, for all boundaries on which this feature occurs.
3. Check Use constraint specified in coefficients and the non-ideal dim constraint radio buttons.
4. Enter the variable name for which the original BC was specified in the constraint variable edit field.

5. Solve problem by simply highlighting all the variables in Solve for variables menu.
6. In order to run this recipe, you need FEMLAB 2.3 (2.2 won't do).

We would recommend this treatment for all nonlinear boundary couplings, not just those in time-dependent models. Why? Because it signals to FEMLAB that a non-standard boundary condition with nonlinear coupling should be treated with symbolic contributions to the assembly of the FEM system of matrices. See the Reference Manual for the command

`[K,N,L,M]=assemble(fem)`

and the description in Chapter 2 (§2.3.1) for the FEMLAB implementation of boundary conditions by Lagrange multipliers for a better understanding of why nonlinearity should be treated this way.

Even though this aside has been focused on the difficulties of nonlinear boundary constraints, nonlinear coupling constraints suffer a related problem. The FEMLAB symbolic engine does not contribute the full dependency (or even any in our case) of the coupling variables on the degrees of freedom. So the Jacobian matrix formed can be incomplete or inaccurate. The numeric Jacobian also lacks this dependency, as we found in the last section. The time-dependent solver lacks it as well, but the nonlinear coupling does manifest itself, with a delay of one time step as the coupling variable is updated at each time step. Since modern time stepping algorithms have quality control built-in, even stiff nonlinearity can be ferreted out by the time dependent solver.

Figure 7.2 shows six frames of the potential contours for times $t=0.01$ through $t=1$. Quantitatively, very little change occurs in the potential lines out to $t=4$. Wait a moment. The ECT problem, (7.1) and BCs defined in Figure 7.1, is time independent. So what is the time scale for t? Answer: completely fictitious. For stationary problems, a pseudo-time is another way of iterative solving. If a time-asymptotic, steady-state emerges, then the time-dependent solver hase done its job.

Another way of thinking of the time-dependent form of (7.1) is as an analogous problem in heat or mass transfer in a heterogeneous medium. The forward problem determines the boundary flux at fixed temperature (concentration) boundary segments interspersed among insulated segments. The inverse problem is to determine the distribution of transport coefficients internally in the heterogeneous medium consistent with the measured boundary fluxes.

In this context, the time scale is that for conduction and is physically meaningful. The potential (temperature or concentration) diffuses into the domain from the source boundary segment. Initially, the domain has uniform potential different from the source. Field lines are warped by the inclusions of non-uniform diffusivity. The field lines largely lead to flux out of the domain from the nearest two "sink" segments. The farthest sink, in the third quadrant, gets about half the flux of the other two sinks.

The time dependent solution for the charges q_1, q_2 and q_3 asymptotes to plateau values near to those computed in the forward problem (see before (7.5)) as shown in Figure 7.3. The arbitrary split between the graphs is due to the solution in three stages, arbitrarily split among (1) $t \in [0, 0.1]$; (2) $t \in [0.1, 1]$; and (3) $t \in [1, 4]$. As the system approaches the time asymptote, i.e. in interval (3), it is not particularly stiff and computes rapidly. The first interval is extremely stiff andonly minor fluctuations in the potential lines with fast animation, and no change to the colour coding.

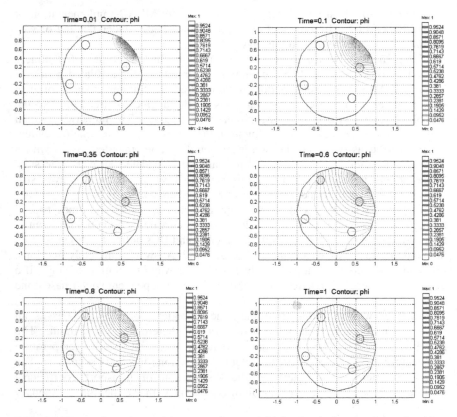

Figure 7.2 Progression of potential contours from an initially electrically neutral domain with a pseudo-time scale imposed. The voltage from the "firing" electrode with outward unit normal (0.707,0.707) diffuses out through the duct, being warped by the inclusions, eventually reaching an asymptotic profile that changes imperceptibly with further time evolution. Computations from time 1—4 show requires small step sizes.

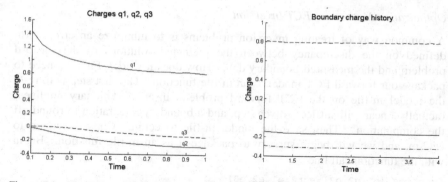

Figure 7.3 Time history of charges on the grounded electrodes q_2 and q_3 and on the fired electrode q_1.

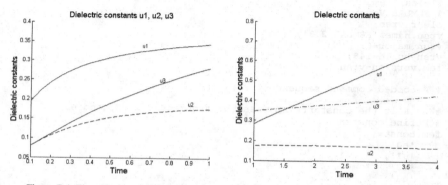

Figure 7.4 Time history of dielectric constants estimated on the inclusions within the duct.

Figure 7.4 shows the computed dielectric constants u1, u2, u3 as functions of time. So the surprise is that even though there is little difference between the computed boundary charges and the "measured values," the steady state has not been found and the dielectric constants inferred are diverging. Nevertheless, the quantitative values of the potential lines (Figure 7.2) are barely changing. The succinct rationale for this pathological behaviour is that there are an infinite family of dielectric constants for which the system outputs (q_1, q_2 and q_3) are flat – insensitive to coordinated variation of the dielectric constants. The final frame of Figure 7.2 is imperceptibly different from the right frame of Figure 7.1 – the inverse problem is badly conditioned. Now, it could be that time asymptotic convergence would occur if we started at close enough to the forward solution. Given the nearly singular nature of the problem, however, a different class of solution altogether is the prescription – optimization techniques.

Optimization Method for ECT Inversion

A common way of treating inversion problems is to minimize an error norm defined on the discrepancy between the predicted solutions of the forward problem and the measured boundary data. How does it work? First we need to package our forward ECT model as an m-file function. The first step is to reset the model m-file on the ECT forward problem, insert a stationary nonlinear (actually linear will suffice) solver step, and a boundary integration to round off the computation. Then save the model m-file to ect.m. Copy a version to ect2.m, and we will begin surgery to package it as an m-file function. Begin with the function call line

```
function [q1,q2,q3]=ect2(e1,e2,e3)
% FEMLAB Model M-file  for ECT forward problem
% Generated 07-Nov-2002 09:53:56 by FEMLAB 2.3.0.145.

flclear fem
% FEMLAB Version
clear vrsn;
vrsn.name='FEMLAB 2.3';
vrsn.major=0;
vrsn.build=145;
fem.version=vrsn;

% Recorded command sequence
...
%WZ: Edit the constants to pass the arguments e1,e2, and e3
% Define constants
fem.const={...
    'e0',      1,...
    'e1',      e1,...
    'e2',      e2,...
    'e3',      e3,...
    'e4',      0.05};

...
%WZ: Manufacture the output data
% Integrate on subdomains
q1=postint(fem,'0.707107*phix+0.707107*phiy',...
        'cont',    'internal',...
        'contorder',2,...
        'edim',    1,...
        'solnum',  1,...
        'phase',   0,...
        'geomnum',1,...
        'dl',      24,...
        'intorder',4,...
        'context','local');
% Integrate on subdomains
q2=postint(fem,'0.707107*phix-0.707107*phiy',...
        'cont',    'internal',...
        'contorder',2,...
        'edim',    1,...
        'solnum',  1,...
        'phase',   0,...
        'geomnum',1,...
```

```
                    'dl',       21,...
                    'intorder',4,...
                    'context','local');
% Integrate on subdomains
q3=postint(fem,'-0.707107*phix-0.707107*phiy',...
            'cont',     'internal',...
            'contorder',2,...
            'edim',     1,...
            'solnum',   1,...
            'phase',    0,...
            'geomnum',1,...
            'dl',       5,...
            'intorder',4,...
            'context','local');
% Integrate on subdomains
q4=postint(fem,'-0.707107*phix+0.707107*phiy',...
            'cont',     'internal',...
            'contorder',2,...
            'edim',     1,...
            'solnum',   1,...
            'phase',    0,...
            'geomnum',1,...
            'dl',       6,...
            'intorder',4,...
            'context','local');
```

We are now ready to give our m-file function a test. Make sure it is saved in the MATLAB current directory, and then execute the function on the MATLAB command line as below:

```
>> [q1,q2,q3]=ect2(0.05,0.05,0.05)
** Several warning messages print here **
Iter     ErrEst       Damping      Stepsize nfun njac nfac nbsu
  1      2e-014     1.0000000         3.1     2    1    1    2
  2      1.2e-016   1.0000000      8.7e-015   3    2    2    4
q1 = 0.7707
q2 =-0.3070
q3 =-0.1654
```

The error messages are a minor nuisance. Recall MATLAB's standard output is rounded to four significant figures. Now we are ready to compute the error norm, with a succinct m-file function:

```
function b=errornm(v);
x=v(1);
y=v(2);
z=v(3);
[q1,q2,q3]=ect2(x,y,z);
x=q1-0.77067;
y=q2+0.30704;
z=q3+0.16538;
[v(1),v(2),v(3),q1,q2,q3,sqrt(x^2+y^2+z^2)]
b=x^2+y^2+z^2;
```

The m-file function errornm.m should also be stored in the MATLAB current directory. Checking on the known "solution" yields an error norm of $O(10^{-5})$. Given the sparsity of the mesh, greater accuracy would not be expected.

```
>> errornm([0.05,0.05,0.05])
** Several warnings print here **
Iter      ErrEst    Damping    Stepsize nfun njac nfac nbsu
   1     2e-014   1.0000000        3.1    2    1    1    2
   2    1.2e-016  1.0000000    8.7e-015    3    2    2    4
ans =   0.00001699
```

Finally, we are ready to create a MATLAB script file to call MATLAB's built-in optimization routine, fminsearch() for scalar valued functions with vectorial arguments:

```
fmin.m contains three simple commands
v=[0.01,0.01,0.01];
a=fminsearch(@errornm,v);
quit
```

The @ preceeding the function name treats it as a pure function argument. The second argument represents the initial condition. fminsearch() provides a simple algorithm for minimizing a scalar function of several variables. It implements the Nelder-Mead simplex search algorithm, which modifies the input arguments "v" to find the minimum of f(v). This is not as efficient on smooth functions as some other algorithms, especially those that compute the derivatives, but on the other hand, costly gradient calculations are not made either. It tends to be robust on functions that are not smooth. If the function to be minimized is inexpensive to compute, the Nelder-Mead algorithm usually works very well.

This m-file script is best executed from the UNIX command line to avoid the GUI overheads. It takes about 10 CPU minutes on a Pentium IV 1.2 GHz processor:

matlab –nojvm <fmin.m >err 2>err &

Figure 7.5 contains the first 131 iterates. Apparently, the error norm has hit a plateau at about 0.0006 and is finding it exceedingly difficult move to smaller error norm. Similarly, the dielectric constants are convergent around the values

v=[0.0517 ,   0.0125,   0.0495].

Given that the "known" solution is

v=[0.05 ,   0.05,   0.05],

the fact that it is not found must be explained. Clearly, given the small error norm, the solution found is nearly as satisfactory as the "known" solution. In fact, there is a wide range of iterates that show nearly identical error norm. This suggests that there are many choices of the dielectric constants that result in nearly identical boundary data – the outputs are weakly sensitive to input variations in this regime. This result is part and parcel of the ill-posedness of the inverse ECT problem in this case. If many sets of dielectric constants are capable of reproducing nearly the same boundary data as the forward problem, the boundary data is insufficient to select among the potential interior configurations.

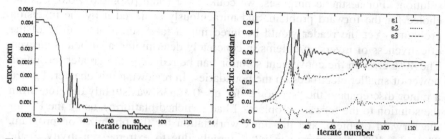

Figure 7.5 Optimization of distance from measured charges q_1, q_2, q_3 by predictions of same with dielectric constants from ε_1, ε_2, ε_3 chosen by Nelder-Mead optimization iterates. Left: error norm (see m-file errornm.m) with increasing iterate number; Right: dielectric constants versus iterate number. Initial dielectric vector is [0.01,0.01,0.01].

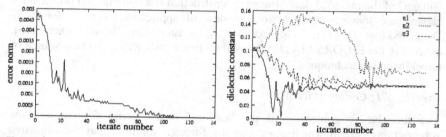

Figure 7.6 Optimization of distance from measured charges q_1, q_2, q_3 by predictions of same with dielectric constants ε_1, ε_2, ε_3 chosen by Nelder-Mead optimization iterates. Left: error norm (see m-file errornm.m) with increasing iterate number; Right: dielectric constants versus iterate number. Initial dielectric vector is [0.1,0.1,0.1].

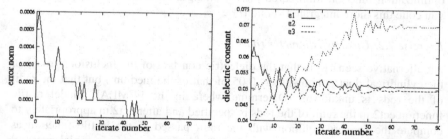

Figure 7.7 Optimization of distance from measured charges q_1, q_2, q_3 by predictions of same with dielectric constants ε_1, ε_2, ε_3 chosen by Nelder-Mead optimization iterates. Left: error norm (see m-file errornm.m) with increasing iterate number; Right: dielectric constants versus iterate number. Initial dielectric vector is [0.06,0.05,0.04].

Figures 7.6 and 7.7 are also convergent to the same dielectric constant solution vector [0.0508, 0.0703, 0.0497] from two substantially different initial guesses. Nevertheless, neither convergent solution is the one found from the forward solution. For aesthetic purposes, we could have taken [0.05,0.07, 0.05] as the choice for the forward problem, and miraculously achieved it by the inversion problem. Yet the reader would be lured into a false sense of security of the effectiveness of inverse problems in concretely determining a sufficient answer. The 40% error in the convergent solution can be reduced with greater resolution power of smaller electrodes on the boundaries. In reviewing this chapter, I noted a minor discrepancy: the "set point" for q3=0.16538 was slightly different from the solution to the forward problem. I had switched platforms, using the PC for the forward problem, and the linux workstation for the optimization program. Thus the 40% discrepancy in u2 is largely due to extreme sensitivity to the measurement error. But if greater resolution in the composition of the inclusions in the domain or their positions or sizes are desired, then the better quality boundary data is diluted across the domain, again possibly obscuring the "image" of the included data. Image reconstruction is a complicated problem for capacitance tomography. A good review of applications can be found by Dyakowski et al. [15]. The work of WRB Lionheart and coworkers [16], especially the EIDORS MATLAB based software package, is the best source of novel inversion techniques.

Exercise 7.1: Coding efficiency

The modular programming of the "calling" m-file script fmin.m and the subprograms ect2.m which computes the forward solution, and the objective function for error minimization errornm.m, is not particularly efficient, though good for pedagogical purposes. To improve the efficiency, the m-file script should have the FEMLAB model set up defined in global variables, and the optimization function fminsearch merely changes the dielectric constants. Can you code this more integrated version?

Exercise 7.2: Unknown diameter rods

An alternative scenario is that the dielectric constant of the inclusions (rods) in the cylindrical duct is known (say 0.05 of that of the medium) but that the radii of the rods is unknown. Determine where in the FEMLAB model m-file function ect2.m the radii of the rods is specified and alter ect2.m appropriately to compute the forward solution with the radii passed to the m-file function as arguments. Repeat the search procedure for the inversion from initial guess of the rod radii. Is this problem any better conditioned than the unknown dielectric constant problem? How could you improve the error in the estimated radii from the boundary data?

Projection Coupling Variables and Line Integrals

The projection coupling variable performs a line integral across a 2-D domain according to a specified coordinate dependent transformation, i.e. a path integral. That this is a useful concept is attested to by its use in formulating quantum electrodynamics [14]. In its simplest form, the path is taken as one of the coordinates (a simple grid line) and thus achieves a reduction in the order of the domain or variable dependence:

$$I(x) = \int_{y_1(x)}^{y_2(x)} f(x, y) \, dy \text{ on } D_2 \tag{7.7}$$

I(x) is the coupling variable, which must be defined on a domain D_1 of dimension one less than $D_2 = (x1,x2) \times (y1(x),y2(x))$ in the case shown above. I: $D_2 \rightarrow D_1$. A more complicated projection can be achieved by local mesh transformation using either the space coordinates (dependent variables x, y, z ...) or local mesh parameters, e.g. s, s1, or s2, which are then used to make a new source mesh either for interpolation or directing the curves on which the line/projection integrals are to be computed. For example

$$I(x) = \int_{C(x)} f(\xi(s), \eta(s)) \, ds \tag{7.8}$$

where the curve C is parametrized by x. So there is one such line integral for each point x in the destination domain. Generally, a projection coupling variable is one order of dimensionality lower than the source domain and therefore must be defined on a new domain, perhaps created explicity to receive the coupling variable as its destination domain. Inherently, a projection coupling requires two distinct domains (though the destination might be a boundary of the source domain) and thus must be planned from the start as at least a two domain (and potentially two geometry) model.

The coupling variable I(x) contains more information than one line integral. So if you are interested in a particular value of the line integral, then you need only click on the point in the destination domain on the post plot of the coupling variable, and the message window will display the interpolated value at that point. Alternatively, you can export the FEM structure and use postinterp to provide numerical value.

Example: Lidar positioning and sizing of a dispersing pollutant cloud

Lidar works on the same principle as several other optical devices, for instance spectrometry and spectroscopy, where light received of a given wavelength is of lower intensity due to absorption by a chemical species. For dilute chemical species, the signal received is proportional to the integrated concentration along the optical path, i.e.

$$proj = \int_C c\big(x(s), y(s)\big)ds \qquad (7.9)$$

where $\underline{C}$ is the curve $(x(s),y(s))$ and s is the coordinate along the length of arc. $c(x,y)$ is the concentration field of the chemical species. Suppose the domain is quasi-2D and an array of lidar are arranged along the x-axis which is the lower bound of the domain which is mapped to $[0,1] \times [0,1]$. Then the lidar array receives the discrete equivalent of the projection coupling variable $proj_1$:

$$proj_1(x) = \int_0^1 c(x, y)dy \qquad (7.10)$$

The curves C in (7.9) are taken here to be vertical lines. This is the standard action for FEMLAB projection coupling variables on a 2-D domain. The projection coupling variable is only a function defined on a 1-D independent variable and the default choices of "local mesh transformation" ($x \leftarrow x$, $y \leftarrow y$) for the source domain and of "evaluation point" for the destination domain ($x \leftarrow x$), produces (7.10) on a unit square. The choices for nonrectangular domains make more sense if one uses local coordinates: (s1, s2) in 2D for domains with curving boundaries, s in 1D for nonlinear curves.

Now for the example. The standard model for an instantaneous release of a dense pollutant gas in the atmosphere is a cloud with an average profile os a Gaussian in 2D. Zimmerman and Chatwin [17] analyze wind tunnel data of such dense gas releases, showing the instantaneous structure of fluctuations is highly intermittent. Yet the ensemble average or windowed time averages approach the 2-D Gaussian profile as the cloud becomes dilute.

So let's suppose that we have an initial profile of concentration of

$$c_0(x, y) = \exp\left(-\frac{(x-x_0)^2}{l_x^2} - \frac{(y-y_0)^2}{l_y^2}\right) \qquad (7.11)$$

Suppose that this profile is subjected to a uniform velocity field $\mathbf{u}=(u_0,v_0)$. It follows that the projections for horizontal and vertical arrays, respectively, of lidar would initially measure:

$$proj_1(x) = c_1 \exp\left(-\frac{(x-x_0)^2}{l_x^2}\right)$$

$$proj_2(y) = c_2 \exp\left(-\frac{(y-y_0)^2}{l_y^2}\right)$$

(7.12)

Let M be the total concentration dose in the domain

$$M = \iint_\Omega c(x,y)\,d\Omega$$

(7.13)

Then these projection coupling variables can be treated as normalized conditional probability density functions, with moments, in the case of $proj_1$:

$$m_{x,1} = \int_0^1 x\frac{proj_1(x)}{M}\,dx$$

$$m_{x,2} = \int_0^1 x^2\frac{proj_1(x)}{M}\,dx$$

(7.14)

For $c = c_0(x,y)$, the initial Gaussian, one can show that the first moment locates the x-coordinate, $m_{x,1} = x_0$. The second moment permits the computation of the standard deviation according to

$$s_x = \sqrt{m_{x,2} - m_{x,1}^2} = \frac{l_x}{\sqrt{2}}$$

(7.15)

Both conclusions obviously hold for x-projections onto the y-axis. So projections can locate and size a cloud of Gaussian shape. For non-Gaussian clouds, they provide at least a notion of centrality and degree of spread.

Now we demonstrate this example in action in FEMLAB. Launch FEMLAB and in the Model Navigator do the following:

Model Navigator
- Select 2-D dimension
- Select ChE⟹Cartesian⟹Mass Balance, Convection and Diffusion, time-dependent (cd mode)>>
- 1 dependent variable c
- OK

Pull down the options menu and select Add/Edit constants. The Add/Edit constants dialog box appears.

Add/Edit Constants

- Name of constant: x0   Expression: 0.4
- Name of constant: y0   Expression: 0.6
- Name of constant: lx   Expression: 0.1
- Name of constant: ly   Expression: 0.12
- Name of constant: u0   Expression: 1
- Name of constant: v0   Expression: 0
- Name of constant: Pe   Expression: sqrt(u0^2+v0^2)
- Apply
- OK

Pull down the Options menu and set the grid to $(-0.1,1.1) \times (-0.1,1.1)$ and the grid spacing to 0.1,0.1. Pull down the Draw menu.

Draw Mode

Select Draw Rectangle. Set $R1 = [0,1] \times [0,1]$
- Apply / OK

Now pull down the Subdomain menu and select Subdomain settings.

Subdomain Mode

- Select domain 1
- Set D=1/Pe; u=u0; v=v0
- Select the init tab and give c(t0) according to (7.11)
- Apply / OK

Now for the boundary conditions. We follow our standard recipe for periodic boundary conditions in both directions. Pull down the Boundary menu and select Boundary Settings.

Boundary Mode

- Check View as Coefficients on the Boundary menu
- Select domain 1 h=1 r=0
- Select domain 4 h=-1 r=0
- Select domain 2 h=1 r=0
- Select domain 3 h=-1 r=0
- Apply
- OK

In Mesh mode, we need to set the symmetry boundaries as 1 4 2 3, which is treated pairwise so that 1 and 4 are symmetry boundaries as are 2 and 3. The combination of symmetry boundaries and boundary condition coefficients achieves doubly periodic boundary conditions. Upon meshing, 417 elements with 772 nodes were created in the 2-D domain.

The major action is the computation of the projection coupling variables. Select Add/Edit Coupling Variables from the Options Menu.

Add/Edit Coupling Variables

Projection add proj1. Source Geom 1, subdomain 1, Integrand: c; int ord 2
Local mesh transformation $(x \leftarrow x, y \leftarrow y)$
Destination Geom 1 bnd 2 Check "Active in this domain" box.
Evaluation point $(x \leftarrow x)$
Projection add proj2. Source Geom 1, subdomain 1, Integrand: c; int ord 2
Local mesh transformation $(x \leftarrow y, y \leftarrow x)$
Destination Geom 1 bnd 1 Check "Active in this domain" box.
Evaluation point $(x \leftarrow x)$
- Apply / OK

Set the Solver Parameters on the Solve menu with output times [0:0.001:0.06] on the time stepping page. Select Apply/OK and hit the Solve = button on the toolbar. After about twenty seconds of overhead computation, the time stepping begins. As the problem is linear, it does not take long per step.

Computation of (7.14) follows as below for t=0.06:

Post Mode

Subdomain integration: domain 1	c (at any time)	I1= 0.037702
Boundary integration: bnd 2	proj1*x/0.037702	I2= 0.49325
Boundary integration: bnd 1	proj2*y/0.037702	I3= 0.51522
Boundary integration: bnd 2	proj1*x^2/0.037702	I4= 0.31825
Boundary integration: bnd 1	proj2*y^2/0.037702	I5= 0.34188

The latter two give s_x=0.2738 and s_y=0.2765, nearly identical spread, but this is expected given the nearly diffused final state. For time t=0., the same contributions result in:

Post Mode

Boundary integration: bnd 2	proj1*x/0.037702	I2= 0.39999
Boundary integration: bnd 1	proj2*y/0.037702	I3= 0.60012
Boundary integration: bnd 2	proj1*x^2/0.037702	I4= 0.165
Boundary integration: bnd 1	proj2*y^2/0.037702	I5= 0.36726

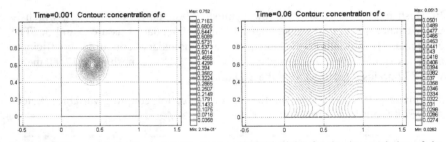

Figure 7.8 Isopycnals for times t=0.001 (left) and t=0.06 (right) for the time evolution of the concentration field from c_0 (7.11) according to the convective-diffusion model with Pe=1 and uniform horizontal flow under doubly periodic boundary conditions.

The latter two give $s_x=0.0707$ and $s_y=0.0872$, consistent with the settings of l_x and l_y as expected.

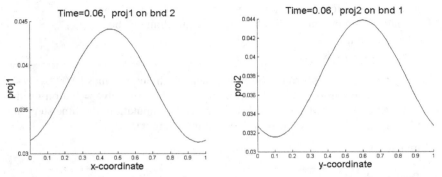

Figure 7.9 Line integral projections for t=0.06 onto the horizontal boundary (left) and vertical boundary (right) for the model of Figure 7.8.

Figure 7.8 shows the extent to which the initial condition diffuses very rapidly. Although the Pe=1 in this simulation, the numerical diffusivity is strong on this mesh resolution. Likely the result is less rapid diffusion on a finer mesh. Figure 7.9 gives the projection coupling variables demonstrating the near Gaussian profiles captured by our synthetic "lidar." Clearly, even the late stage evolution where periodic boundary conditions obscure the usual "long tails" of the Gaussians, exhibit a central peak and spread captured by the central moments according to (7.14) and (7.15).

Exercise 7.3: Artificial (numerical) diffusivity

Repeat the lidar example with a refined mesh. Does the Gaussian cloud disperse slower (less spread) with a refined mesh. How could you use this computation to quantify the numerical diffusivity that is artificially created?

Extrusion Coupling Variables

An extrusion coupling variable was named after one of its most common uses; it maps information from a domain of dimension n to one of higher dimension $n+1$. Yet extrusion is only one of its potential uses, which are generalized as interpolation, projection, or mapping, depending on the information passed. The other two coupling variable types – scalar and projection – perform integrations over their source domains (or subdomains) and are thus able to be incorporated in integral equations. Extrusion coupling variables map detailed or distributed information from one domain to another, with the destination position selected by the local mesh transformation. So extrusion variables are useful intermediaries in models with multi-domain coupling. Yet they need not be defined on domains of different geometries. In FEMLAB seminars, the common example given of extrusion coupling variables is for aesthetic reasons. Frequently, given the symmetry in a physical configuration, the model can be solved over only part of the domain or even a lower dimension, yet the real physical configuration is required to visualize the solution. So, for instance, in a cylindrical duct, axi-symmetry may only require solution in the r-z plane, yet visualization on the cylinder may be desirable. Extrusion over the θ-coordinate of the r-z solution will permit the desired visualization. Suppose placement of baffles with hexagonal symmetry in the domain permitted solution over a wedge of $\theta \in [0,\pi/3]$ with r and z bounded. Yet, if visualization is required over the whole duct, extrusion of the wedge to the other fiver wedges would permit this. So extrusion coupling variables may merely extend information for postprocessing into other domains.

Integral Equations

Integral equations are distinguished by containing an unknown function within an integral. As with differential equations, linear systems are the best characterized and therefore most commonly occurring. The classification system is straightforward.

If the integration limits are fixed, the equation is termed of Fredholm type. If one limit is a variable, it is termed of Volterra type.

If the unknown function appears only under the integral sign, it is labeled as of the "first kind." If it appears both inside and outside the integral, it is labeled of the "second kind."

Here are the four combinations symbolically:

Fredholm integral equations of the first kind:

$$f(x) = \int_a^b K(x,t)g(t)\,dt \qquad (7.16)$$

Process Modelling and Simulation with Finite Element Methods

Fredholm integral equations of the second kind:

$$g(x) = f(x) + \lambda \int_a^b K(x,t)g(t)dt \tag{7.17}$$

Volterra integral equations of the first kind:

$$f(x) = \int_a^x K(x,t)g(t)dt \tag{7.18}$$

Volterra integral equations of the second kind:

$$g(x) = f(x) + \lambda \int_a^x K(x,t)g(t)dt \tag{7.19}$$

In all four cases, g is the unknown function. $K(x,t)$, called the kernel, and $f(x)$ are assumed to be known. When $f(x)=0$, the equation is said to be homogeneous.

One might reasonably ask why we bother with integral equations. The answer is the theme of this chapter – integral equations are fundamentally nonlocal. Some physical phenomena are inherently nonlocal in character, so their description leads to integral or integro-differential equations. For instance, Shaqfeh [18] derived a theory for transport properties of composite materials that naturally leads to a nonlocal description of effective properties. Many systems are "elliptical" in nature – the boundary data diffuses everywhere, say steady state heat transfer or mass transfer – which results in the solution at a point depending on the solution everywhere. Such nonlocal systems can be conveniently described in terms of a Green's function, which then leads to an integral equation description for inhomogeneous systems. Finally, some processes are conveniently described in a phase space (Fourier space, Laplace space, size, volume or mass distribution) that involve nonlinear coupling of the variables in phase space. When described in physical space, these phase space couplings manifest as convolution integrals which are both nonlocal and nonlinear. Rarely, transform methods, for instance the Abel transform, through a clever change of variables, permits the restatement of an integral equation as an equivalent differential equation, at least for smooth functions. Howison et al. [19] give an example that was cited with regard to film drying in Chapter 6. Otherwise, either discretization or power series expansion are the preferred analysis techniques.

It is not the intention of this chapter to teach integral equation theory. An introduction worth reading is given in Arfken's book [20] and a thorough grounding can be found in Stakgold [21] or Lovitt [22]. Here we intend only to explore some aspects of FEMLAB's ability to compute solutions to integral and integro-differential equations.

Solving a Fredholm Integral Equation of the Second Kind

Zimmerman [23] gives the derivation of a Fredholm integral equation of the second kind as an intermediate in the solution for the drag on a thin disk in broadside motion in a cylindrical duct. The variation on (7.17) is slight:

$$g(x) = 1 + \varepsilon \int_0^1 K(x,t) g(t) \, dt \qquad (7.20)$$

where $\varepsilon < 1$ is a small parameter. The kernel K was bounded, so a theorem in integral equation theory [21] ensures that a solution for g(x) can be found by iteration, with each iterate improving in accuracy by at least one order of correction in ε. Zimmerman [23] demonstrated a solution by series expansion in powers of x and ε, albeit relying on numerical computation of the series coefficients. As the kernel of that problem is not particularly tractable (it too was expanded in powers of x and t), a simpler kernel will be selected here for demonstration. The FEMLAB implementation is a *tour de force* in projection and extrusion coupling variables.

As alluded to in section 7.4, projection variables are the variable of choice for a line integration that returns a function. Although we wish to achieve a line integration of the form

$$f_1(x_1) = \int_0^1 K(x_1, x_2) g_2(x_2) \, dx_2 \qquad (7.21)$$

it is easier to achieve

$$f_1(x_1) = \int_0^1 K(x_1, x_2) g_2(x_1, x_2) \, dx_2 \qquad (7.22)$$

where $g_2(x_1,x_2) = g_1(x)$, with the mapping $(x_2 \leftarrow x)$ and extruded along the x_2 coordinate. Figure 7.10 shows this graphically for the kernel $K(x,t) = sin(2\pi x t)$. The left figure is extruded along the horizontal coordinate after being mapped to the vertical. Alternatively, think of the left figure as the projection of the right 2-D domain onto its vertical axis. The computation (7.22) has the intermediate swelling by one dimension of the domain in order to preserve the functionality of the line integral though the projection coupling variable. If this concept is clear, then the FEMLAB implementation is merely "turning the crank."

Launch FEMLAB and in the Model Navigator do the following:

Model Navigator

- Select 1-D dimension, Geom 1
- Select PDE modes⇒General⇒Stationary nonlinear model, weak form (mode g1). Dependent variable u1, independent variable x>>
- Multiphysics Tab. Add Geom 2, Select 2-D dimension
- Select PDE modes⇒General⇒Stationary nonlinear form (mode g2) Dependent variable u2, independent variables x1, x2>>
- Apply/OK
---

down the **Options** menu and set the grid to (-0.1,1.1) × (-0.1,1.1) on Geom 2 and the grid spacing to 0.1,0.1. Pull down the **Draw** menu.

Draw Mode

Geom 1: Specify geometry; interval [0,1]
Geom 2: Select Draw Rectangle. Set R1 = [0,1] × [0,1]
- Apply / OK

Pull down the **options** menu and select **Add/Edit constants**. The Add/Edit constants dialog box appears.

Add/Edit Constants

- Name of constant: eps   Expression: 0.05
- Apply / OK

The major action is the computation of the extrusion and projection coupling variables. Select **Add/Edit Coupling Variables** from the **Options Menu**.

Add/Edit Coupling Variables

extrusion add f2. Source Geom 1, subdomain 1, Expression: u1
Local mesh transformation (x ← x)
Destination Geom 2 subdomain 1, Check "Active in this domain" box.
Evaluation point transformation (x ← x2)
projection add f1. Source Geom 2, subdomain 1, Integrand: sin(2*pi*x1*x2);
integration order 2
Local mesh transformation (x ← x1, y ← x2)
Destination Geom 1 subdomain 1Check "Active in this domain" box.
Evaluation point transformation (x ← x)
- Apply / OK

By comparison, the Subdomain settings are pedestrian:

Subdomain Mode

- Select mode g1 (geom1 domain 1)
- Set Γ=0, da=0, F=u1-1-eps*f1
- Apply/OK
- Select mode g2 (geom2 domain 1)
- Set Γ=0 0, da=0, F=u2-f2
- OK

Now for the boundary conditions. Neutral are needed. Pull down the Boundary menu and select Boundary Settings.

Boundary Mode

- Mode g1: geom1 domain 1,2 Select Neumann, G=0
- Mode g2: geom2 domain 1,2,3,4 Select Neumann, G=0
- Apply
- OK

In Mesh mode, accept the standard mesh for mode g2 (417 nodes, 772 elements) and in mode g1, refine to 61 nodes, 60 elements. Solve. The solution should appear as in Figure 7.10.

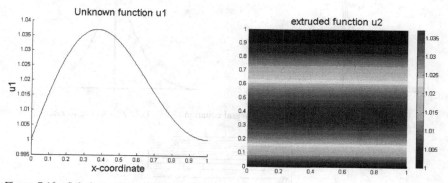

Figure 7.10   Solution g(x) to (7.20) with $K(x,t)=sin(2\pi \, x \, t)$.   Left: 1-D solution.   Right: 2-D extrusion of g(x).

Solving a Volterra Integral Equation of the Second Kind

In searching for a Fredholm integral equation of the second kind as an example from the literature for the last section, I hit upon Shaqfeh's [18] equation (7.23) for the edge effect near an impermeable wall for characterization of effective boundary conditions for thermal conduction in a fiber composite medium, where the fibers are better conductors than the fluid matrix:

$$g(z)+1+N\int_{-2}^{2}K(z,x)g(z+x)dx=0 \qquad (7.23)$$

Here, $g(z)$ is the gradient of the ensemble average temperature at a distance z from the edge of the wall in scaled coordinates. Shaqfeh's theory derives the non-local contributions for average extra flux due to the presence of randomly positioned fibers. N is the dimensionless parameter expressing number density and slenderness of the fibers. The kernel is given here in MATLAB notation

K(z,x)=((x>0)*(z^2*(6-3*x-2*z)+(2*z-x+2)*(x+z-2)^2*(x+z>2))+
(x<0)*((2*z+3*x+2)*(z-2)^2*(z>2)+(x-2*z+6)*(z+x)^2*(z+x>0)))/12;

It should be noted that this expression corrects equation (83b) of [18] for a typographical error. The proof of this is that with the correction, Shaqfeh's assertion that the kernel is homogeneous for $z\geq2$ is borne out. Figure 7.11 (left frame) shows the invariant kernel profile in this regime. The contours of $K(z,x)$ are shown in Figure 7.12, with the regime of parallel lines at the top consistent with this assertion. Essentially, in this regime, the heat flux sees the same environment whether in the direction of the edge or away from it, statistically,

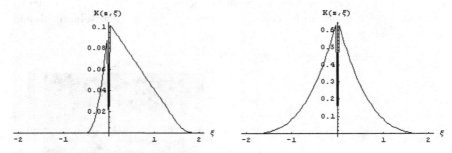

Figure 7.11 Kernel K(z,x), of the integral equation (7.23) Left: z=0.5. Right: z≥2.

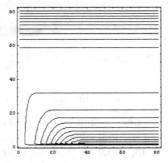

Figure 7.12 Contour plot of the kernel K(z,x), of the integral equation (7.23) Abcissa and ordinate are in the range $(x,z)\in[0,4]\times[0,4]$. Homogeneity of K for $z\geq2$ is apparent.

and therefore, the driving force for non-local heat flux is lost. Furthermore, with the correction, we find that $K(z,x)$ is continuous at the origin, but has a discontinuous slope, typical of Green's functions in 1-D [21]. Finally, we shall see that we reproduce results consistent with Shaqfeh's finite difference solution of (7.23).

It turns out that (7.23), however, is not a Fredholm integral equation at all. Why? Because the $g(z+x)$ dependency in the integrand is not the standard form for a Fredholm equation. Change of variable leads to a Volterra integral equation. Let $x_2 = z + x$. Then $dx_2 = dx$. Re-writing (7.23) yields

$$g(z) + 1 + N \int_{z-2}^{z+2} K(z, x_2 - z) g(x_2) dx_2 = 0 \qquad (7.24)$$

Since how one writes the kernel is not at issue, (7.24) clearly has the dependent variable in the limits of integration, so can be identified as a Volterra integral equation. Nevertheless, this a second alteration to the kernel, we can re-write (7.24) in a form that is treatable by our recipe for Fredholm integral equations in FEMLAB, using extrusion and projection coupling variables on an abstract intermediate domain with coordinates (z, x_2):

$$g(z) + 1 + N \int_a^b K'(z, x_2) g(x_2) dx_2 = 0 \qquad (7.25)$$

where $K' = (x_2 > z - 2) * (x_2 < z + 2) * K(z, x_2 - z)$. With this kernel, (7.25) is of the same form as (7.20), so the same strategy should suffice to a large extent. The one major modification is that (7.23) is ill-posed as it stands for any finite interval in z. Simply, for $z \in [0, l]$, (7.25) shows that g must be defined for $x \in [-2, l+2]$. Shaqfeh posited that to regularize the problem, the homogeneous behavior of K for $z > 2$ leads to the asymptotic solution that $g \rightarrow -1/(1 + 2/3\ N)$. This can be taken as the solution in the regime $x \in [l, l+2]$. For $x \in [-2, 0]$, there position is within the wall, so the homogeneous conductivity there must match the flux at the wall, $g = -1$. So in our abstract 2-D domain of coupling variables, we impose these two limiting behaviors outside the solution domain $z \in [0, l]$.

Launch FEMLAB and in the Model Navigator do the following:

Model Navigator
• Select 1-D dimension, Geom 1
• Select PDE modes⇒General⇒Stationary nonlinear model, weak form (mode g1). Dependent variable u1, independent variable x>>
• Multiphysics Tab. Add Geom 2, Select 2-D dimension
• Select PDE modes⇒General⇒Stationary nonlinear form (mode g2) Dependent variable u2, independent variables x1, x2>>
• Apply/OK

Pull down the Options menu and set the grid to (-0.1,5.1) × (-2.1,7.1) on Geom 2 and the grid spacing to 0.5,0.5. Pull down the Draw menu.

Draw Mode

Geom 1: Specify geometry; interval [0,5]
Geom 2: Select Draw Rectangles.
Set R1 = [0,5] ×[-2,0], R2=[0,5] ×[0,5] , R3=[0,5] ×[5,7].
- Apply
- OK

Pull down the options menu and select Add/Edit constants. The Add/Edit constants dialog box appears.

Add/Edit Constants

- Name of constant: eps   Expression: 1
- Apply
- OK

The major action is the computation of the extrusion and projection coupling variables. Select Add/Edit Coupling Variables from the Options Menu.

Add/Edit Coupling Variables

extrusion add f2. Source Geom 1, subdomain 1, Expression: u1
Local mesh transformation (x ← x)
Destination Geom 2 subdomain 1, Check "Active in this domain" box.
Evaluation point transformation (x ← x2)
projection add f1. Source Geom 2, subdomain 1,2,3; Integrand
```
u2*(x2>x1-2)*(x2<x1+2)*((x2>x1)*(x1^2*(6-3*(x2-x1)-2*x1)+(x1+(x2-
x1)>2)*(2*x1-(x2-x1)+2)*(x1+(x2-x1)-2)^2)/12+(x2<x1)*((2*x1+3*(x2-
x1)+2)*(x1-2)^2*(x1>2)+((x2-x1)-2*x1+6)*(x1+(x2-x1)))^2*(x1+(x2-
x1)>0))/12);
```
integration order 2
Local mesh transformation (x ← x1, y ← x2)
Destination Geom 1 subdomain 1, Check "Active in this domain" box.
Evaluation point transformation (x ← x)
- Apply / OK

By comparison, the Subdomain settings are pedestrian:

Subdomain Mode
• Select mode g1 (geom1 domain 1)
• Set $\Gamma=0$, da=0, F=u1+1+eps*f1
• Apply/OK
• Select mode g2 (geom2 domain 1) Set Γ=-u2x1 0, da=0, F=u2+1
• Select mode g2 (geom2 domain 2) Set Γ=-u2x1 0, da=0, F=u2-f2
• Select mode g2 (geom2 domain 3) Set Γ=-u2x1 0, da=0, F=u2+1/(1+2*eps/3)
• Apply
• OK

The horizontal diffusive flux in our second geometry (abstract 2-D domain) is merely a numerical convenience to help insure stability. Since the model, by construction, is horizontally homogeneous in this space, with no flux BCs (see below), no amount of horizontal diffusion can alter the solution theoretically. Yet stronger diffusion will damp out any horizontal numerical errors which might creep in due to truncation. With vertical diffusion, however, this is not true, so it is excluded. Now for the boundary conditions. Neutral are needed. Pull down the Boundary menu and select Boundary Settings.

Boundary Mode
• Mode g1: geom1 domain 1,2 Select Neumann, G=0
• Mode g2: geom2 all domains Select Neumann, G=0
• Apply / OK

In Mesh mode, set max edge size general to 0.35 for geom2, which gives mesh for mode g2 (691 nodes, 1296 elements) and in mode g1, set max edge size to 0.1, to give 251 nodes, 250 elements. Solve. The solution should appear as in Figure 7.13.

Note that our 2-D abstract domain gives u2 horizontally as fairly homogeneous. So vertical line integrals traducing all three subdomains experience u1(x2) when integrating u2(x1,x2) along the x2 coordinate.

Figure 7.13 (left) shows that the physically important region over which the temperature gradient moves from edge value (-1) to asymptotic value -1/(1+2N/3), is not more than about a dimensionless length of 2.5. This limiting behavior matches the theoretical predictions of Shaqfeh [18], and is a consistency check on the kernel. An interesting feature of this profile is the internal maximum of temperature *gradient*. Shaqfeh graphed the *temperature* profile itself, so given the relatively smooth transition, the integration of g might have such a modest internal maximum discernable, but here, in the gradient, it

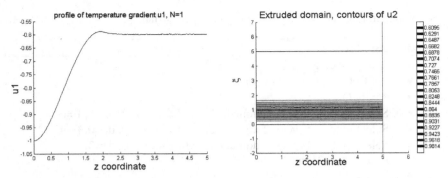

Figure 7.13 Temperature gradient, solution g(z) to (7.23). Left: 1-D solution. Right: 2-D extrusion of g(x). N=1.

clearly manifests. I am curious if this feature is an intrinsic aspect of the edge conduction near an impermeable (to the fibers) wall bounding a composite. From the description of the calculation in [18], it is not clear which conditions are applied for the region $x2 \in [-2,0]$. The original integral equation (7.23) clearly shows that the integral is computed over that region. The last term of the kernel, with the factor $(x+z>0)$, clearly has a contribution from $x \in [-2,0]$ when $z \in [0,2]$. So what is the consistent value of the temperature gradient? I argued that $g(x)=-1$ for $x \in [-2,0]$. Perhaps this choice influences the prediction of an internal maximum in temperature gradient.

Figure 7.14 gives a parametric study over N (eps in our FEMLAB model) for the same eight values given for the profiles of temperature in Figure 6 of [18]. The internal maximum in temperature gradient seems to be a persistent feature for N>1, but is not apparent for N<0.5. The computation given here is self-consistent, leading to confidence in using FEMLAB to compute the solutions to canonical linear 1-D integral equations of either the Fredholm or Volterra type, of either kind. Although not particularly envisaged by the software developers themselves, this feature has its own niche among software packages for general engineering/mathematical physics productivity. As an experienced, Mathematica, Macsyma, and Matlab user, I can confidently claim that solution to integral equations by other means is a *tour de force* in difference equations, matrix assembly, and sparse matrix solvers for linear integral equations. As we will see in the next subsection, our FEMLAB recipe for integral equations extends to nonlinear integral equations, even of the convolution type, in a straightforward manner.

As a coda to this subsection, one notes that the problem considered here is a variant on the electrical capacitance models of §7.3.2 and §7.3.3, particularly as there is a direct analogue to heat conductance in a fluid medium with solid inclusions. The difference is that the ECT models were of nonhomogeneously placed rods and thus the relative positions dominated the flux calculations. Here, the homogeneity of the fibrous inclusions simplifies the conductance model.

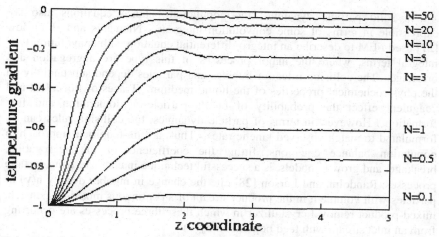

Figure 7.14 Solution g(z) to (7.23) for parameters N=0.1,0.5,1,3,5,10,20,50. Shaqfeh's Figure 6 solves for the integral $\langle T \rangle = \int_0^z g(z')dz'$ where the reference temperature is at the wall.

Convolution Integrals and Integral Equations

Convolution integrals are typically nonlinear and nonlocal, viz.

$$I(z) = \int_a^b K(z,x)g(z)g(z+x)dx \qquad (7.26)$$

They arise naturally in turbulence theory as two point correlation functions – statistics of the turbulence [24]. There is also a well known duality with nearly all linear transforms – convolutions in physical space transform to quadratic products of the individual transforms in transform space, and vice versa – known as the convolution theorem [25]. Since quadratic nonlinearity is fairly common in transport phenomena (inertia and convective terms), convolutions in transform space are just as common. Another important class where convolutions occur is in phase space descriptions of combination processes. In liquid-liquid (droplets) and gas-liquid (bubble) flows, the population changes due to coalescence [26] are expressible as convolutions. Fragmentation mechanisms can be partially treated by collision rules. The kinetics of some mechanisms, like vibration breakup, bag breakup, bag-and-stamen breakup, sheet stripping, wave crest stripping, and catastrophic breakup can only be estimated by rate and probability laws for isolated bubbles/droplets for given local conditions. Nevertheless, collision-based processes are inherently represented in a size phase space as a convolution integral for the population change.

In solid particle dynamics, the population balance equations also are expressible in terms of some convolution integrals. Nicmanis and Hounslow [27] used FEM to describe an integro-differential equation with convolution-type integral terms, where the major processes of this type are aggregation and breakage. The collision rules for bubbles and particles depend substantially on the physicochemical properties of the liquid medium. Traces of flocculent and coagulent effect the probability of bubble-particle agglomeration and floc formation. However, in terms of particle dynamics, the collision rules can be formulated to match observed kinetic rates. Thus, a semi-empirical approach to population balance equations, fitting the coefficients of the aggregation, breakage, and growth models, is a successful technique in characterizing particle processes. Randolph and Larson [28] cite the change in number density $n(v)$ of particles with volume n in the product stream of a continuous mixed-suspension, mixed-product removal crystallizer in which these three processes are occurring from an inlet stream with feed population $n_{in}(v)$:

$$\frac{n(v)-n_{in}(v)}{\tau}+\frac{d}{dv}\big(G(v)n(v)\big)=b(v)-d(v) \qquad (7.27)$$

where τ is the residence time in the crystallizer, $G(v)$ is a volume-dependent growth function and the number density of nuclei is incorporated into the equation as a boundary condition $n(0)=n_0$. $b(v)$ and $d(v)$ are suggestively denoted as the birth and death terms for the volume fraction of size v. In general, there are contributions to both terms from both aggregation and breakage. Case 1 considered by [27] is a purely aggregation model, so we will cite only the forms derived by Hulburt and Katz [29] for aggregation.

$$b(v) = \int_a^{v/2} \beta(v-w,w)n(v-w)n(w)dw$$

$$\qquad\qquad\qquad\qquad\qquad (7.28)$$

$$d(v) = n(v)\int_0^\infty \beta(v,w)n(w)dw$$

Succinctly, birth by aggregation is due to the probability of combining particles with volumes which sum to v (and sticking). Death is by the probability of a particle of volume v participating in a collision (and sticking).

FEMLAB Model

Case 1 of Nicmanis and Hounslow is for aggregation only, characterized by the assignments $\beta(v,w)=$beta0, $G=0$, and no breakage contributions to b and d source/sink terms. The idealized feed is the exponential inlet condition

$$n_{in}(v) = \exp(-v) \qquad (7.29)$$

Launch FEMLAB and in the Model Navigator do the following:

Model Navigator
- Select 1-D dimension, Geom 1
- Select PDE modes⇒General⇒Stationary nonlinear model, weak form (mode g1). Dependent variable n1, independent variable v>>
- Multiphysics Tab. Add Geom 2, Select 2-D dimension
- Select PDE modes⇒General⇒Stationary nonlinear form (mode g2) Dependent variable n2, independent variables v1, v2>>
- Apply/OK

Nicmanis and Hounslow [27] suggest that the domain $v \in [0,2500]$ has a suitable ceiling for convergence. Pull down the Options menu and set the grid to $(-50,2550) \times (-50,2550)$ on Geom 2. Pull down the Draw menu.

Draw Mode
Geom 1: Specify geometry; interval [0,2500]
Geom 2: Select Draw Rectangles
Set R1 = [0,2500] ×[0,2500]
- Apply
- OK

Pull down the options menu and select Add/Edit constants. The Add/Edit constants dialog box appears.

Add/Edit Constants
- Name of constant: tau   Expression: 1
- Name of constant: n0   Expression: 1
- Name of constant: beta0   Expression: 1
- Name of constant: G0   Expression: 1
- Name of constant: v0max   Expression: 2500
- Apply/OK

The major action is the computation of the extrusion and projection coupling variables. Select Add/Edit Coupling Variables from the Options Menu.

Add/Edit Coupling Variables

---

extrusion add N1. Source Geom 1, subdomain 1, Expression: n1
Local mesh transformation (x ← v)
Destination Geom 2 subdomain 1, Check "Active in this domain" box.
Evaluation point transformation (x ← v2)
extrusion add N2. Source Geom 1, subdomain 1, Expression: n1
Local mesh transformation (x ← v)
Destination Geom 2 subdomain 1, Check "Active in this domain" box.
Evaluation point transformation (x ← v1-v2)
projection add ba. Source Geom 2, subdomain 1;
Integrand: `(v2>0)*(v2<v1/2)*N1*N2`
integration order 2
Local mesh transformation (x ← v1, y ← v2)
Destination Geom 1 subdomain 1, Check "Active in this domain" box.
Evaluation point transformation (x ← v)
projection add da. Source Geom 2, subdomain 1;
Integrand: `N1`
integration order 2
Local mesh transformation (x ← v1, y ← v2)
Destination Geom 1 subdomain 1, Check "Active in this domain" box.
Evaluation point transformation (x ← v)
  • Apply / OK

It should be noted that projection coupling variable ba computes the convolution integral for the birth term in (7.28), with the awkward offset coordinate (v1-v2) treated neatly by the evaluation point transformation in the extrusion variable N2. The independent variable in the limits of integration are catered for by the MATLAB binary logic factors (v2>0)*(v2<v1/2), in the same fashion as the treatment of the Volterra integration limits in the last section. da is far more pedestrian, only requiring the projection coupling variable for the line integral to be computed. Although da is the same for all v (a constant), it must be computed by a coupling variable. On reflection, its source could be Geom 1, subdomain 1, with integrand n1 to save computer labor in this case due to the assignment of $\beta(v,w)=beta0$. The treatment here is more general to accommodate potentially greater complexity of β.

By comparison, the Subdomain settings are still pedestrian:

Subdomain Mode
• Select mode g1 (geom1 domain 1)
• Set $\Gamma=0$, da=1, F=(n1-exp(-v))/tau-ba+n1*da
• Apply/OK
• Select mode g2 (geom2 domain 1) Set $\Gamma=0$ 0, da=0, F=n2-N1*N2
• Apply
• OK

Now for the boundary conditions. Neutral are needed. Pull down the Boundary menu and select Boundary Settings.

Boundary Mode
• Mode g1: geom1 domain 1,2 Select Neumann, G=0
• Mode g2: geom2 domain 1,2,3,4 Select Neumann, G=0
• Apply
• OK

In Mesh mode, accept the standard mesh for geom2, which gives for mode g2 (415 nodes, 768 elements) and in mode g1, set number of elements per subdomain to 1 100, to give 251 nodes, 250 elements. Solve using the Stationaty Nonlinear solver. The solution should appear as in Figure 7.15.

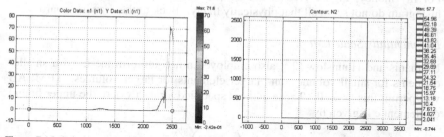

Figure 7.15 Solution n1(v) to (7.27). Left: 1-D solution. Right: 2-D solution for extrusion variable N2=n1(v2,v1-v2).

Parametric continuation will permit us to persevere out to $\tau=1.175$ before convergence is lost. Yet the solution in Figure 7.15 hardly seems plausible – nearly all the particles "gang up" in the maximum volume of the truncated infinite domain – hardly likely to have infinitesimal truncation error. These particles would aggregate up to the next higher "bin" if they were permitted. So this is not likely a stationary solution to the full problem.

"Time Dependent" Solution

The stationary nonlinear solution just doesn't work. It apparently becomes ill-posed at $\tau=1.175$. A check of the eigenvalues suggests that the Jacobian matrix is becoming singular – the condition number is large. It is possible that the coupling variables are not contributing significantly to the assembly of the stiffness matrix, which would then become singular. Iteration worked for Nicmanis and Hounslow [27]. One way to iterate is to specify a pseudo time scale and use the time-dependent solver. We anticipated this by putting da=1 in mode g1. For time integration, use the fldaspk solver, as it turns out the computation is stiff. The time integration out to t=0.3 is shown in Figure 7.16.

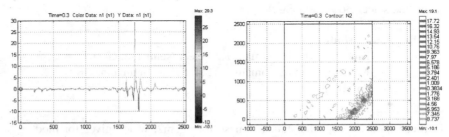

Figure 7.16 Pseudo time-dependent model solution n1(v) to (7.27) at time t=0.3. Left: 1-D solution. Right: 2-D solution for extrusion variable N2=n1(v2,v1-v2).

For my money, Figure 7.16 is not a winning solution. Negative counts in a histogram do not seem all that physically likely. Back to the drawing board.

Exponentially Scaled Mesh

Nicmanis and Hounslow [27] also employed a non-uniform mesh, with smaller elements for small volumes, and larger elements for larger volumes. FEMLAB will permit this as well. Those authors specified a mesh where the upper bound of element e is given by

$$x_e = v_b \left(\frac{v_{0\max}}{v_b} \right)^{e-1/N-1} \tag{7.30}$$

where N is the number of elements and v_b is the bin volume size for the first element. After some arithmetic, the mesh size h can be deduced as a function of position only

$$h = x_e - x_{e-1} = x_{e-1} \left(\left(\frac{v_{0\max}}{v_b} \right)^{1/N-1} - 1 \right) \tag{7.31}$$

and it is seen to be linear. FEMLAB will not permit a zero size element at the origin, so you need to specify an affine term. In mesh mode, specify, for instance:

Mesh Mode: g1

Mesh size expression: 0.08+(x+0.08)*0.08

Which yields 98 nodes and 97 elements. Similarly,

Mesh Mode: g2

Mesh size expression: 0.08+(x+0.08)*0.08+(y+0.08)*0.08

Which yields 2426 nodes and 4624 elements.

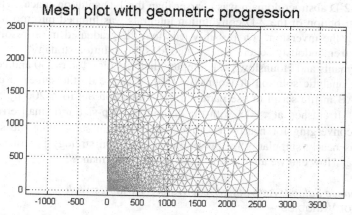

Figure 7.17 Mesh in abstract 2-D extrusion domain used for computing convolution integrals.

Figure 7.17 shows the result of element spacing being a linear function of position – mesh is packed in near the origin, and is sparse near the truncated corner of the doubly infinite domain. So what effect does this have on the solution?

The solution is found to be slowly convergent for $\tau=10$.

Iter	ErrEst	Damping	Stepsize	nfun	njac	nfac	nbsu
1	11	1.0000000	14	2	1	1	2
2	0.82	0.4967352	1.4	3	2	2	4
3	0.1	1.0000000	0.4	4	3	3	6
4	0.0023	1.0000000	0.077	5	4	4	8
5	5.1e-07	1.0000000	0.0016	6	5	5	10

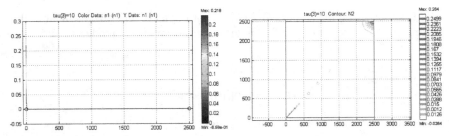

Figure 7.18   Solution n1(v) to (7.27) with graduated mesh of Figure 7.17 for tau=10.   Left: 1-D solution.   Right: 2-D solution for extrusion variable N2=n1(v2,v1-v2).

However, given that it was non-convergent after $\tau=1.175$ the last time we tried a stationary nonlinear solution, the graduated mesh is an improvement.   The difficulty with the standard uniform mesh that leads to ill-posedness could be that the 2-D abstract domain was coarser than the 1-D solution space.   Thus, the full contribution of the 2-D convolution integrals was lumped into too few bins to permit the inversion.   Or possibly it is that the graduated mesh resolves the solution tremendously better.   But how good is this solution strategy?

Nicmanis and Hounslow [27] compute their solution for $\tau=200$ and compare with the analytic solution of Hounslow [30].   Can we do the same?   I coded a MATLAB m-file script to use parametric continuation out to $\tau=200$ by steps of $\Delta\tau=0.5$.   It crashed at $\tau=18$ with the ubiquitous step-size too small error after slowly converging for nearby τ.   Eigenanalysis again shows that the stiffness matrix is nearly singular.   So this promising solution strategy is still not fully effective.   How can we alter it to achieve better performance?

Last Chance Saloon: Activating/Deactivating Variables With Solve for Variables

Perhaps you noticed that we are solving in the abstract domain for n2, which at steady state should be the trivial solution of F=0, i.e. n2=N1*N2.   n2 is pretty useless to us, but as it is a diagonal system at steady state, it should not be hard to solve, right?   And we do have to solve for something in our fictitious domain, don't we?   Wrong!   Even the trivial diagonal solution for F=0 uses sparse matrix solvers with somewhere around 4600 back substitutions.   Eventually, this work will lead to an ill-conditioned numerical solution due to round-off error alone. Furthermore, we do not need to *solve* for anything in our fictitious 2-D domain. We can disable the solution for n2.

Multiphysics: Solve for Variables

Select and highlight only Geom1: 1 variable gen form (g1): n1

This will disable the solution for n2. Since it is superfluous, computing n2 can only harm us. To test how good the solution is now, we will compare the analytic and FEM computed moments. Moments are defined on the distribution as

$$m_j = \int_0^\infty v^j n(v)\, dv \qquad (7.32)$$

Moments are computed on our truncated domain by subdomain integrations:

τ	m_0	m_1	m_2
1	0.73082	0.99701	2.9864
2	0.35748	0.99379	11.867
30	0.22653	0.99573	31.884
70	0.15503	0.99768	72.533
140	0.11237	1	143.11
200	0.094934	1.0001	201.72

Hounslow [30] gives the analytic values for $\tau = 200$ to 3 significant figures as

200	0.0951	1	202

In the above computations, parameter space continuation was done with old solutions taken as the new guess. This is a more complex version of iteration than used in [27], since the FEMLAB standard stationary nonlinear solver assembles the Jacobian matrix. Even with only 98 elements, the solution is rather good for the moments at this level of τ. Greater refinement is necessary for higher τ values.

Exercise 7.4: An integro-differential equation

(7.17) is an integro-differential equation when $G \neq 0$. Set up a variation of our stationary nonlinear model for PBE with $G=1$ and boundary condition n1=1 at v=0. Solve for the steady solution with residence time $\tau=200$. Since the pde is first order, only one boundary condition can be applied sensibly. The recipe with FEMLAB is to impose a "non-condition" at v=v0max, i.e. the Neumann BC that the derivative of n1 vanishes at the top volume. Since this is the natural BC in FEM, no Lagrange multiplier equation is augmented to the system. Does this natural boundary condition make physical sense in the case of PBE?

Summary

This chapter has a plumbed the depths of coupling variables of all three varieties to solve an array of inverse and integral equations. We encountered several features of FEMLAB not previously explored – coupling to optimization tools through MATLAB, extended meshes, using the time-dependent solver as an iterative tool for stationary nonlinear models, and the ability to selectively activate/deactivate multiphysics modes in coupled models. The latter is particularly useful if there is only one-way coupling (as in the hydrodynamics around the catalyst supported on the pellet in Chapter 3). In the case of the integral equations treated here, a fictitious dependent variable on an auxiliary domain is set up. The domain is used by coupling variables for various operations, but the dependent variable is never needed itself. So deactivating it results in better conditioning the FEM approximation to the integral equation. Although we implemented this procedure only with the convolution integral in our last model of the PBE, this is a generically useful technique for all the integral equations posed here.

My contacts at COMSOL have led me to believe that coupling variables and extended multiphysics were an addition to FEMLAB 2.2 "because they could" without necessarily a vision of how they might prove to be practically useful. With the wide survey of applications shown here, and earlier in Chapter 4, my impression is that coupling variables and extended multiphysics are the features of FEMLAB most likely to lead to rapid growth in its usage. Complex system modeling and simulations that are envisaged for the biological systems, micromachines (MEMs), and generic networked systems are readily modeled by these features of FEMLAB. Process simulation packages such as HYSYS and Aspen have long had the capability of simulating networks of coupled units comprising ODEs and nonlinear, algebraic constraints. Computational fluid dynamics packages such as FLUENT and FIDAP and finite element solvers like ANSYS contain the elements of PDE solver engines. FEMLAB, through extended multiphysics and coupling variables, have made the combination appear seamless to the user.

References

1. RA Gingold and JJ Monaghan, "Smooth particle hydrodynamics: theory and application to nonspherical stars," *Mon. Not. Roy. Astr. Soc.* 181:375–289, 1977.
2. AV Potapov, ML Hunt, and CS Campbell, "Solid-liquid flows using smoothed particle hydrodynamics and the discrete element method." *Powder Technology* 116:204–213, 2001.

3. DC Rapaport, <u>The Art of Molecular Dynamics Simutation</u>, Cambridge University Press, 1995.
4. BJ Adler and WG Hoover, in <u>Physics of simple liquids</u>, ed. HNV Temperly, JS Rowlinson, and GS Rushbrooke. Amsterdam: Horth-Holland Publishing Co., 1968.
5. S Kim and SJ Karrila, <u>Microhydrodynamics. Principles and Selected Applications</u>. Stoneham, MA: Butterworth-Heinemann series in chemical engineering, 1991.
6. D Coca and SA Billings, "A Direct Approach To Identification of Differential Models from Discrete Data", *J Mechanical Systems and Signal Processing*, 13:739–755, 1999; D Coca, Y Zheng, JEM Mayhew and SA Billings, "Nonlinear system identification and analysis of complex dynamical behaviour in reflected light measurements of vasomotion", *Int J Bifurcation and Chaos*, 10:461–476, 2000.
7. MM Mansour and F Baras, "Microscopic simulation of chemical systems" *Physica A* 188:253, 1992.
8. IE Grossman, "Mixed integer nonlinear programming techniques for the synthesis of engineering systems," *Res. Eng. Des.* 1:205, 1990.
9. M Mitchell, An Introduction to Genetic Algorithms, MIT Press, Cambridge, Massachusetts, 1996.
10. SC Roberts, D Howard, JR Koza, "Evolving modules in genetic programming by subtree encapsulation", *Genetic Programming, Proceedings Lecture Notes In Computer Science* 2038: 160–175, 2001.
11. SM Huang, AB Plaskowski, CG Xie, and MS Beck, "Tomographic imaging of two-component flow using capacitance sensors." *J.Phys. E, Sci. Instrum.* 22:173–177, 1989.
12. SL McKee, T Dyakowski, RA Williams, T Bell, T Allen, "Solids flow imaging and attrition studies in a pneumatic conveyor." *Powd. Tech.* 82:105–113, 1995.
13. T. Dyakowski, "Process tomography for multiphase measurement." *Measurement Science & Technology* 7:343–353, 1996.
14. RP Feynman, RB Leighton, and M Sands, The Feynman Lectures on Physics, vol II. Addison-Wesley, 1964.
15. T Dyakowski, LFC Jeanmeure, AJ Jaworski, "Applications of electrical tomography for gas-solids and liquid-solids flows - a review," *Powder Technology* 112 (3): 174–192, 2000.
16. N. Kerrouche, CN McLeod, WRB Lionheart, Time series of EIT chest images using singular value decomposition and Fourier transform, *Physiol. Meas.* 22(1): 147–157, 2001; M Vauhkonen, WRB Lionheart, LM Heikkinen, PJ Vauhkonen and JP Kaipio, A MATLAB package for the EIDORS project to reconstruct two-dimensional EIT images, *Physiol. Meas.* 22(1): 107–111, 2001. http://www.ma.umist.ac.uk/bl/eidors/.

17. WB Zimmerman and PC Chatwin, "Fluctuations in dense gas concentrations measured in a wind-tunnel." *Boundary Layer Meteorology* 75:321–352, 1995.

18. ESG Shaqfeh, "A nonlocal theory for the heat transport in composites containing highly conducting fibrous inclusions" *Phys. Fluids* 31(9):2405–2425, 1988.

19. SD Howison, JA Moriarty, JR Ockendon, EL Terrill, and SK Wilson, "A mathematical model for drying paint layers." *J. Engineering Math.* 32:377–394, 1997.

20. G Arfken, <u>Mathematical Methods for Physicists</u>, 3rd edition, Academic Press, New York, 1985.

21. I Stakgold, <u>Green's Functions and Boundary Value Problems</u>, Wiley, New York.

22. WV Lovitt, <u>Linear Integral Equations</u>, McGraw-Hill, New York, 1924, reprinted 1950, Dover, New York.

23. WB Zimmerman, "The drag on sedimenting discs in broadside motion in tubes." *International Journal of Engineering Science*, 40:7–22, 2002.

24. SB Pope, <u>Turbulent Flows</u>, Cambridge University Press, 2000.

25. RN Bracewell, "The fast Hartley transform". *Proc. IEEE*, 72(8):1010, 1984.

26. NL Kolev, "Fragmentation and coalescence dynamics in multiphase flows." *Experimental Thermal and Fluid Science*, 6:211–251, 1993.

27. M Nicmanis and MJ Hounslow, "Finite-element methods for steady-state population balance equations," AIChE J., 44(10):2258–2272, 1998.

28. AD Randolph and MA Larson, Theory of Particulate Processes, 2nd edition, Academic Press, New York, 1988.

29. HM Hulburt and S Katz, "Some problems in particle technology. A statistical mechanical formulation." *Chem. Eng. Sci.* 19:555, 1964.

30. MJ Hounslow, "A discretized population balance for continuous systems at steady state." *AIChE J.* 36(1):106, 1990.

Chapter 8

MODELING OF MULTI-PHASE FLOW USING THE LEVEL SET METHOD

K. B. DESHPANDE and W.B.J. ZIMMERMAN

*Department of Chemical and Process Engineering, University of Sheffield,
Newcastle Street, Sheffield S1 3JD United Kingdom*

E-mail: w.zimmerman@shef.ac.uk

Multiphysics, the feature of FEMLAB that allows coupling of different types of physics, is demonstrated in this chapter for the level set method for modeling multiphase flow, illustrating various scenarios for the coalescence of drops. In the level set method for biphasic fluid systems, one fluid has strictly positive phase function ϕ, the other strictly negative ϕ, so the interface is tracked by the zero level set of ϕ. The transport of ϕ is computed by solving an advection-diffusion equation for ϕ and the incompressible Navier-Stokes equations simultaneously. The level set method is extensively applied here to study the coalescence of drops in biphasic flows for different configurations such as drops under influence of gravity, an acoustically suspended drop, drops approaching one another and the interaction among three drops. The curvature analysis here shows the power of FEMLAB's post integration tools for statistical analysis of evolving fields, capturing the occurrence of coalescence by a distinguished feature – cusp formation.

8.1 Introduction

Multiphase flows are often difficult to model computationally, especially because of the difficulty in tracking the fluid-fluid interface. Furthermore, there is a steep change in physical properties such as density, viscosity etc., which makes the computation yet more stiff. There are various computational methods available to solve incompressible two-phase problems such as the front tracking method [1], the boundary integral method [2], the volume of fluid method [3], the Lattice Boltzmann method [4], diffuse interface modeling [5], and the level set method [6][11]. We use the level set method in this chapter, illustrating its use to compute the coalescence of two drops.

All the above mentioned methods have their advantages and disadvantages. In the front tracking method, marker particles are explicitly introduced to keep track of the front that reduces the resolution needed to maintain the accuracy. However, re-gridding algorithms should be employed with front tracking method to prevent marker particles from coming together, especially at the points of larger curvature.

The volume of fluid method (VOF) is based on discretization of the volume fraction of one of the fluids. The motion of the interface is captured by solving a conservation law for volume fraction and the Navier-Stokes equations simultaneously. Since the interface is represented in terms of volume fraction, mass is always conserved, while maintaining a sharp representation of the interface. The VOF method needs to have accurate reconstruction algorithms to solve for the advection of volume fraction. A disadvantage of the VOF method is that it is difficult to compute accurate local curvature from volume fraction. This is due to the sharp transition in volume fraction near the interface.

The Lattice Boltzman method (LBM) is a mesoscopic approach to the numerical simulation of fluid motions based on the assumption that a fluid consists of many particles whose repeated collision, translation, and distribution converge to a state of local equilibrium, yet always remaining in flux. LBM has advantages such as implementation on a complex geometry, very efficient parallel processing, and ease of reproduction of the interface between the phases. However, LBM is not yet a widely used computational method to track the fluid motion in multiphase systems, due to its computational intensity.

The level set approach is another potential numerical method to solve incompressible two-phase flow incorporating surface tension term. In the level set method, the interface is represented as the zero level set of a smooth function. This has the effect of replacing the advection of physical properties with steep gradients at the interface with advection of level set function that is smooth in nature. Although level set method does not have the same conservation properties as of VOF method or front tracking method, the major strength of level set method lies in its ability to compute curvature of the interface easily. Furthermore, level set method does not require complicated front tracking regridding algorithms or VOF reconstruction algorithms. Level set method is based on continuum approach in order to represent surface tension and local curvature at the interface as a body force. This facilitates the computations in capturing any topological change due to change in surface tension.

The diffuse interface method is a kindred notion to the level set method and VOF in that it computes the transport of another function that varies between the phases – the chemical potential. As is well known (see [7]), the surface tension between two fluids is also the excess partial molar Gibbs free energy per unit surface area, so that the change of chemical potential across an interface between immiscible fluids is treated by the notion of surface tension as infinitely steep. The diffuse interface method permits this condition to be merely relaxed to be steep, and then a field equation for chemical potential is tracked, rather than the imposition of topology and stress balance equations implied by the notion of surface tension. The latter method still requires grid adaption, which in state of the art computational models (see [8] and references therein) employ auxiliary equations for elliptic mesh diffusion, but are fragile in the face of topological change, e.g. coalescence or breakage phenomena [9]. Whether greater accuracy at the same computational intensity is available by the topological method of monitoring the free surface position with mesh re-gridding or by solving an auxiliary transport equation for a field variable (VOF, diffuse interface, or level set methods) is arguable. The latter auxiliary equation methods have ease of coding in their favor, which will be illustrated in this chapter with the level set method.

The level set method is used in this chapter to illustrate the coalescence of two axisymmetric and non-axisymetric drops. Computations are performed using FEMLAB. This FEM approach simplifies the level set method by eliminating all the complexities in grid discretization required for free surface/interface tracking methods. The governing equations for the level set method are described in following section.

8.2 Governing Equations of the Level Set Method

In the level set method, a smooth function called a level set function is used to represent the interface between two phases. The level set function is always positive in the continuous phase and is always negative in the dispersed phase. The free surface is implicitly represented by the set of points in which level set function is always zero. Hence we have,

for the continuous phase $\qquad \phi(x, y, t) > 0$ $\qquad$ (8.1a)

for the interface $\qquad \phi(x, y, t) = 0$ $\qquad$ (8.1b)

for the dispersed phase $\qquad \phi(x, y, t) < 0$ $\qquad$ (8.1c)

From such a representation of the free surface, the unit normal on the interface pointing from dispersed phase to continuous phase and curvature of the interface can be expressed in terms of level set function as,

$$\mathbf{n} = \frac{\nabla \phi}{|\nabla \phi|} \qquad (8.2)$$

$$\kappa = \nabla \cdot \frac{\nabla \phi}{|\nabla \phi|} \qquad (8.3)$$

The motion of the interface can be captured by advection of the level set function,

$$\frac{\partial \phi}{\partial t} + \mathbf{u} \cdot \nabla \phi = 0 \qquad (8.4)$$

The governing equation for the fluid velocity and pressure can be written in terms of the Navier-Stokes equations which is the equation of motion for incompressible flow:

$$\rho \frac{\partial \mathbf{u}}{\partial t} - \nabla \cdot \mu \left(\nabla \mathbf{u} + \left(\nabla \mathbf{u} \right)^T \right) + \rho \left(\mathbf{u} \cdot \nabla \right) \mathbf{u} + \nabla p = \mathbf{F} ; \tag{8.5}$$

$$\nabla \cdot \mathbf{u} = 0 \tag{8.6}$$

where F is body force which includes gravitational force and, due to the level set treatment of interfacial stresses, the surface tension term. The two components of the F term can be represented as,

$$F_x = \sigma \kappa \delta \left(\phi \right) n_x \tag{8.7}$$

$$F_y = \sigma \kappa \delta \left(\phi \right) n_y + \rho g \tag{8.8}$$

The delta function treats the surface tension term at the interface which is determined by the position of the zero level set—which can be as many fluid-fluid interfaces as necessary to demarcate the dispersed phase. The Heaviside function, incorporated in order to describe the steep change in physical properties, is represented in terms of level set function such as,

if $\phi < 0$ $\qquad\qquad\qquad H(\phi) = 0$ $\qquad\qquad\qquad$ (8.9a)

if $\phi = 0$ $\qquad\qquad\qquad H(\phi) = \dfrac{1}{2}$ $\qquad\qquad\qquad$ (8.9b)

if $\phi > 0$ $\qquad\qquad\qquad H(\phi) = 1$ $\qquad\qquad\qquad$ (8.9c)

The density and viscosity are constant in each fluid and are represented in terms of Heaviside function as,

$$\rho = H(\phi) + \frac{\rho_d}{\rho_c} \left(1 - H(\phi) \right) \tag{8.10}$$

$$\mu = H(\phi) + \frac{\mu_d}{\mu_c} \left(1 - H(\phi) \right) \tag{8.11}$$

We solve the above set of equations using FEMLAB. Smoothed approximants to the Heaviside function are used to avoid Gibbs phenomena resulting in poor convergence.

8.3 Curvature Analysis: Methodology

In the present simulations of multi-phase modeling, the coalescence phenomenon is demonstrated for the various scenarios where the motion of the interface is significant, particularly at the time of coalescence. The curvature analysis is an attempt to capture the rupture of the interface during the coalescence event.

In the level set method, the curvature of the interface is represented as shown in the equation (8.3). The mean value of the curvature can be estimated by integrating $|\kappa|$ over the interface as,

$$\kappa_{mean} = \frac{\int\limits_{\Omega} |\kappa| d\Omega * (\phi = 0)}{\int\limits_{\Omega} d\Omega * (\phi = 0)} \tag{8.12}$$

Similarly, standard deviation of $|\kappa|$ can also be evaluated by first calculating variance as,

$$var = \frac{\int\limits_{\Omega} \kappa^2 d\Omega * (\phi = 0)}{\int\limits_{\Omega} d\Omega * (\phi = 0)} - (\kappa_{mean})^2 \tag{8.13}$$

The standard deviation, σ is,

$$\sigma = \sqrt{var} \tag{8.14}$$

Thus, the first and the second moments of $|\kappa|$ can be evaluated at different time steps to study the behaviour of $|\kappa|$ at the time of coalescence. The numerical results are shown for coalescence, followed by the curvature analysis with the associated MATLAB m-file script in the next section.

8.4 Results and Discussion

The numerical simulation presented here demonstrates the power of FEMLAB in the modeling of multi-phase flow. We use multiphysics, the basic versatility of FEMLAB that enables us to incorporate as many modes (physics) as we wish to include.   The level set method, which requires two application modes: Incompressible Navier-Stokes and ChEM: Convection and Diffusion modes, respectively, has been applied extensively here to capture the coalescence of two drops in a two-phase system. Since the interface can be tracked by setting the zero level set at the interface, this permits the study of the evolution of the interface after the merging of two drops. In coalescence phenomena, the contact point between interfaces of two drops is very important and hence the approach of two drops is treated by various means in this computational study and is described in detail in following sections.

8.4.1 Coalescence of two axisymmetric drops

The simplest way to start the numerical simulation is to assume a symmetrical 2-D domain, which would significantly reduce the computational time. The system can be physically described as a rectangular domain with one boundary acting as an axis of symmetry and all other boundaries are insulated. Two equally sized drops are initially separated by axial distance equal to two times their diameter. It is a tricky task to initiate two drops of the same physical properties. This is accomplished using following initial condition which is used in the sub-domain settings.

$$\phi(t=0) = \min(\sqrt{(x^2 + (y-2)^2)} - 0.25, \min((6-y), \sqrt{x^2 + (y-1)^2} - 0.25)) \quad (8.15)$$

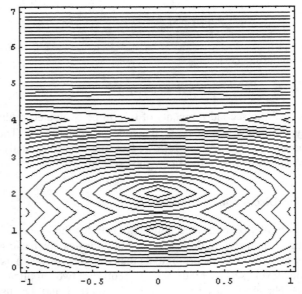

Figure 8.1 Contour plot (level sets) of the initial condition for ϕ described in terms of the MATLAB function min in (8.12).

The above system of equations (8.2)—(8.11) with initial condition (8.12) in an initially quiescent fluid (no motion) can be solved using level set method as described below.

Start up FEMLAB and enter the Model Navigator:

Model Navigator
- Select Multiphysics
- Select 2-D dimension
- Select Incompressible Navier-Stoke and
- Select ChEM: Convection and diffusion. Dependent variable: phi
- Element: Lagrange – quadratic
- OK

These application modes give us four dependent variables u, v, and p for Incompressible Navier-Stoke mode and one dependent variable called 'phi' for ChEM: convection and diffusion mode. The 2-D space coordinates are x and y.

Now we can define our domain by pulling down the Draw menu and select a rectangle.

Draw Mode
- Draw a rectangle by clicking left mouse button
- Edit the dimensions of a rectangle by double clicking
- Set X_{min} to 0, X_{max} to 1, Y_{min} to 0 and Y_{max} to 6
- Apply / OK
- Select Axes/Grid Settings from Options menu
- Set Axis Equal
- Apply / OK

We need to specify the boundary conditions for both the modes separately in the Multiphysics menu. Select the mode from Multiphysics. Pull down the Boundary menu and select Boundary Settings.

Boundary Mode
- Select Incompressible Navier-Stokes (ns mode) from Multiphysics
- Set domain 1 to symmetry and domain 2-4 to no-slip
- Apply / OK
- Select ChEM: Convection and diffusion (cd mode) from Multiphysics
- Set domain 1-4 to Insulation/symmetry
- Apply / OK

Before setting up the problem, we need to define all the constants to be used in the simulation. Pull down the Options menu and select Add/Edit constants. The Add/Edit constants dialog box appears.

Add/Edit Constants

- Name of constant: rhod   Expression: 1
- Name of constant: rhoc   Expression: 10
- Name of constant: nu   Expression: 1
- Name of constant: gy   Expression: -10
- Name of constant: dadd   Expression: 0.005
- Name of constant: n   Expression: 100
- Name of constant: sigma   Expression: 1
- Apply / OK

It is useful to define some expressions, i.e. intermediate variables, to make the FEM data entry more concise. Pull down the Options menu and select Add/Edit expressions. Expressions can be functions of the dependent variables (u,v,p, and phi). Contant expressions can be defined in the Add/Edit Constants dialogue box.

Add/Edit Expressions

- Variable name: **smhs**
- Variable type: subdomain then Add
- Now click on the definition tab
- Enter expression: (1+tanh(n*phi))/2
- Variable name: **smdelta**
- Variable type: subdomain then Add
- Now click on the definition tab.
- Enter expression: n/sqrt(pi)*exp(-n^2*phi^2)

- Variable name: **kappa**
- Variable type: subdomain then Add
- Now click on the definition tab
- Enter expression: (phiy^2*phixx-2*phix*phiy*phixy +phiyy*phix^2)/(phix^2+phiy^2)^(3/2)
- Variable name: **r0**
- Variable type: subdomain then Add
- Now click on the definition tab
- Enter expression: rhod + rhoc*smhs
- Apply
- OK

smhs (a hyperbolic tangent) is a common smooth approximant to the Heaviside function. **smdelta**, similarly, is a smooth approximant to the Dirac delta function. The prefactor on the Gaussian is for normalization – the quadrature over the real line must be unity. It is potentially the case that weak terms could be used to define point forces along the zero level set of phi, but the smooth approximants are easier to code. **kappa** is the major component of the curvature defined in (8.3). **r0** is the expression of the density as in (8.10).

For subdomain specifications, select the mode form Mulitiphysics and then pull down the Subdomain menu. Select Subdomain settings.

Subdomain Mode
<ul><li>Select Incompressible N-S from Multiphysics</li><li>Select the Coefficient tab</li><li>Set ρ =r0, η = nu</li><li>Set F_x =sigma*kappa*smdelta*phix/sqrt(phix^2+phiy^2)</li><li>Set F_y =sigma*kappa*smdelta*phiy/sqrt(phix^2+phiy^2)+r0*gy</li><li>Click Stream line diffusion on</li><li>Select Init tab</li><li>Set u=0, v=0 and p=0</li><li>Click Apply and then OK</li><li>Select ChEM: Convection and diffusion from Multiphysics</li><li>Select 'phi' tab</li><li>Set D $_i$ = dadd</li><li>Set R $_i$ = 0, u=u and v=v</li><li>Click Stream line diffusion on</li><li>Select Init tab</li><li>Set phi = $\phi(t = 0)$, see (8.12).</li><li>Click Apply and then OK</li></ul>

Now pull down the Solve menu and select the Parameters option.

Solver Parameters
<ul><li>Select Time dependent solver</li><li>Select Timestepping tab</li><li>Enter in output times: 0:0.025:3</li><li>Select fldaspk Timestepping algorithm</li><li>Define tolerance limits to 0.01</li><li>Click Apply, OK and then Solve</li></ul>

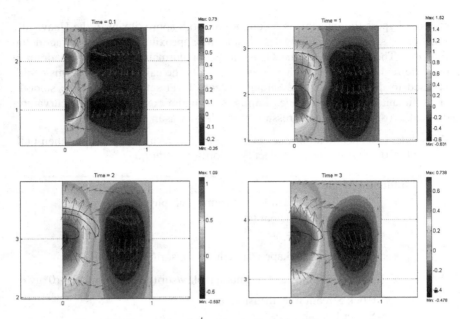

Figure 8.2 Contour plot of ϕ with velocity field at different time steps.

The contour plots were generated by the "Copy to Figure" option on the Edit Menu for different time steps. Figure 8.2 captures the rise of two drops in a column that ultimately results in coalescence. The interface of the two drops is represented by the contour plot at ϕ =0. The velocity field is also represented in the above figure by activating the surface field for v velocity and the arrows field. Two drops are initially separated by a distance equal to two times diameter of drops and their motion under gravity is captured at different time steps. The lower drop travels faster than the upper one, although two drops are of same density and of uniform size. This can be explained by wake formation for the upper drop. The lower drop becomes entrapped into the wake region of the upper one and experiences the velocity field of the upper drop, thereby lowering the effective velocity of the upper drop. Thus, two drops suspended freely rise in a column, eventually coalesce and the subsequent coalesced drop rises again. In this way, the motion of the interface of two drops can be monitored readily using level set method in FEMLAB. Various other configurations of the approach of the drops are discussed in the following sections.

Curvature Analysis: An Application

The results of a FEMLAB simulation that is run in the GUI can be exported to MATLAB workspace by using "Export FEM structure as fem" from File Menu.

The model m-file script shown below was saved as 'analysis.m', edited, and executed from the MATLAB command line to generate the plot of standard deviation and mean of |κ| as shown in Figure 8.2.

```
    t=[1:121];
m1int=postint(fem,'abs(kap)*(phi<0.00015 & phi >-0.00015)',...
    'cont',    'off',...
    'contorder',2,...
    'edim',    2,...
    'solnum',  t,...
    'phase',   0,...
    'geomnum',1,...
    'dl',      1,...
    'intorder',4,...
    'context','local');
v=postint(fem,'1*(phi<0.00015 & phi >-0.00015)',...
    'cont',    'off',...
    'contorder',2,...
    'edim',    2,...
    'solnum',  t,...
    'phase',   0,...
    'geomnum',1,...
    'dl',      1,...
    'intorder',4,...
    'context','local');
m1=m1int./v;
plot(0.025*t,m1);
hold on
m2int=postint(fem,'(abs(kap)*(phi<0.00015 & phi >-0.00015))^2',...
    'cont',    'off',...
    'contorder',2,...
    'edim',    2,...
    'solnum',  t,...
    'phase',   0,...
    'geomnum',1,...
    'dl',      1,...
    'intorder',4,...
    'context','local');
m2=m2int./v;
var=m2-(m1).^2;
sd=sqrt(var);
plot(0.025*t,sd);
hold off
save collisionlong.dat t m1 sd -ascii;
```

The variables m1int, m2int and v approximate the numerators and denominator of (8.12) and (8.13). The range of tolerances surrounding the $\phi=0$ contour were selected to weight the subdomain integration by contributions in a narrow band surrounding the interface. It is to be noted that the variable 't' used in above calculation is the solution number and varies from 0 to 121, since we ran the simulation with time range 0:0.025:3.

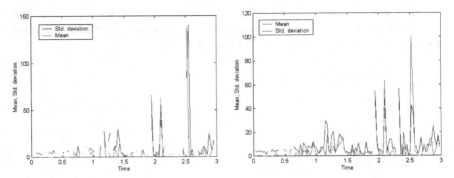

Figure 8.3 Plot of standard deviation and mean value of |κ| with respect to time for the coalescence of two drops under gravity. Left: ε (solution tolerance)=0.0001. Right: ε =0.0002.

The plots of standard deviation and mean of |κ| are shown in Figure 8.3 for two different values of tolerance (commonly used in MATLAB as a variable 'eps') 0.0001 and 0.0002 respectively. The smaller the value of tolerance, the greater the depth of the spike observed. Both the mean and standard deviation of the curvature as estimated are sensitive to topological changes in the connectivity of the domains. The greatest spike was observed at the time of coalescence that can be attributed to the rupture of the interface or cusp formation.

8.4.2 Coalescence of acoustically suspended drops

The technique of acoustic levitation, using tuned sonic fields to oppose the drag force on a droplet and levitate it, has been known for many years. The group of Sadhal at USC have studied the phenomenon and its implications for droplet dynamics for years (see [10] and references therein). The coalescence of acoustically suspended drops where the lower drop is rising and upper drop is held stationary is simulated. Unlike the previous simulation where only half of the domain was considered by assuming symmetry of the domain, the present simulation is performed over the entire domain. Hence, the no-slip boundary condition is applied to all the boundaries in all the modes of Multiphysics. Change the initial condition to generate two drops as follows,

$$\phi(t=0) = \min(\sqrt{(x^2 + (y-2)^2)} - 0.25, \min((3-y), \sqrt{x^2 + (y-1)^2} - 0.25))$$

$$(8.16)$$

The only other change would be in the body force (gravity term) in the Navier-Stokes equations which is modified in such a way that upper drop does not experience any gravitational force. This is the bare effect of the acoustic levitation, without consideration of capillary-gravity waves induced on the free surface by acoustic interactions. But, lower drop is rising in a column due to buoyancy. The above mentioned changes can be incorporated by changing F_y term in sub-domain settings Menu as follows:

Subdomain Mode
• Select Incompressible N-S from Multiphysics Menu
• Select the Coefficient tab
• Set F_y = sigma*kappa*smdelta*phiy + ro*gy*(tanh(-(y-yc))>0)
• Apply / OK

The new constant used yc is set to 2, i.e. y co-ordinate of the center of upper bubble. The force term used in this way applies no gravity to upper drop whereas lower drop experiences gravitational force equal to ρg.

Numerical results are shown in Figure 8.3 in terms of contour plot of level set function at ϕ =0 and surface plot of velocity field. The two drops initially separated by a distance equal to two times their diameter approach quite faster than the previous simulation where both the drops were rising. Eventually, two drops coalesce quickly and evolution of the interface of two drops after the coalescence event has been brought out through this simulation. Cusp formation is observed at time *t*=2 sec. The coalesced drop regains its original shape as it rises in a column. The different shapes of two drops before collision can be attributed to the fact that pressure is continuously decreasing along the length of the column and hence radii of curvature would increase according to the Young-Laplace equation. This can be validated by changing the configuration so that pressure change is uniform as described in the following section.

Curvature Analysis

The procedure outlined for the curvature analysis of the coalescence of two drops under gravity is followed for the coalescence of acoustically suspended drops. The fem structure is exported to the MATLAB workspace after the simulation is over and MATLAB model m-file analysis.m is run to study the standard deviation and mean of |κ| as shown in the Figure 8.5.

Both the first and second moments of |κ| show a sharp peak at the time of the coalescence, attributed to the rupture of the interface.

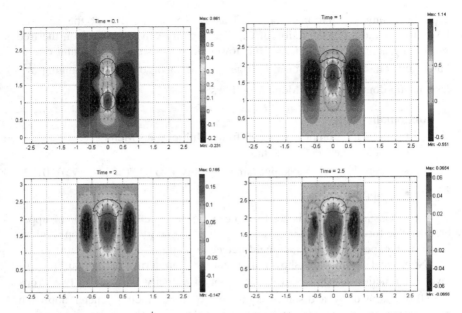

Figure 8.4   Contour plot of ϕ with velocity field at different time steps for the coalescence of acoustically suspended drops.

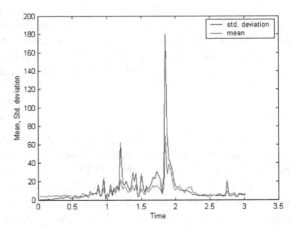

Figure 8.5 Plot of standard deviation and mean value of |x| with respect to time for the coalescence of an acoustically suspended drop.

8.4.3 Coalescence between two drops approaching each other

The coalescence between two drops approaching each other in opposing directions can be achieved by defining a driving force that attracts two drops to each other. This can be simulated by incorporating a driving force term in sub-domain settings for Incompressible Navier-Stoke application mode.

Subdomain Mode
• Select Incompressible N-S from Multiphysics Menu • Select the Coefficient tab • Set F_y = sigma*kappa*smdelta*phiy + ro*gy*tanh(-(y-yc)) • Apply / OK

The new constant used yc is a midpoint of the line of centres of two drops and set to 1.5. The force term defined in this manner applies force equal to ρg when *(y-yc) < 0* and *-ρg* when *(y-yc)>0*. Thus, upper and lower drop experiences exactly equal force but in the opposite direction.

Computational results are represented in terms of a contour plot of the level set function where ϕ =0, a surface plot for pressure field and arrows for velocity field as shown in Figure 8.3. Two drops separated by a distance equal to two times their diameter attract to each other, ultimately resulting in coalescence at time *t=2* sec. Cusp formation has been clearly brought out at that time step. The coalesced drop regains its original shape at later time steps. The important feature of this simulation is that symmetry is observed at the midplane between the two drops.

The velocity field is also found to be symmetrical for both the drops which retains after the coalescence event as well. Another important feature is that both the drops are identical in their shape and size. This can be explained on the basis of a surface plot of pressure that is found to be symmetrical around the midplane between the two drops. Since the two drops experience same pressure force, they follow the same change in radii of curvature. Also, less droplet deformation is observed for the present simulation as compared to the earlier two cases. This can also be attributed to lower magnitude of the pressure force.

Curvature Analysis

The curvature analysis performed for two drops approaching one another is shown in Figure 8.7. The peak in standard deviation and mean value of |κ| is confirmed at the time of the coalescence. No other spikes were observed for the present simulation because the deformation of the drops was found to be smaller than that in the earlier two cases considered. Thus, it can be concluded that peaks observed in the mean and standard deviation value of |κ| are indeed due to the rupture of the interface.

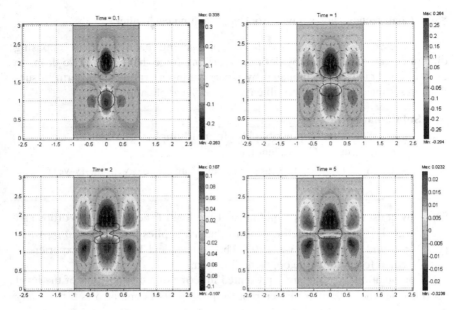

Figure 8.6 Contour plot of ϕ with velocity field at different time steps.

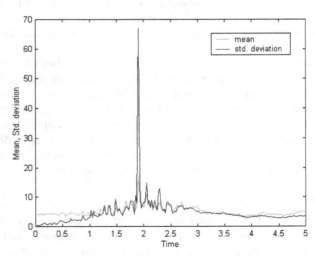

Figure 8.7 Plot of standard deviation and mean value of |κ| with respect to time for the coalescence between two drops approaching one another.

8.4.4 Multi-body coalescence

The coalescence between three drops is simulated assuming the symmetry of the domain. Hence, only half of the domain is simulated in the present case defining one boundary as an axis of symmetry.

The above system brings out the effect of horizontal offset amongst interacting drops. In the present case, centers of drops are more than one radius apart from each other. Two drops are initiated using following initial condition,

$$\phi(t=0) = \min(\sqrt{(x^2 + (y-3)^2)} - 0.5, \min((4-y), \sqrt{(x-0.75)^2 + (y-1)^2} - 0.5)) \ (8.17)$$

Above initial condition generates two uniform sized drops of radius 0.5 whose centers are separated by a distance equal to 0.75. The present simulation is similar to earlier one where two drops are traveling towards each other except the horizontal offset. Hence, there is no other change in the formulation of the problem than the initial condition. The contour plot of level set function at $\phi = 0$, surface plot and arrows of velocity field are represented in Figure 8.8.

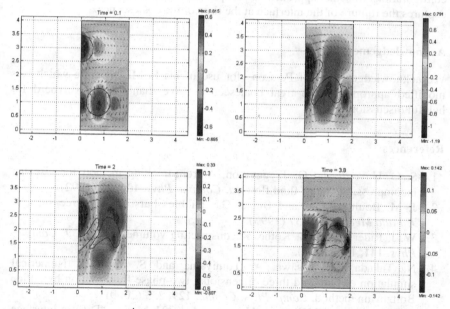

Figure 8.8 Contour plot of ϕ with velocity field at different time steps for three-body coalescence.

The interaction between drops with horizontal offset is found to be very different from that without offset. The velocity field of the lower drop is found to be diverted due to the influence of the velocity field of upper drop that is traveling downwards. Hence, the shape of rising drop changes drastically. The lower drop almost skids downward traveling drop and changes the contact point between the interacting drops. Finally, coalescence takes place at 3.8 seconds.

Thus, it is found that the contact point of the interacting drops is very significant in the coalescence phenomenon. Different types of contacting schemes are observed for different approaches considered in the present set of simulations merely by changing the initial condition.

8.5 Summary

The level set method is extensively used in the present simulations to study computationally the coalescence of droplets in a two-phase system. FEMLAB handles computationally intensive multi-phase modeling with ease using its multiphysics utility that allows the coupling of different physics into one problem as demonstrated here for the level-set method. In the present simulations, the coalescence phenomenon has been extensively studied using various configurations for the approach of drops. The curvature analysis performed captures the rupture of the interface at the time of the coalescence.

Acknowledgements

We acknowledge Per-Olof Persson for use of his FEMLAB code which was modified appropriately to treat coalescence. We thank Peter Norgaard for helpful discussions.

References

1. S. O. Unverdi and G. Tryggvason, A front-tracking method for viscous, incompressible, multi-fluid flows, *J. Comput. Phys.* **100,** 25 (1992).
2. J. M. Boulton-Stone and Blake, Gas bubbles bursting at a free surface, *J. Fluid Mech.* **254,** 437 (1993).
3. W. J. Rider and D. B. Kothe, Reconstructing volume tracking, *J. Comput. Phys.* **141,** 112 (1998).
4. N. Takada, M. Misawa, A. Tomiyama and S. Fujiwara, Numerical simulation of two- and three-dimensional two-phase fluid motion by lattice Boltzmann method, *Comp. Phys. Comm.*, **129,** 233 (2000).
5. M. Verschueren, F.N. van de Vosse, and H.E.H. Meijer, "Diffuse interface modeling of thermocapillary flow instabilities in a Hele-Shaw cell," *J. Fluid Mech.* **434**:153–166, 2001.

6. M. Sussman, A. S. Almgren, J. B. Bell, P. Colella, L. H. Howell and M. L. Welcome, An adaptive level set approach for incompressible two-phase flows, *J. Comput. Phys.* **148,** 81 (1999).

7. W. Shyy, <u>Computational modeling for fluid flow and interfacial transport</u>, Elsevier, (1994).

8. W.B. Zimmerman, "Excitation of surface waves due to thermocapillary effects on a stably stratified fluid layer." Submitted to *Journal of Fluid Mechanics*.

9. NL Kolev, "Fragmentation and coalescence dynamics in multiphase flows." *Experimental Thermal and Fluid Science*, **6**:211–251, 1993.

10. H Zhao, SS Sadhal, Eh Trinh, "Singular perturbation analysis of an acoustically levitated sphere: Flow about the velocity node" *J Acoust Soc Am* **106** (2): 589–595, 1999.

11. Sethian, J. A., <u>Level set methods and fast marching methods</u>, Cambridge University Press, 1999.

Chapter 9

ELECTROKINETIC FLOW

W. B. J. ZIMMERMAN and J. M. MACINNES

*Department of Chemical and Process Engineering, University of Sheffield,
Newcastle Street, Sheffield S1 3JD United Kingdom*

E-mail: j.m.macinnes@shef.ac.uk

This chapter explores the multiphysics modeling appropriate for electrokinetic flow in microchannel networks. In setting up our case study, we learned how FEMLAB's weak boundary constraints are needed for coupled boundary conditions that incorporate non-tangential boundary conditions. To illustrate the utility of weak boundary conditions in accurate flux computations, we revisit the electrical capacitance tomography forward problem defined in §7.3.2. After this simple example, we move on to implementing more complicated weak boundary constraints in the electrokinetic flow model. The latter explores FEMLAB's guidelines for when to use a weak boundary constraint and when they fail.

9.1 Introduction

The purpose of this chapter is to demonstrate the facility of setting up a model for electrokinetic flow in FEMLAB. A cutting edge application for electrokinetic flow is microfluidics, wherein small quantities of chemicals (nanoliters) are transported "just-in-time" for complicated switching and sequencing in a network of microchannels to achieve high reproducibility of chemical reactions and compositional changes by tight control. Moving fluids by physicochemical phenomena is especially important since it involves fast response times and no moving mechanical parts that can become damaged. There is a strong overlap between microfluidics and micromechanical machines (MEMs). For instance, moving macromolecules adjacent to walls and side channels as soft actuators is considered microfluidics, but these are also molecular machines, but at a scale too small to be considered conventional moving parts.

In order to set up even our simplest electrokinetic model, however, multiphysics is essential – coupling electric potential, chemical transport, and momentum transport (Navier-Stokes). Furthermore, a first approach introduces some coupling through boundary conditions to approximate the electrochemical boundary layer motion. Although this coupling is linear, we found that to get an acceptable model in FEMLAB, the set up requires weak boundary constraints.

Thus, in preparation for our model, a simple example illustrating how weak boundary constraints can be used, in this case to compute boundary fluxes more accurately, is shown in §9.2, with the example drawn from the forward ECT problem of §7.3.2. In section §9.3, our "building block" 2-D electrokinetic flow model is set out.

9.2 Weak Boundary Constraints: Revisiting ECT

The FEMLAB 2.3 User's Guide and Introduction [1] has an excellent section on modeling approaches, including a discussion of weak boundary constraints. The section gives three primary reasons for using weak boundary constraints:

- Very accurate flux computations
- Handling nonlinear constraints
- Implementing constraints using derivatives.

In this section, we discuss how to use the weak boundary constraint to compute fluxes very accurately.

Before we do this, we need to re-visit the role of Lagrange multipliers in boundary and auxiliary constraint satisfaction through the finite element method. This was done in its full glorious detail in §2.3.1. Now, if you have Neumann boundary conditions, on a boundary B, then there is nothing to do – these are naturally computed using the FEM weak formulation as described in Chapter 2. You can think of Neumann conditions as being "neutral" in that unless you specify a constraint, they happen by default. So we will put it here simplistically that on a given boundary B we have a nonlinear constraint $r(\phi)=0$, where ϕ is the dependent variable. A second quantity that FEMLAB utilizes is h, which is the derivative of the constraint, i.e. $h=-r'(\phi)$. If there are more than one dependent variable, then h is a vector valued quantity (the gradient). The simplest form that can be taken for the constraint is a linear function: $r(\phi)=\phi_0 - \phi$, which is the Dirichlet condition. In this case, $h=1$. You might have been wondering for some time about what h and r were in specifying Dirichlet conditions.

If you select the Dirichlet radio button in general mode (boundary setting), for instance, specify r=1-phi (in general form, h is automatically computed by symbolic differentiation), then FEMLAB implements ideal boundary conditions for that boundary, which adds two more contributions to the weak formulation of the problem. These are subsequently discretized by the Galerkin method on the finite element basis functions as described in §2.3:

$$\int_B \phi_{test} r \, ds - \int_B \phi_{test} h \lambda \, ds \qquad (9.1)$$

where ϕ_{test} is the test function (conjugate to ϕ), ds is the increment of arc length along the curve B, and λ is the Lagrange multiplier. The Galerkin method for finite elements chooses the optimal value of λ to balance the constraint $r(\phi)=0$ so that the error in satisfying it is minimized "in the energy" sense. Physically, the Lagrange multiplier should have a meaning – inspection of (9.1) shows that it has the same units as r/h. In optimization theory, for instance, the Lagrange multiplier conjugate to a constraint on supply of a commodity is the "shadow price" for that supply – the price that would identically balance the supply and demand. Since $r(\phi)$ is a generalized Dirichlet condition, the Lagrange multiplier is a generalized boundary flux of the field (dependent) variable.

In heat and mass transfer problems, the Lagrange multiplier conjugate to fixed temperature or concentration on the boundary is the heat flux or mass flux across the boundary. So the Lagrange multiplier λ is that value for each individual element. In fluid dynamics, the flux of momentum across a boundary is a force. But which force? It depends on the quantity Γ expressed by the user in general form: $\lambda = \hat{n} \cdot \Gamma$ for a Dirichlet condition with $G=0$. For the Navier-Stokes equations, it is the viscous momentum dissipation in the PDE, so the boundary flux is the viscous force on the boundary. In FEMLAB 3.0, there are plans to include the pressure term in the Γ-vector as the default and to leave the current arrangement (viscous stress) as an option in the application mode. In general form, the PDE which we compute is the divergence of $\Gamma(\phi)$, so the flux computed by λ is the normal component of Γ. To illustrate this, we will revisit the ECT forward problem shown in §7.3.2.

Example: Very accurate flux computations in the ECT forward problem

If you recall, this problem computes the boundary fluxes (charges) across the electrodes held at given voltages on the boundary of a cylinder with badly conducting rods placed axially when the duct is full of a much better conducting substance. The heat transfer analogue is that the electrode surfaces are held at fixed temperatures and we compute the heat flux across these surfaces. All other external boundaries are insulated (no flux) and internal boundaries have continuous temperature and flux. Rather than set up the whole problem again, we start by reading in the MAT file ect.mat with the old solution (see Figure 7.1). We will investigate the accuracy of the original computations of flux with two different meshes (coarse and fine) and then compare with the flux calculation by the weak boundary constraint using the Lagrange multiplier. Figure 9.1 shows the original coarse mesh used to compute the boundary fluxes across the electrode surfaces.

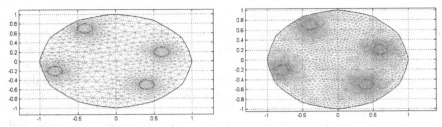

Figure 9.1 Coarse (2073 elements) and fine (8292) meshes used for ECT flux computations.

In FEMLAB, load ect.mat (file menu, Open). Pull down the Multiphysics menu from the menu bar and select Add/Edit modes. With ect.mat, we should already have a Geom1: PDE general form in place on the right hand side list.

Add/Edit Modes
- Select "Weak boundary constraint"
- \>\>
- Apply / OK

This now defines the mode wc1 (default name) and dependent variable lm (default name).

Back in the FEMLAB GUI main window, select Boundary Mode and Boundary Settings for mode wc1.

Boundary Mode and Boundary Settings (wc1)
- Select domain 5, check active in this subdomain, type 'phi' into the constraint variable entry box, and Apply
- Select domain 6, check active in this subdomain, type 'phi' into the constraint variable entry box, and Apply
- Select domain 21, check active in this subdomain, type 'phi' into the constraint variable entry box, and Apply
- Select domain 24, check active in this subdomain, type 'phi' into the constraint variable entry box, and Apply
- OK

Boundaries 5,6, and 21 are held at ground potential (phi=0) and boundary 24 is the source electrode, held at unit potential (phi=1). Note that you have accepted the non-ideal dim constraint as the default for all four boundary segments.

Equation (9.1) defined the *ideal* dim constraint. So what is this? The analogous contributions to the weak formulation are:

$$\int_B \phi_{test}\, r\, ds - \int_B \phi_{test}\, \lambda\, ds \qquad (9.2)$$

The derivative of the constraint function (h) is now missing from the second contribution in (9.2). It is argued that (9.2) better "balances" the constraint $r(\phi)=0$ in the case when r is nonlinear or contains derivatives of ϕ, which are not as accurately estimated by FEM for their contribution to h.

Now use the triangle on the toolbar (mesh) then select solve (=). ect.mat was a linear problem, so the linear solver is the default. Ours computed so rapidly that we did not notice the solution time. Enter post mode and compute the following:

Post Mode
Boundary integration: bnd 24,21,5,6   nx*phix+ny*phiy
Boundary integration: bnd 24,21,5,6   lm

Since writing Chapter 7, we have learned that nx and ny are symbols available on the boundary to compute the components of the normal vector in the coordinate directions. Thus the first calculation is equivalent to the standard formula

$$\frac{\partial \Phi}{\partial n} = \hat{n} \cdot \nabla \Phi \qquad (9.3)$$

where the unit outward pointing normal is used. We did this "by hand" since we had defined the normal vector as a constant, even though the sector boundary was slightly curved in defining the geometry. Now refine the mesh using the standard toolbar – inverted triangle in the triangle. Recompute the boundary integrations. Table 9.1 gives the summary data:

	1m coarse	7.3.2 coarse	(9.3) coarse	1m fine	7.3.2 fine	(9.3) fine
bnd 24	-0.92145	0.77067	0.77269	-0.91705	0.81237	0.81467
bnd 21	0.37001	-0.30704	-0.30786	0.36776	-0.32426	-0.3252
bnd 5	0.19415	-0.16518	-0.16559	0.19374	-0.17347	-0.17394
bnd 6	0.35729	-0.30096	-0.30182	0.35555	-0.31617	-0.31713
Σ	0	-0.00251	-0.00258	0	-0.00153	-0.0016

Table 9.1   Comparison of the coarse and fine mesh computations of boundary fluxes by three methods.

Boundary 24 is the "source" boundary. Since the lm computation gives it a negative boundary integral, we should interpret it as the flux out of the domain across that boundary. The other two methods oppose the sign of the lm method in every instance, they have the interpretation of the flux into the domain across that boundary. Slight numerical differences occur due to the *(nx,ny)* method and "by hand" (7.3.2) having errors on the order of the grid scale. The final row is the sum of all the boundary fluxes and is consistent only for the lm method giving exactly zero to five decimal places. The others have cumulative errors on the order of the grid scale. Conservation of electric charge should give flux in equals flux out, or net flux is zero. Clearly the percentage change upon refining the mesh by approximately four-fold the number of elements results in an order of magnitude less change in the lm method flux estimates than in the direct computation. As the values of the different methods are approaching upon refining the mesh, it is clear that the Lagrange Multiplier estimate of the flux is substantially better than the direct computation. The Lagrange multiplier is an "integrated balance" for the constraint, and in FEM, integrated quantities are better approximated than differentiated quantities in general. This is a feature of the weak formulation of the PDE.

The FEMLAB 2.3 User's Guide and Introduction [1, p. 1-400] gives a laundry list of caveats for the use of weak boundary constraints. We reproduce them here for completeness, and, on advice from COMSOL, update them, now that we have a concrete example for discussion:

- Strong and weak constraints should not be mixed on adjacent boundaries, i.e. those sharing common nodes.
- You must always have a constraint on boundaries when you enable the weak boundary constraint. N.B. only Dirichlet-type boundary conditions count as a constraint. Neumann conditions, being natural to FEM, even if inhomogeneous, do not count as a constraint.
- Scale your equations so that all coupled quantities are the same order, to avoid convergence difficulties. Automatic scaling of variables, a solver parameters option, does this by default.
- Discontinuous constraints are only satisfied by theoretically infinite Lagrange multipliers. In practice, this leads to large oscillations.
- Be careful not to use different element shape types between boundary and application modes. Derivative only boundary conditions should have *lower* order elements (same shape) than the "bulk."
- Iterative solvers do not like the structure of the matrices (not sparse enough) so use incomplete LU factorization as the preconditioner for the iterative solver.

Now for our clarifications of this list in light of our example:

- Strong and weak constraints should not be mixed on adjacent boundaries. *But Neumann boundary conditions do not count as strong (automatically weak), so they can be mixed on adjacent boundaries.* Our "electrodes" were surrounded by Neumann BC segments with no apparent difficulty. **This is consistent with the policy that Neumann boundary conditions do not count as a constraint for the purposes of a weak boundary constraint.**

- You must always have a *constraint* on boundaries when you enable the weak boundary constraint. We used three **zero** boundary conditions for boundaries bnds 5,6, 21 and still got the correct answer. So even a homogeneous constraint still counts as a constraint.

9.3 Electrokinetic Flow

9.3.1 Background

Electrokinetic flow is produced by the interaction of an electric field and charged (ion) species in a liquid. Two distinct interactions are present: the electric force on the liquid in the double layer region adjacent to wall surfaces where there is a net charge and the movement of individual ions in the bulk of the flow (outside the double layer region) where there is generally no net charge. The double layer may be taken as infinitesimal for channel sizes of interest (say greater than about 1 μm) and its effect on the flow is then equivalent (MacInnes, 2002) to application of the boundary conditions for velocity, u_i, electric field, ϕ, and mass fraction of a relevant chemical species, Y :

$$u_i = \zeta \frac{\partial \phi}{\partial x_i} \qquad n_i \frac{\partial \phi}{\partial x_i} = 0 \qquad n_i \frac{\partial Y}{\partial x_i} = 0 \qquad (9.4)$$

where n_i is the unit normal vector to the wall surface.

The system of equations that must be solved comprises the momentum equation, the continuity of mass equation, the charge continuity equation and a species equation. A simplest case may be expressed in non-dimensional form by

Momentum transport and continuity (Navier-Stokes):

$$\frac{\partial u_i}{\partial t} + u_j \frac{\partial u_i}{\partial x_j} = -\frac{\partial p}{\partial x_i} + \frac{\partial}{\partial x_j}\left(\frac{1}{\text{Re}}\frac{\partial u_i}{\partial x_j}\right) \qquad (9.5)$$

$$\frac{\partial u_j}{\partial x_j} = 0$$

Species transport including electrophoresis:

$$\frac{\partial Y}{\partial t}+u_j\frac{\partial Y}{\partial x_j}=\frac{\partial}{\partial x_j}\left(\frac{1}{\text{Pe}}\frac{\partial Y}{\partial x_j}+\beta z Y\frac{\partial \phi}{\partial x_j}\right) \tag{9.6}$$

Charge balance:
$$\frac{\partial}{\partial x_j}\left(\sigma\frac{\partial \phi}{\partial x_j}\right)=0 \tag{9.7}$$

The electric field satisfying Eq 9.7 must also satisfy Gauss' law (c.f. equation 7.1), which becomes an equation determining charge density as a function of position in the flow. In the typical conditions of electrokinetic flow, the charge density may be taken as negligible for purposes of both charge conservation (Eq 9.7) and the momentum balance (Eq 9.5). The electrical conductivity and zeta potential may depend on concentration of species Y and linear relations are assumed here: $\sigma =1+\sigma_r(1-Y)$ and $\zeta =-1-\zeta_r(1-Y)$, where subscript 'r' indicates the ratio of the property in the two pure solutions involved in the flows considered.

Boundary conditions at the flow inlets are that electric potential, pressure and species concentration must be specified, and at flow outlets electric potential and pressure must be specified. Species concentration is not known at the boundary and an approximation regarding species diffusion, the only term that connects the species field within the domain to the species distribution on the outflow boundary, is required. As usual, the species diffusion is neglected at the outflow boundary, i.e. a Neumann boundary condition just on the diffusion part of the flux term Γ is used.

The electric field is taken as quasi-steady, that is the electric field adjusts practically instantly to changes in the velocity and concentration. The above equations represent a generic problem providing a test of the numerical implementation which when verified may allow computation of any particular electrokinetic flow conditions. For the test implementation, suggested coefficient values are $1/\text{Re}$ = 30, $1/\text{Pe}$ = 0.03, ζ_r = 1 (no variation in wall zeta potential), $z = 0$ (no charge on species Y) and σ_r = 1 (no variation in electrical conductivity).

9.3.2 Problem set up

The basic problem one can solve is the propagation of a concentration front along a channel. Initially, a sharp front is placed at mid channel and the evolution of the front is then computed in time. The test problem is two-dimensional and the channel width can be taken as 1 unit, with the length equal to 6 units.

There are a number of distinct steps in problem complexity. (1) With the parameter values suggested above, the electric field will be uniform and in the direction of the channel. The concentration will move with a uniform flow with the front thickening from diffusion. (2) Setting $\zeta_r \neq 1$ gives a non-uniform wall zeta potential with walls exposed to full concentration of the computed species having zeta potential ζ_r and those exposed to zero concentration $\zeta = -1$, giving variation of slip boundary velocity through the first of boundary conditions 9.4. The electric field remains uniform and in the channel direction, but the velocity field will be altered. The concentration front will be modified from the pure diffusion case by the non-uniform velocity field. (3) Setting $z = \pm 1$ and $\beta = 1$ will introduce electrophoresis. The computed species will translate in the channel direction in addition to being moved by the liquid velocity. (4) Finally, setting $\sigma_r \neq 1$ introduces non-uniform electrical conductivity. This leads to changes in the electric field associated with changes in concentration (Y) so the electric field is no longer uniform or, where concentration gradient is not everywhere in the direction of the channel, in the channel direction.

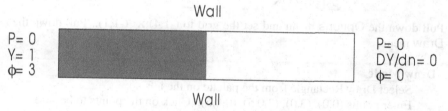

Figure 9.2 Problem definition in a nutshell.

9.3.3 FEMLAB implementation

There are application modes for conductive media, convection and diffusion, and the Navier-Stokes equations. To have best knowledge of what the computation entails, we start with the Navier-Stokes equations and add two general modes for (9.6) and (9.7).

Load up FEMLAB and in the Model Navigator, Multiphysics tab, set up as follows:

Model Navigator

- Select Multiphysics tab, 2D
- Select Incompressible Navier-Stokes variables u v p; ns   mode >>
- Select PDE general form,   variable Y; species   mode >>
- Select PDE general form,   variable phi; potential mode >>
- Solver Type: time dependent
- OK

Accept the default element shape and order for each mode.

Pull down the options menu and select Add/Edit constants. The Add/Edit constants dialog box appears.

Add/Edit Constants

- Define as follows:

Pec	30
Re	0.03
zel	1
betael	1
zetar	1
sigr	1

- Apply / OK

Pull down the Options menu and set the grid to $(-1,4) \times (-1,1)$. Pull down the Draw menu.

Draw Mode

Select Draw Rectangle from the palette on the left.
Enter points (0,0), (3,0), (3,0.5), (0,0.5). Click on the points to be sure that you snapped to the grid or to edit them.
- Apply / OK

Now pull down the Subdomain menu and select Subdomain settings.

Subdomain Mode/Settings	Γ	F	d_a
Subdomain 1 (ns mode)	$\rho = 1$; $\eta = 1/Re$; Fx = 0; Fy = 0		
Subdomain 1 (species mode)	-(1/Pec)*Yx-betael*zel*Y*phix (space delimiter) -(1/Pec)*Yy-betael*zel*Y*phiy	-u*Yx-v*Yy	1

Init tab	Y(t0) = 1		
Subdomain 1 (potential mode)	-sig*phix -sig*phiy	0	0
•	Apply / OK		

Note the space delimiter is necessary to specify Γ as a vector in species mode and potential mode. There are few loose ends to tie up. Two expressions above are undefined. As we are already in the **Subdomain Mode**, it is convenient to define them now. Pull down the **options** menu and select **Add/Edit expressions**. The Add/Edit expressions dialog box appears.

Add/Edit Expressions

- Define two expressions at geometry level
- Name: zeta   Definition: - (Y+zetar*(1-Y))
- Name: sig   Definition: Y+sigr*(1-Y)
- Apply / OK

Now for the boundary conditions. Pull down the **Boundary** menu and select **Boundary Settings**.

Boundary Mode / Boundary Settings

- ns mode

bnd 1	bnd 2	bnd 3	bnd 4
outflow/pressure p=0	u = zeta*phix v = zeta*phiy	u = zeta*phix v = zeta*phiy	outflow/pressure p=0

- species mode

bnd 1	bnd 2	bnd 3	bnd 4
Dirichlet G = 0; R = -Y	Neumann G = 0	Neumann G = 0	Neumann G = betael*zel*Y*phix

- potential mode

bnd 1	bnd 2	bnd 3	bnd 4
Dirichlet G = 0; R = 3 - phi	Neumann G = 0	Neumann G = 0	Dirichlet G = 0; R = -phi

- Apply / OK

Accept the standard mesh parameters and hit the mesh button on the toolbar (triangle). 87 nodes and 136 elements are created for us. Note that to produce the desired Neumann boundary condition at the outlet boundary for species, G is set to the expression that precisely eliminates the electrophoretic term from Γ in the species equation.

Pull down the Solver menu and select Solver Parameters.

Solver Parameters

- Click on the Settings button under "Scaling of variables." Check the None option. OK
- Select the timestepping tab. Set output times to 0:0.1:1 (default)
- Set the timestepping algorithm to fldaspk
- Set absolute tolerance: u 0.01 v 0.01 p Inf Y 0.001 phi 0.001
- Apply / OK

Now a warning. This is not going to work. Just to verify it, try solving and take a break for five minutes. You will come back to find little solution progress has been made. Cancel the computation.

Where does the model go off the rails? We tried the usual suspects – system is stiff? So we tried the stiff solver (ode23s)! System is differential-algebraic, so we tried a dae solver (fldaspk). No good. Reduce the time step. No good. When reduced to a ridiculously small time step, we did get a converged solution after ages of waiting. Remember, this is a small mesh (compare with ECT problem in §9.2!).

We foreshadowed the problem in Chapter 7 and in §9.1, so no prizes for guessing it involves weak boundary constraints. The problem is that the standard Multiphysics couplings are not picking up the boundary couplings, even though they are linear or pseudo-linear, in (9.4). The top and bottom velocity boundary conditions couple to the electric field (gradient phi) and the outlet species concentration does as well. The latter, though pseudo-linear, is eventually a nonlinear term, feeding back both species and field strength quadratically.

Our prescription, following the ECT example in §9.2, is to add two weak boundary constraints; one for each velocity.

Pull down the Multiphysics menu from the menu bar and select Add/Edit modes. We should already have ns, species, and potential in place on the right hand side list.

Add/Edit Modes

- Select "Weak boundary constraint"
- Name the mode wcu and the variable lmu>>
- Select "Weak boundary constraint"
- Name the mode wcv and the variable lmv>>
- Apply / OK

This now defines the mode wcu and wcv and dependent variables lmu and lmv. We could do this slightly tidier with one weak boundary constraint with two variables. Back in the FEMLAB GUI main window, select **Boundary Mode** and **Boundary Settings** for mode wcu and then mode wcv.

Boundary Mode and Boundary Settings
- **Mode wcu.** Select domain 2 and 3, check active in this subdomain, type 'u' into the constraint variable entry box, and Apply
- **Mode wcv.** Select domain 2 and 3, check active in this subdomain, type 'v' into the constraint variable entry box, and Apply
- OK

Now it is safe to click on the Solve (=) toolbar button. It still takes some substantial time to make progress in this model – the coupling does not help the sparseness of the matrix assembled – but timestepping does proceed to solution in 7 minutes. A Pentium III 866Hz produced the first output time step in 4 minutes.

Figure 9.3 shows all the information rolled up into one plot for the final time $t=1$. By this time, all streamlines are parallel and velocity vectors uniform – flat profile. The spreading of concentration and speeding up of the flow are all driven by the electric field, which is now apparently uniform in magnitude. A few cross plots (see Figure 9.4) show that the steady state electric field relaxes its transients within the first output time and remains constant thereafter (phi is linear for all times after t=0.1). As expected, electrokinetic flow is dragged along by its boundary layer coupling to the electric field.

But why did this recipe work? Of course we tried everything we could think of. For instance, we tried adding an additional time dependence in the electrostatic potential equation, da=0.001, as an attempt to overcome the stiffness of instantaneous relaxation to electrostatic equilibrium. But the final result uses weak boundary conditions for the side wall Navier-Stokes velocities which are linearly coupled to the electric field, but not for the outlet species condition which is nonlinearly coupled to concentration and electric field. We tried some variations on the species mode:

Trial 1: No weak boundary constraint (general form) – apparently fine
Trial 2: Weak boundary constraint (general form) - does not work
Trial 3: Weak boundary constraint (weak form) - does not work

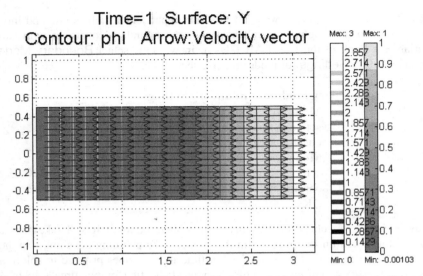

Figure 9.3 Combined concentration (Y, color), electric potential (phi, contour), and velocity vector (u, v, arrow) plot. Coupling on the boundary of species electrophoresis/diffusion with electric field drags the fluid along.

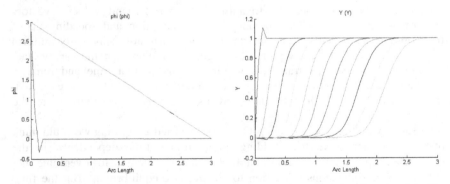

Figure 9.4 Histories of electric potential (phi) and species concentration Y along boundary 2 (wall).

So the first question is why does our recipe work? §9.2 supposes three reasons for implementing weak boundary constraints. Our application satisfies the final two:

(1) Handling nonlinear constraints

The nonlinear solver in FEMLAB handles linear or nearly linear standard constraints. Note well, however, that Neumann conditions are not considered a constraint in this context. Weak constraints can include nonlinearities because they contribute to the stiffness matrix and residual instead of the constraint matrices [1]. *So one expects that the outlet species concentration, though nonlinear, may be treated satisfactorily by the standard handling of constraints. Very simply, as a Neumann condition, it does not count as a weak boundary constraint – it is naturally in FEM (see chapter 2) so it automatically is treated correctly.*

(2) Implementing constraints using derivatives

In FEMLAB 2.2 and later, the derivatives of the dependent variables are available also on the boundary. Constraints on only the tangential component of the derivative work when using standard constraints, whereas here it is necessary to use a weak constraint to be able to handle *non-tangential* constraints (the velocity BCs on the walls).

Condition (3) is clearly satisfied, yet condition (2) is not violated with the caveat that Neumann constraints do not count. We should not need to use a weak boundary constraint on the outlet boundary (bnd 4) for the species transport equation, we did not, and it works. When we tried a weak boundary constraint, it failed.

From (9.6), we defined for our species general mode

$$\Gamma = \nabla Y / Pe + \beta z Y \nabla \phi \tag{9.8}$$

as the straightforward way of dealing with the electrophoretic term. Consequently, our boundary condition on the outlet takes the form

$$n \cdot \Gamma = \beta z Y \left(n \cdot \nabla \phi \right) \tag{9.9}$$

Equation (9.9) is a *non-zero Neumann* condition with regard to the flux Γ. But since Neumann conditions do not count as constraints, the standard BC works fine.

Figure 9.5 shows convincingly that the expected value with uniform Y and phix on the outlet boundary is achieved by the model at all times.

9.3.4 Links to physical boundaries

Current microchannel devices may consist of many distinct channel segments joined at several junctions. Future ones may well comprise hundreds of segments joined at a similar number of junctions. Detailed computation of the flow in such a system is unlikely to be feasible for some time to come and, indeed, is probably not desirable. Rather, an approach in which a particular junction of interest or perhaps an evolving mixing zone such as that considered in MacInnes et al. (2003) is probably appropriate. The approach emerges from

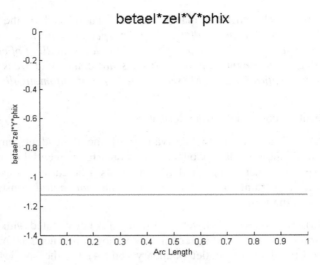

Figure 9.5   Neumann boundary term for all output times (identical) along the outlet boundary (bnd 4).

the fact that electrokinetic flows in microchannel networks virtually always are characterized by very low Reynolds number, $\mathrm{Re} \ll 1$. In channel segments of uniform section and liquid and wall properties, the flow is developed along the entire segment length except for a region within about one channel width of junctions or other disturbances to uniformity. If the segment is many channel widths in length, it is a good approximation to neglect the junction effects and one can write linear relations between pressure and electric potential differences and the liquid volume flow rate, Q, and the charge flow rate, I :

$$-\frac{\mathrm{Re}\,R^2 A}{f \Delta s} \Delta p + \frac{\zeta A}{\Delta s} \Delta \phi = Q \quad \text{and} \quad -\frac{\sigma A}{\Delta s} \Delta \phi = I \qquad (9.10)$$

These equations are coupled to the detailed flow solution through the liquid and charge flow rates. We will consider the specific example of an electrokinetic switching at a 'Y' junction in the arrangement shown in Figure 9.6. By changing voltages at reservoirs A, B and C in an alternating pattern, 'slugs' of the liquid fed in at A interspersed with the liquid fed in at B will be formed in the channel leading to C. No property non-uniformity will be present so the zeta potential and conductivity are uniform over each channel segment. We wish only to compute the flow in the vicinity of the junction where slug formation takes place.

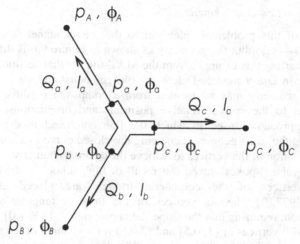

Figure 9.6. The simple network considered.

The 'Y' junction domain is computed in detail with algebraic relations linking to the known boundary conditions at reservoirs A, B and C to provide the boundary conditions at a, b and c. In the case of non-uniform composition in channel segment C, a one-dimensional pie must be solved for the species composition to provide the link to downstream conditions.

The boundary conditions at the junction flow boundaries a, b and c come from the network flow equations linking junction boundary conditions to the known conditions of the reservoirs connected up to the network. In the simple case considered here (where $R = A = 1$), the linking equations for boundary a are

$$-\frac{\text{Re}}{f\Delta s_a}(p_A - p_a) + \frac{\zeta}{\Delta s}(\phi_A - \phi_a) = Q_a$$

and

$$-\frac{\sigma A}{\Delta s_a}(\phi_A - \phi_a) = I_a \qquad (9.11)$$

where Q_a and I_a are the flow rates at boundary a given in terms of computed conditions on that boundary by

$$Q_a = \int_A u_i n_i dA \qquad \text{and} \qquad I_a = -\int_A \sigma \frac{\partial \phi}{\partial x_i} n_i dA \qquad (9.12)$$

Precisely similar relations result at boundaries b and c.

Example: Y-junction electrokinetic valve

The physics of this problem is identical to the microchannel flow in §9.3.1, equation (9.4)—(9.7), but the geometry as shown in Figure 9.6 is different. We recommend starting this example from the MAT-file for that section, and to edit the geometry in Draw mode. Delete the old geometry and start with a fresh plane. The geometry may be formed from a composite of three rectangular solids rotated to the correct relative positions and orientations. Here, the alternative approach is used in which the line command to draw segments approximating the exact geometry, coercing the closed curve to a solid, and then editing the location of the vertices to achieve the desired symmetry. The vertices of the rectangular "spokes" taxed our recall of high school analytic geometry. Once the vertices of the equilateral triangle are placed at $\{(1,-0.5),$ $(1,0.5),(0.133977,0)\}$, the six vertices of the three rectangular sections are determined from requiring that the slopes be appropriate – $\{+1,-1,0\}$, The last is the easiest, with vertices at $(3,-0.5)$ and $(3,0.5)$ for a channel of length two. All other channels should have the same length at least initially. So, for example, the lowest vertex is found by satisfying the distance formula and the slope constraint simultaneously:

$$\sqrt{(x_1+1)^2+(y_1+1)^2}=2$$

$$\frac{y_1+0.5}{x_1-1}=1 \tag{9.13}$$

There are two solutions to this quadratic system. The one we are after is $(x_1,y_1)=(-0.414214,1.91421)$. We coded this system of equations in MATLAB since the algebra is straightforward, though tedious. Follow these steps in Draw Mode:

Pull down the Draw menu.

Draw Mode

- Set axis/grid settings to the domain [-3,3]×[-2,2].
- Select Draw Line from the palette on the left.
  Enter points buy clicking on them and dragging the line to the next point in the list:
  (1,-0.5), (3,-0.5), (3,0.5), (1,0.5),(-0.414214,1.91421), (-1.28024,1.41421), (0.133977,0),(-1.28024,-1.41421),(-0.414214,-1.91421), (1,-0.5)
  It is OK to enter points near these. After the figure is drawn, click on the points to edit them.
- Under the drawn menu, select Coerce Objects to Solid
- Apply / OK

We now need to reconfigure our boundary conditions and PDE coefficients to reflect the new geometry. In our model, the boundary segment number assignments are 1,3 (inlet), 2,4,5,6,7,8 (side walls), 9 (outlet). Pull down the Boundary menu and select Boundary Settings. In parallel with the previous set up, make the following assignments:

Boundary Mode / Boundary Settings

- ns mode

bnd 1,3	bnd 2,4,5,6,7,8	bnd 9
outflow/pressure p=0	u = zeta*phix v = zeta*phiy	outflow/pressure p=0

- species mode

bnd 1,3	bnd 2,4,5,6,7,8	bnd 9
Dirichlet G = 0; R = -Y	Neumann G = 0	Neumann G = betael*zel*Y*phix

- potential mode

bnd 1,3	bnd 2,4,5,6,7,8	bnd 9
Dirichlet G = 0; R = 5-phi	Neumann G = 0	Dirichlet G = 0; R = -phi

- Apply / OK

The Subdomain mode/settings are identical as before, but still need to be made, since we threw out the previous subdomain, and with it, its subdomain settings:

Subdomain Mode/Settings	Γ	F	d_a
Subdomain 1 (ns mode)	$\rho = 1$; $\eta = 1/Re$; Fx = 0; Fy = 0		
Subdomain 1 (species mode)	-(1/Pec)*Yx-betael*zel*Y*phix (space delimiter) -(1/Pec)*Yy-betael*zel*Y*phiy	-u*Yx-v*Yy	1
Init tab	Y(t0) = 1		
Subdomain 1 (potential mode)	-sig*phix -sig*phiy	0	0
	Apply / OK		

The constants and expressions were made geometry-wide through geom1, so they are inherited. We also inherit the weak boundary modes wcu and wcv, but need to make the new assignments.

Boundary Mode and Boundary Settings
• **Mode wcu.** Select domain 2,4-8, check active in this subdomain, type 'u' into the constraint variable entry box, and Apply
• **Mode wcv.** Select domain 2,4-8, check active in this subdomain, type 'v' into the constraint variable entry box, and Apply
• OK

Before solving the time evolution, we are going to try a new trick to improve convergence. The biggest problem that was making the previous simulation slow was the rapid change in the velocity and electric fields required in the first few instances from the initial condition (no field, no flow) to practically pseudosteady flow and field. The viscous and dielectric response times are much faster than the diffusive and convective time scales, so this problem is inherently stiff. In the computational modeling of the Navier-Stokes equations for *incompressible* flow, this problem is encountered for the pressure. Since in the incompressible approximation, sound speed is infinite, the pressure field adjusts instantaneously to changes in velocity. Computationally, time stepping over such widely different time scales leads to problems with stiffness and slow convergence, requiring miniscule time steps. The SIMPLE algorithm (Patankar, 1980) circumvented this pitfall by staging the time stepping of the velocity with rapid solution to the pressure field consistent with mass conservation and the current velocity field by solving a separate elliptic equation for the pressure – the Lighthill Poisson equation. The difference is that the corrections to the last pressure field are not found – small changes on the order of the time step -- but rather the pressure can be wholly different from that at the previous time step. Instantaneous changes in the pressure that depend on the field everywhere in the domain are thus catered for, and the Navier-Stokes simulation is no longer stiff. The FEMLAB Navier-Stokes application mode has this fast Poisson solver for pressure built in. But our electrostatic potential mode does not. So even though ϕ should, in principle, change instantaneously to applied alterations in voltages, which should change the slip velocity instantaneously, the ns mode will respond on the incompressible time scale, but the electric field needs to be relaxed. So to mimic the SIMPLE algorithm, we need to implement a fast elliptic solver for flow and electric field while freezing the concentration profile. Once the velocity field and electric field have been established, we can release the mass transport. Our no electrokinetic relaxation time in the potential mode is necessary for the model to advance steadily with only small changes to u,v,p and phi at each time step. The fast elliptic step overcomes the rapid relaxation time needed for abrupt changes in the electric field.

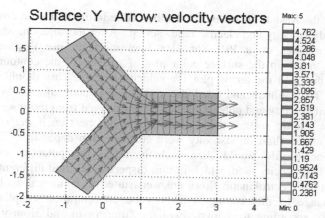

Figure 9.7 Stationary solution for flow field and potential lines from the initial conditions (Y=1 everywhere) and boundary potentials imposed, such that the pseudosteady flow and electric fields are achieved instantaneously after imposition of the boundary potentials.

So how do we make this initial fast elliptic step? -- By selective solution of the equations. Pull down the Solver menu and select Solver Parameters.

Solver Parameters

- Click on the Multiphysics tab
- Select all modes but species and Apply.
- Click on the general tab.
- Select stationary nonlinear. Apply / Solve

If you have difficulty with selecting the combination of modes, use variations of holding down the shift key or the control key, which have their standard Windows effect. Note that the MATLAB window has shown that it takes three Newton iterations to arrive at the converged profile, shown in Figure 9.7, for contours of phi and flow along velocity vector arrows as shown. We can now turn on the mass transport equations and let the transient solution begin. Pull down the Solver menu and select Solver Parameters.

Solver Parameters

- Click on the Multiphysics tab
- Select all modes and Apply.
- Click on the general tab.
- Select time dependent solver.
- Apply / OK

Now select the Restart button on the toolbar. The initial condition now has the pseudosteady velocity and electric fields set up. After the solution, which takes about six minutes on a PIII 866MHz Intel machine, use Post mode/Plot parameters to set up the surface color plot of species Y, the contour plot for potential lines, and the arrow plot for velocity vectors. This should resemble Figure 9.8. Then animate and watch the streams merge and flow out. We can see that the final velocity and potential fields are unchanged from those set up in the initial "fast elliptic solver" step. This would not be true if the conductive properties of the fluid or viscosity were concentration dependent. The local concentration profile would then globally affect the potential field, thus modifying the velocity field, which in turn disperses the fluid differently. Such highly coupled electrokinetic flows are well suited to numerical analysis by a modest change to this coding – adding the concentration coupling to the conductivity according to a mixture rule. Zimmerman and Homsy treat the mixture rule for concentration dependency of viscosity, for instance, in the instability of viscous fingering in porous media [2].

Y-Junction Switcher: An Application of Linkages Through Coupling Variables

As described in Figure 9.6, by including relatively long channels to the reservoirs of the component species where electric potentials are applied, the flow and the electric potential can be described by algebraic relations for the inlet and outlet dependencies. We implement these physical linkages between nodes *a* and A, for instance, using the appropriately named coupling variables. Originally, we set up a pseudo 0-D geometry (geom2 as in Chapters four and seven), but on advice from COMSOL consultants, realized that the conceptual 0-D domain was not needed – it can all be done in coupling variables. Using the model MAT-file of the last section, set up the following twelve coupling variables:

Select Add/Edit Coupling Variables from the Options Menu.

Add/Edit Coupling Variables

scalar add Qa. Source Geom 1, subdomain 1, boundary 3.
Integrand: zeta1*(PHIA-volta)/dsa-Re*(PA-bara)/f/dsa
Integration order: 2
Destination Geom 1 bnd 3 Check "Active in this domain" box.
scalar add Qb. Source Geom 1, subdomain 1, boundary 1.
Integrand: zeta1*(PHIB-voltb)/dsb-Re*(PB-barb)/f/dsb
Integration order: 2
Destination Geom 1 bnd 1 Check "Active in this domain" box.
scalar add Qc. Source Geom 1, subdomain 1, boundary 9.
Integrand: zeta1*(voltc-PHIC)/dsc-Re*(PC-barc)/f/dsc

Integration order: 2
Destination Geom 1 bnd 9 Check "Active in this domain" box.
scalar add Ia. Source Geom 1, subdomain 1, boundary 3.
Integrand: sigr*(PHIA-volta)/dsa
Integration order: 2
Destination Geom 1 bnd 3 Check "Active in this domain" box.
scalar add Ib. Source Geom 1, subdomain 1, boundary 1.
Integrand: sigr*(PHIB-voltb)/dsb
Integration order: 2
Destination Geom 1 bnd 1 Check "Active in this domain" box.
scalar add Ic. Source Geom 1, subdomain 1, boundary 9.
Integrand: sigr*(PHIC-voltc)/dsc
Integration order: 2
Destination Geom 1 bnd 9 Check "Active in this domain" box.
scalar add bara. Source Geom 1, subdomain 1, boundary 3.
Integrand: p
Integration order: 2
Destination Geom 1 bnd 3 Check "Active in this domain" box.
scalar add barb. Source Geom 1, subdomain 1, boundary 1.
Integrand: p
Integration order: 2
Destination Geom 1 bnd 1 Check "Active in this domain" box.
scalar add barc. Source Geom 1, subdomain 1, boundary 9.
Integrand: p
Integration order: 2
Destination Geom 1 bnd 9 Check "Active in this domain" box.
scalar add volta. Source Geom 1, subdomain 1, boundary 3.
Integrand: phi
Integration order: 2
Destination Geom 1 bnd 3 Check "Active in this domain" box.
scalar add voltb. Source Geom 1, subdomain 1, boundary 1.
Integrand: phi
Integration order: 2
Destination Geom 1 bnd 1 Check "Active in this domain" box.
scalar add voltc. Source Geom 1, subdomain 1, boundary 9.
Integrand: phi
Integration order: 2
Destination Geom 1 bnd 9 Check "Active in this domain" box.
- Apply / OK

That was pretty long-winded to evaluate (9.11) in a straightforward way. Now we need to implement the appropriate boundary conditions from the "flux type variables": Q's (volume flux) and I's (current flux): Pull down the **Boundary** menu and select Boundary Settings. In parallel with the previous set up, make the following assignments:

Boundary Mode / Boundary Settings

- ns mode

bnd 1	bnd 3	bnd 9	bnd 2,4,5,6,7,8
u=0.707107*Qb v=0.707107*Qb	u=0.707107*Qa v=-0.707107*Qa	u=Qc v=0	unchanged

- species mode: unchanged
- potential mode

bnd 1	bnd 3	bnd 9	bnd 2,4,5,6,7,8
Neumann G=Ib/sigr	Neumann G=Ia/sigr	Neumann G=Ic/sigr	unchanged

- Apply / OK

There is an underlying assumption in the above formulation. Use of a uniform velocity at each flow boundary is only possible if pressure gradient can be neglected. In 'pure' electrokinetic flow, that is where conductivity, zeta potential, viscosity are each uniform, the approximation of uniform velocity at the flow boundaries is excellent. The total pressure in each reservoir must also be the same (Cummings et al., 2000). However, when liquid properties are not uniform or a differential of dynamics pressure between reservoirs is present, pressure gradients arise within the network and the assumption implicit in the above treatment that velocity is uniform at each boundary is not appropriate. The generally correct treatment would be to determine I and Q from the flow boundaries and use relation 9.11 for the uniform pressure and potential at the boundary. That pressure or potential are uniform at each boundary follows rigorously when the boundary is at a position where the flow is developed, that is sufficiently far (say, a channel width) from a disturbance region such as a junction.

The formulation used does avoid the need for further weak boundary constraint modes – only wcu and wcv are needed. Although there is an analogy between pressure and electric potential, current and velocity, these quantities are treated fundamentally differently with regard to the need for weak boundary constraints. The velocity boundary conditions now require weak boundary constraints on all boundaries (not just the wall surfaces). So we will need to alter the Boundary Settings for wcu and wcv to include *all* boundaries. This is because velocities are implemented as Dirichlet boundary conditions. The Neumann BCs for the current in potential mode, however, do not require and are incompatible with weak boundary constraints as we learned earlier. Neumann conditions, since they are the default for FEM, are non-constraints even when they are inhomogeneous as here. The inhomogeneous Dirichlet conditions for u and v, however, are constraints and so require the weak boundary treatment to bring out the full nonlinearity and coupling.

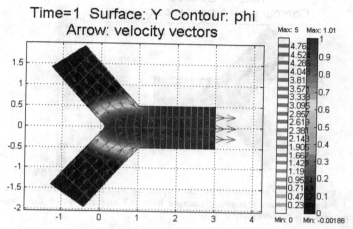

Time=1 Surface: Y Contour: phi
Arrow: velocity vectors

Figure 9.8 Merging concentration profile after unit time -- each leg has a forming front that merges in th Y-junction.

Now we are prepared for implementing the microfluidic Y-junction switcher in two stages:

A. <u>Merging two input streams</u>

All as before, but uniform properties ($\zeta_r = \sigma_r = 1$) with an uncharged species ($z = 0$). $\Delta s_a = \Delta s_b = 10$, $\Delta s_c = 6$ and $p_A = p_B = p_C$. Set the voltages to $\phi_A = 24$, $\phi_B = 12$, $\phi_C = 0$.

B. <u>Alternation between voltages</u>

$\phi_A = 24$, $\phi_B = 12$, $\phi_C = 0$ and $\phi_A = 12$, $\phi_B = 24$, $\phi_C = 0$ with a period of around 6 time units. This is a square wave signal for A and a complementary one for B.

A. Merging two input streams

We are all set up for merging two input streams. The constant potentials are set for nodes A, B, and C according to the values of three constants PHIA, PHIB, PHIC. Use Add/Edit Constants to introduce these three values.

As before, we need to stage our solution to set up the pseudosteady velocity and potential fields initially, then turn on the species transport.

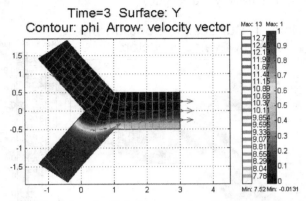

Figure 9.9   Developed flow of species Y=0 along upper leg with inhibited flow of species Y=1 in the lower leg for t=3. Hardly anything changes from t=3 onwards within the domain. The concentration profile is pseudosteady.

The result is shown in Figure 9.9. The fully developed flow of species Y=0 along upper leg with inhibited flow of species Y=1 in the lower leg for t=3. Hardly anything changes from this time onwards within the domain. The concentration profile is near its steady distribution. It is prudent to check the consistency of the calculation of the velocity and potential solutions. Using Boundary Integration under Post Menu, we find the following values:

Table 9.2 gives the summary data:

	$\bar{I} = \int \hat{n} \cdot \nabla \phi \, ds$	$\bar{\phi} = \int \phi \, ds$	$\bar{u} = \int u \, ds$	$\bar{v} = \int v \, ds$	$\bar{p} = \int p \, ds$
bnd 3 (a)	1.1184	12.816	0.83909	-0.8391	19.537
bnd 1 (b)	0.13518	10.649	0.10208	0.1021	2.6554
bnd 9 (c)	-1.2535	7.5213	1.2499	0	-0.64633
Σ	0.00008				

Table 9.2 Boundary fluxes across the three open boundaries.

The conservation of charge is satisfied to 10^{-4}. The conservation of mass does not hold so well. You can verify that $\sqrt{2} \times (0.83909 + 0.10201) = 1.33 \neq 1.25$. This discrepancy suggests that the velocity flow field is not spatially well resolved at this level of meshing. To improve the result, it is likely that greater mesh density is required in the "Y" vertex which clearly has discontinuity in velocity from the upper leg to the lower leg.

The algebraic relations hold roughly for the electric potential, pressure, current, and velocity, for instance: $u_a \approx \dfrac{I_c}{\sqrt{2}} = 0.790828$. Some are spot on --

$\dfrac{0-\phi_c}{\Delta s_c} = \dfrac{0-7.5213}{6} = -1.2535$. In general, the electric potential conditions hold exceedingly well, but the velocity field/pressure fields do not hold as well. Again, this is a strong indicator that the solution is not mesh resolved.

Plot (see Figure 9.9) and Animate the solution. The animation shows clearly how the front evolves to set up a fully developed upper flow, bypassing the lower leg. The intermediate voltage in the "b-leg" of the Y-junction tends to hold back the flow of species Y=1 from the merging into the main flow along the (a)-(c) open switch – only slow diffusion out and in, along with some modest convection, occur. With concentration dependent conductivity (and viscosity [2]), it is possible to counteract the diffusion to some extent which sharpens some fronts.

B. Alternation between voltages

Now for this application to be a Y-junction switch, we need to be able to replace the constant voltages ϕ_a and ϕ_b with $\phi_a(t)$ and $\phi_b(t)$, respectively. The signals could be arbitrary, however, in practice they are discontinuous level adjustments, which can be idealized as a square wave. A suitable choice is the alternation between values ϕ_{a0} and ϕ_{b0} (PHIA and PHIB), with the signals 90° out of phase. We coded the square wave in coupling variables in two different ways:

B1. Logical functions

Under Add/Edit Coupling Variables, we made the following changes: switching

```
Qa
zeta1*(PHIA-PHIA*(sin(2*pi*t)<0)+PHIB*(sin(2*pi*t)<0)-volta)/
dsa-Re*(PA-bara)/f/dsa
Qb
zeta1*(PHIB-PHIB*(sin(2*pi*t)<0)+PHIA*(sin(2*pi*t)<0)-voltb)/
dsb-Re*(PB-barb)/f/dsb
Ia
sigr*(PHIA-PHIA*(sin(2*pi*t)<0)+PHIB*(sin(2*pi*t)<0)-volta)/dsa
Ib
sigr*(PHIB-PHIB*(sin(2*pi*t)<0)+PHIA*(sin(2*pi*t)<0)-voltb)/dsb
```

The logical statement `(sin(2*pi*t)<0)` takes the value of 1 when true (during the second half period) and 0 when false (during the first half period), which is the essence of the square wave that makes a discontinuous switch. It is easy to code. We substituted this code into the above FEMLAB model without success. The numerical analysis proceeds smoothly until the end of the first half period, at which time FEMLAB crashes. We tried fldaspk and ode15s time integration schemes without success. The discontinuity makes the convergence criteria unattainable. Perhaps with fixed time step it is possible to "ram through" the discontinuity, accepting the large error, but even then the abrupt change is likely to lead to oscillatory artifacts. There are some schemes, like total variance diminishing and essentially non-oscillatory methods, which might alleviate this difficulty, but they are not implemented in MATLAB or finite element methods to our knowledge. So we abandoned this approach.

B2. Smoothed square waves

Perhaps the nonconvergence was due to the instantaneous switch which could be alleviated by smoothing the signal. We coded a Fourier Cosine Series representation of the square wave:

$$Sq(t;\tau) = \frac{4\tau}{\pi}\left[\cos\left(\frac{2\pi t}{\tau}\right) + \frac{1}{3}\cos\left(\frac{6\pi t}{\tau}\right) + \frac{1}{5}\cos\left(\frac{10\pi t}{\tau}\right) + \ldots + \frac{1}{2n+1}\cos\left(\frac{(2n+1)\pi t}{\tau}\right) + \ldots\right]$$

This was coded as a MATLAB m-file function, square.m, and placed in the MATLAB current directory:

```
function a=square(t,tau)
sum=0;
for n=1:10
sum=sum+4*tau*cos(pi*(2*n-1)*t/tau)/(2*n-1)/pi;
end
a=sum;
```

Figure 9.10 shows the square wave form approximating the first ten terms (n=9). Although the jumps are no longer infinitely steep, the price paid is non-physical oscillations and overshooting the steady levels. These are historical difficulties for electronic circuits used as function generators for square waves, overcome by sophisticated filters. The coding in coupling variables as below was successful to a greater extent than the logical function coding:
switching cosine series

```
Qa
zeta1*((PHIB-PHIA)*square(t,1)+PHIA-volta)/dsa-Re*(PA-bara)/f/dsa
Qb
zeta1*((PHIA-PHIB)*square(t,1)+PHIB-voltb)/dsb-Re*(PB-barb)/f/dsb
Ia
sigr*((PHIB-PHIA)*square(t,1)+PHIA-volta)/dsa
Ib
sigr*((PHIA-PHIB)*square(t,1)+PHIB-voltb)/dsb
```

The success was that this method actually integrates, yet exceedingly slowly. Why? Because the time integration must resolve all the non-physical oscillations in the square wave, which slows down the within half-period integrations, and then the jumps are inordinately slow, but eventually the new flow configuration is found, an improvement over the logical function method, but at an inordinate price.

By the way, you might have found our choice of how to code the coupling variables in terms of either the logical functions or square wave as long-winded. The reason for not using the coding with greatest algebraic simplicity was to insure that our initial step of the fast elliptic solution finds the correct initial conditions. By trial and error, we learned that any MATLAB function with t as an independent variable evaluates to zero when using the stationary nonlinear solver. It does not substitute $t=0$ into the formula, but rather chucks out the function altogether. By coding as we did, the correct $t=0$ conditions are found (either PHIA or PHIB) in spite of this quirk.

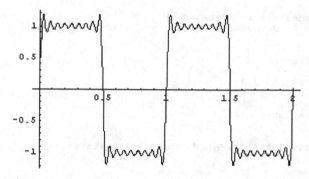

Figure 9.10 Square wave approximation from ten terms of the Fourier cosine series.

B3. A MATLAB wrapper for individual half periods

So we have found that programming the square wave as signals in the coupling variables did not work. The smoother the signal, the greater likelihood that coding the time dependence succeeds in an efficient time integration. The discontinuity in the ideal square wave is the enemy of convergence. For our third attempt, we recognized that we have already used an excellent strategy to overcome the effects of the discontinuity in the initial conditions – staging the fast elliptic step without the species transport, and then turning on the transient solution with the species now mobile. We could simply implement the square wave by successively manually swapping the values of the constants PHIA and PHIB, restarting the stationary nonlinear solver to find the fast elliptic switch of the flow and potential fields, then let the species transport continue under the new flow conditions by restarting with this initial condition and the transient solver. The MATLAB code we wrote merely puts a loop around this operation to continue as long as specified. Since we have put loops around a number of FEMLAB model m-files generated from the GUI, this is not a new technique.

However, the coding warrants a look for a crafty way of patching together the array of solution vectors (fem.sol.u) and the array of solution times (fem.sol.tlist) so that the FEMLAB GUI can be used to animate the movie. Using the manual technique in the FEMLAB GUI described above, the user would only have the most recent half-period available for animation. The m-file script outline below should be run in MATLAB, and thereafter the fem.sol structure imported into the FEMLAB GUI from the "Insert from Workspace" option on the file menu, specifying "Compatible solution" as fem.sol. Then the animation proceeds smoothly. Curiously, loading the m-file script from the FEMLAB GUI as a model m-file still only leaves the last half-period available for post-processing. No doubt this is due to our imperfect understanding of FEMLAB's variable assignments.

```
% FEMLAB Model M-file (Ystaged.m)
% Generated 05-Jun-2003 21:11:46 by FEMLAB 2.3.0.148.

flclear fem
% FEMLAB Version
clear vrsn;
vrsn.name='FEMLAB 2.3';
vrsn.major=0;
vrsn.build=148;
fem.version=vrsn;


%%%%%%%%%%%%%%%%%%%%%%%%%%%WBJZ contants %%%%%%%%%%%%%%%%%%%%
tau=3;
cycles=4;
phia=24;
phib=12;

% Recorded command sequence    (set up first half cycle before the
loop)
...
% Define constants
fem.const={...
        'Pec',     35.399999999999999,...
        'Re',      0.035400000000000001,...
        'zel',     1,...
        'betael',  1.1200000000000001,...
        'zetal',   1,...
        'zetar',   1,...
        'sigr',    1,...
        'f',       1,...
        'PHIA',    phia,...
        'PHIB',    phib,...
        'PHIC',    0,...
        'PA',      0,...
        'PB',      0,...
        'PC',      0,...
        'dsa',     10,...
        'dsb',     10,...
        'dsc',     6};
```

```
...
% Solve dynamic problem
fem.sol=femtime(fem,...
        'tlist',   0:0.10000000000000001:tau,...
        'atol',
{'u',0.01,'v',0.01,'p',Inf,'Y',0.001,'phi',0.001,'lmu',0.001, ...
'lmv',0.001},...
        'rtol',    0.01,...
        'jacobian','equ',...
        'mass',    'full',...
        'ode',     'fldaspk',...
        'odeopt',
struct('InitialStep',{[]},'MaxOrder',{5},'MaxStep',{[]}, ...
'maxiter',{6},'estrat',{0},'complex',{0}),...
        'out',     'sol',...
        'stop',    'on',...
        'init',    init,...
        'report',  'on',...
        'timeind','auto',...
        'index2v',[],...
        'index1v',[],...
        'consistent','bweuler',...
        'krylov',  'direct',...
        'context','local',...
        'sd',      'off',...
        'nullfun','flnullorth',...
        'blocksize',5000,...
        'solcomp',{'p','u','v','Y','phi','lmu','lmv'},...
        'linsolver','matlab',...
        'uscale',  'auto');

% Save current fem structure for restart purposes

%%%%%%%%%%%%%%%%%%%%%%%%%WBJZ%%%%%%%%%%%%%%%%%%%%%%%%%%%%%%%%%%%%%%%%%%%
%%%
fem0=fem;
fem1=fem;

for k=1:2*cycles-1
swap=phia;
phia=phib;
phib=swap;
...
% Solve nonlinear problem
fem.sol=femnlin(fem,...
        'out',     'sol',...
        'stop',    'on',...
        'init',    init,...
        'report',  'on',...
        'context','local',...
        'sd',      'off',...
        'nullfun','flnullorth',...
        'blocksize',5000,...
        'solcomp',{'p','u','v','phi','lmu','lmv'},...
        'linsolver','matlab',...
        'bsteps',  0,...
        'ntol',    9.9999999999999995e-007,...
        'hnlin',   'off',...
```

```
                'jacobian','equ',...
                'maxiter',25,...
                'method', 'eliminate',...
                'uscale', 'auto');

% Save current fem structure for restart purposes
fem0=fem;
...
% Solve dynamic problem
fem.sol=femtime(fem,...
          'tlist',   (k*tau):0.10000000000000001:((k+1)*tau),...
          'atol',
{'u',0.01,'v',0.01,'p',Inf,'Y',0.001,'phi',0.001,'lmu',0.001, ...
'lmv',0.001},...
          'rtol',    0.01,...
          'jacobian','equ',...
          'mass',    'full',...
          'ode',     'fldaspk',...
          'odeopt',
struct('InitialStep',{[]},'MaxOrder',{5},'MaxStep',{[]}, ...
'maxiter',{6},'estrat',{0},'complex',{0}),...
          'out',     'sol',...
          'stop',    'on',...
          'init',    init,...
          'report',  'on',...
          'timeind', 'auto',...
          'index2v', [],...
          'index1v', [],...
          'consistent','bweuler',...
          'krylov',  'direct',...
          'context', 'local',...
          'sd',      'off',...
          'nullfun', 'flnullorth',...
          'blocksize',5000,...
          'solcomp',{'p','u','v','Y','phi','lmu','lmv'},...
          'linsolver','matlab',...
          'uscale', 'auto');

% Save current fem structure for restart purposes

u=[fem1.sol.u, fem.sol.u];
tlist=[fem1.sol.tlist, fem.sol.tlist];
sol=fem.sol;
sol.u=u;
sol.tlist=tlist;
fem.sol=sol;
fem0=fem;
fem1=fem;

clear sol u tlist;

end
```

For clarity, our alterations to the FEMLAB GUI generated model m-file are shown. The model m-file is found from running the first half period with the initial fast elliptic step without species, then a time dependent solution with species transport, then swapping the phia and phib values, then repeating the fast elliptic step and time dependent restarts. The loop is then placed around the second set of solutions. The final part is to append the fem.sol structure with the current set of solution vectors and tlist. Now run the animation to appreciate the speed and electrokinetic switching in action. Figure 9.11 shows the configuration in the "lower" pseudosteady state (second half period). The striking feature of the animation is how reproducible each cycle is – diffusion does not smooth anything out cumulatively.

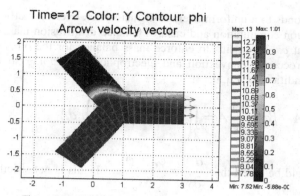

Figure 9.11 Lower flow pattern in the second half-period.

Monitoring the Output Concentration

In the case of species, the variation of concentration along the outlet channel segment C may be of interest and one must solve at least a one-dimensional differential equation for species distribution. The equation for this case is:

$$\frac{\partial Y}{\partial t} + U \frac{\partial Y}{\partial s} = \frac{\partial}{\partial s} \left(\frac{1}{\text{Pe}} \frac{\partial Y}{\partial s} + \beta z Y \frac{\partial \phi}{\partial s} \right) \qquad (9.14)$$

where $U = Q_c / A$ and s is distance along channel C. The boundary condition at reservoir C is the same as that used at the junction domain outflow boundary c, i.e. $\partial Y / \partial s = 0$ at $s = s_C$. At the upstream boundary of the segment one could take the simple route of setting the boundary condition $Y = Y_c$ at $s = s_c$ where

$$Y_c = \frac{1}{A} \int_A Y dA \qquad (9.15)$$

However, imposing this average value of mass fraction will not in general satisfy conservation of species. Care must be taken with the species boundary condition at the connection between the junction domain and the downstream segment.

First of all, now that the downstream spatial variation of species is available from the segment c solution, an improved boundary condition for the junction domain at the outflow boundary is possible. One can impose the computed segment species gradient over the entire outflow boundary:

$$\frac{\partial Y}{\partial x_i} n_i = \left(\frac{\partial Y}{\partial s}\right)_c \qquad (9.16)$$

This corresponds to a uniform average diffusion flux over the outflow boundary for the junction computation and ensures balance of diffusion mass flow rate of the species at c. It remains to enforce mass balance for the other two species mass flux processes, convection and electrophoresis. Doing so leads to the boundary condition for Eq. 9.14:

$$UYA - \beta z Y \frac{\partial \phi}{\partial s} A = \int_A \left(u_i Y - \beta z Y \frac{\partial \phi}{\partial x_i} \right) n_i dA \quad \text{at} \quad s = s_c \quad (9.17)$$

Here, it should be noted that $\dfrac{\partial \phi}{\partial s} = \dfrac{\phi_c - \phi_c}{\Delta s_c}$. If this is not strictly correct, as

will be the case for non-uniform electrical conductivity, one must solve the one-dimensional equation for potential along with 9.14.

Now to the next FEMLAB coding. All as before, but solving for Y in the downstream segment.

Multiphysics Add/Edit Modes
- Add geometry geom2 1D
- PDE general form, mode name outlet, variable c>>
- Apply / OK

In Draw Mode, using Specify Geometry, set the name to outlet and the range [0,6].

Now pull down the **Subdomain** menu and select **Subdomain settings**.

Subdomain Mode/Settings	Γ	F	d_a
Geom 2: Subdomain 1(outlet mode)	-(1/Pec)*cx	-Qu*cx+betael*zel*cx*delphi	1
Init tab	c(t0) = 1		
•		Apply / OK	

Now for the boundary conditions. Pull down the **Boundary** menu and select **Boundary Settings**.

Boundary Mode / Boundary Settings

- outlet mode

bnd 1	bnd 2
Dirichlet	Neumann
$G = 0$; $R = Yc - c$	$G = 0$

- Apply / OK

We have implicitly used three additional scalar coupling variables: delphi, Qu, and Yc. Now we need to define them:

Add/Edit Coupling Variables

scalar add delphi. Source Geom 1, subdomain 1, boundary 9.
Integrand: (PHIC-voltc)/dsc
Integration order: 2
Destination Geom 2 subdomain 1 Check "Active in this domain" box.
scalar add Qu. Source Geom 1, subdomain 1, boundary 9.
Integrand: Qc
Integration order: 2
Destination Geom 2 subdomain 1 Check "Active in this domain" box.
scalar add Yc. Source Geom 1, subdomain 1, boundary 9.
Integrand: Y
Integration order: 2
Destination Geom 2 bnd 1 Check "Active in this domain" box.

- Apply / OK

Recall that the cross-section is unity for channel c, and thus for a 2-D model, $A=1$ for the averages Qu and Yc. For convenience, we have used (9.15) as the boundary condition. Now you are ready to mesh. Set the max element size to 0.1 in geom2 and **Remesh**. Then we can solve. Twice as usual. First solve the fast elliptic step with the stationary nonlinear solver – be careful to de-select species and outlet modes. Then solve with all modes and the time dependent solver. Do not forget to turn off the automatic scaling of variables (Solver Parameters/Settings).

After some calculation time, we arrive at a final state in the outlet geometry (geom2) of **uniform** concentration $c=1$. Upon inspection, we find that it never changed. Since the concentration profile in the outlet does not couple back to the Y-junction dynamics, that was unaffected. But why was there no change in the outlet concentration from the initial condition? A bit of reflection on the theme of this chapter leads to the suspicion that we need a weak boundary constraint on the inlet (bnd 1 of geom2). There is a coupled boundary constraint of the Dirichlet type, which requires treatment with a weak boundary constraint.

Multiphysics Add/Edit Modes
- Select "Weak boundary constraint"
- Mode wcc and variable name lmu>>
- Apply / OK

Back in the FEMLAB GUI main window, select Boundary Mode and Boundary Settings for mode wc1.

Boundary Mode and Boundary Settings (wcc)
- Select domain 1, check active in this subdomain, type 'c' into the constraint variable entry box
- Apply / OK

Repeating the two solver steps – fast elliptic and time dependent – yields for instance Figure 9.12, a cross plot at position x=0.5 for all times, clearly showing the passage of a c=1 slug being displaced by a c=0 diffusive-convective front.

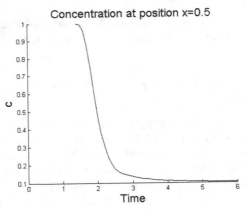

Figure 9.12 Concentration at a fixed point x=0.5 in the outlet geometry over time.

There are clearly many more variations on this theme that can be made. Coupling variables and extended multiphysics enable the connections to be made for entire microfluidic switching networks, if patiently constructed. Electrokinetic microchannel flows have a rich texture of physicochemical coupling for which this chapter has only touched the surface, hinting at the power of control available through combined pressure and voltage application.

The design issues of microfluidic switches can be profitably explored by such 2D/1D networks – our Y-junction is an idealized geometry which could be tailored to achieve greatest segregation among the switching slugs.

9.4 Summary

This chapter explored the multiphysics modeling appropriate for electrokinetic flow in microchannel networks. In setting up our case study, we discovered how to use FEMLAB's weak boundary constraints for coupled boundary conditions that incorporate non-tangential boundary conditions. We showed the utility of weak boundary conditions in accurate flux computations by revisiting the electrical capacitance tomography forward problem defined in §7.3.2. An order of magnitude improvement in convergence rate was found for little extra cost. After this simple example, we moved on to implementing more complicated weak boundary constraints in the electrokinetic flow model. The latter illustrated FEMLAB's guidelines for when to use a weak boundary constraint and when they fail. One typically has the connotation that constraints on the function should be implemented as Dirichlet-type boundary conditions, but that constraints on the derivative should be implemented as Neumann-type boundary conditions. In the context of a weak boundary condition, if you can implement it as a Neumann boundary condition, it does not count as a constraint. Thus, only Dirichlet-type boundary conditions can be treated with weak boundary constraints, and the distinction between constraints on the function and on its derivatives does not apply.

Acknowledgements

We are indebted to Johan Sundqvist and Niklas Rom of COMSOL for consultation on the handling of weak boundary constraints in this chapter.

References

1. FEMLAB 2.3 User's Guide and Introduction, p. 1-398ff (2002).
2. W.B. Zimmerman and G.M. Homsy, "Nonlinear viscous fingering in miscible displacement with anisotropic dispersion." *Physics of Fluids A* 3(8) 1859 (1991).
3. J. M. MacInnes, "Computation of Reacting Electrokinetic Flow in Microchannel Geometries", *Chemical Engineering Science*, 57 (21), 4539–4558 (2002).

4. J. M. MacInnes, X. Du and R. W. K. Allen, "Prediction of Electrokinetic and Pressure Flow in a Microchannel T-Junction", *Physics of Fluids*, in press (2003).
5. S.V. Patankar, S.V., <u>Numerical heat transfer and fluid flow</u>. Hemisphere Publishing Corporation, New York, 1980.
6. E. B. Cummings, S. K. Griffiths, R. H. Nilson and P. H. Paul, "Conditions for Similitude between the Fluid Velocity and Electric Field in Electroosmotic Flow", *Analytical Chemistry*, 72 (11), 2526–2532 (2000).

Appendix

A MATLAB/FEMLAB PRIMER FOR VECTOR CALCULUS

W.B.J. ZIMMERMAN

Department of Chemical and Process Engineering, University of Sheffield, Newcastle Street, Sheffield S1 3JD United Kingdom

and

J.M. REES

Department of Applied Mathematics, University of Sheffield, Hicks Building, Sheffield

Vector calculus underpins partial differential equations and their numerical approximation. Modelers must have a good working knowledge of the basics of vector calculus to use finite element methods effectively. Perhaps because undergraduate engineers are not confronted with realistic applications of vector calculus, but rather learn it as a mathematical discipline, their ability to apply vector calculus in engineering modeling is limited. In this appendix, all the basics of vector calculus are introduced with reference to MATLAB/FEMLAB utility and implementation. So the other way of reading this appendix is as a primer for MATLAB/FEMLAB basics with regard to multivariable differential calculus. When we wrote this appendix, we debated whether or not to augment Chapter One (basics of numerical analysis) with the material directly, as numerical approximation of derivatives is fundamental to the solution of PDEs – a FEMLAB "primitive" operation. Indeed, in learning spectral methods for solving PDEs, the fundamental theorem is the "derivative theorem" – how to use the spectral transform method to approximate derivatives. So by analogy, the fundamental utility of FEM is numerical differentiation. The debate was lost in that Chapter One aims to solve basic problems with FEMLAB straightaway. Approximating derivatives, no matter how useful, is still an intermediate step in modeling, rarely the objective itself. The only MATLAB basics we consider essential that are not used in making vector calculus the point of this appendix are eigenvalue analysis and logical expressions. These are sprinkled throughout the textbook anyway.

A.1 Review of Vectors

A.1.1 Representation of vectors

Since FEMLAB deals with scalar, vector, and matrix quantities, if only as input coefficients, a brief review of the representation of vectors (as a special case of MATLAB's matrix data type) is in order here. Scalar quantities can be represented by a single number, but vector quantities have magnitude and direction. Given a righthanded coordinate system as shown in Figure A1, any vector **a** is expressible in the form

$$\mathbf{a} = a_1\mathbf{i} + a_2\mathbf{j} + a_3\mathbf{k}$$
$$\mathbf{a} = \left(a_1, a_2, a_3\right)$$
(A1)

where $\mathbf{i}, \mathbf{j}$ and $\mathbf{k}$ are unit vector in the coordinate directions, a_1, a_2, a_3 are the components of $\mathbf{a}$ relative to this set of axes. They are the projections of $\mathbf{a}$ on to the unit vectors $\mathbf{i}, \mathbf{j}$ and $\mathbf{k}$. For a point P with coordinates (x,y,z), the position vector of P relative to the origin of the coordinate system, O, is

$$\mathbf{r} = x\mathbf{i} + y\mathbf{j} + z\mathbf{k}$$
$$= \left(x, y, z\right)$$
(A2)

MATLAB represents vectors in component form as either column (countervariant) or row (covariant) vectors:

```
>> a = [1; 2; 3];          % column vector
>> a = [1  2  3];          % row vector
```

In the row vector, the white space (any number of contiguous spaces) serves as the delimiter. The column vector is delimited by semicolons, or alternatively, by newlines:

```
>> a = [1
   2
   3];
```

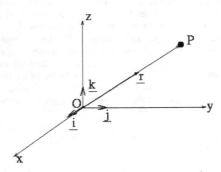

Figure A1. Position vector of a point P with respect to cartesian coordinate axes.

A.1.2 Scalar products, matrix multiplication, unit vectors, and vector products
Typically, scalar products (or dot products) are defined by

$$\mathbf{a} \cdot \mathbf{b} = |a||b|\cos\theta = a_1b_1 + a_2b_2 + a_3b_3 = \sum_{i=1}^{3} a_i b_i$$
(A3)

where θ is the angle between the vectors **a** and **b**. To achieve the same result in MATLAB, we use the * operator

```
>> a = [1; 2; 3];
>> b = [-3 2 -1];
>> b*a
ans =
   -2
```

This is a special case of a row vector (1×3 matrix) multiplying a column vector (3×1 matrix). As the first dimension of the latter and the second dimension of the former are the same, these matrices are compatible and can be multiplied according to the general rule for matrix multiplication

$$\left(AB\right)_{ik} = \sum_{j=1}^{n} A_{ij}B_{jk} \tag{A4}$$

If A is an $m \times n$ matrix and B is an $n \times l$ matrix, then AB is an $m \times l$ matrix. If the common size is not respected, then the matrices are incompatible and the product is not defined. MATLAB can compute scalar products as the special case of matrix multiplication, but care must be taken to respect compatibility of the vectors. For instance,

```
>> a*b
ans =
   -3      2     -1
   -6      4     -2
   -9      6     -3
```

What happened? Simply, **a** is a 3×1 matrix multiplying a 1×3 matrix, **b**. The product, **ab**, is a 3×3 matrix, viz.

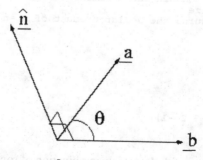

Figure A2. $\mathbf{b} \times \mathbf{a}$ is in the direction of $\hat{n}$.

$$(ab)_{ik} = \sum_{j=1}^{n} a_{ij}b_{jk} = a_{i1}b_{1k} \tag{A5}$$

In the case of vectors, the matrix (ab)$_{ik}$ is called the dyadic product of **a** and **b**, or a dyad. It is a special case of the matrix outer product, where the scalar product is also termed the inner product.

The scalar product of two row vectors or two column vectors can be computed in MATLAB using the transpose operator ', which is a unary operator and deceptively easy to mistake as a single quote of a character string, for instance.

```
>> a=[1; 2; 3]; b=[-3; 2; -1]; b'*a
ans =
      -2
```
but
```
>> a*b'

ans =
     -3      2     -1
     -6      4     -2
     -9      6     -3
```

still yields the dyad. Care must still be taken to respect the matrix compatibility. If **a** and **b** were row vectors, which combination, b'*a or a*b' yields the inner and outer products? MATLAB provides a special function dot for this purpose that blurs the distinction about compatibility:

```
>> help dot
 DOT   Vector dot product.
    C = DOT(A,B) returns the scalar product of the vectors A and B.
    A and B must be vectors of the same length.  When A and B are both
    column vectors, DOT(A,B) is the same as A'*B.
    DOT(A,B), for N-D arrays A and B, returns the scalar product
    along the first non-singleton dimension of A and B. A and B must
    have the same size.
    DOT(A,B,DIM) returns the scalar product of A and B in the
    dimension DIM.
    See also CROSS.
```

Example.
```
> dot(a,b)
ans =
      -2
>> dot([1; 2; 3],[-3 2 -1])
ans =
      -2
```

It simply does not matter with dot which combination of row/column vectors is used.

Vector Magnitude

The norm or magnitude of a vector is found by the formula

$$\|\mathbf{a}\| = \sqrt{\mathbf{a} \cdot \mathbf{a}} = \left(\sum_{j=1}^{n} a_j^2 \right)^{1/2} \tag{A6}$$

MATLAB will compute the norm of a vector with the formula

```
>> sqrt(a'*a)
ans =
    3.7417
```

or with the built-in command `norm`

```
>> norm(a,2)
ans =
    3.7417
```

where `sqrt( )` is the built-in square root function.

Unit Vector

A unit vector is a vector whose norm is one. Unit vectors can be constructed by normalization, i.e.

$$\hat{\mathbf{a}} = \frac{\mathbf{a}}{\|\mathbf{a}\|} \tag{A7}$$

For example,

```
>> ahat=a/norm(a,2)
ahat =
    0.2673
    0.5345
    0.8018
```

The division above is scalar division, which divides each element of the vector by the scalar.

Cross Product

The vector or cross product is defined

$$\mathbf{a} \times \mathbf{b} = |\mathbf{a}||\mathbf{b}| \sin \theta \hat{n} = \sum_{i=1}^{3} \varepsilon_{ijk} a_j b_k \hat{e}_i \tag{A8}$$

where ε_{ijk} is the permutation tensor, which takes the value +1 when indices ijk are a positive permutation of 123, -1 if they are a negative permutation of 123, and zero otherwise. $\hat{n}$ is the unit normal vector to the plane containing **a** and **b**. $\hat{e}_i$ is the unit vector in the i-th coordinate direction.

MATLAB provides a special function to compute cross products

```
>> help cross
CROSS   Vector cross product.
  C = CROSS(A,B) returns the cross product of the vectors
  A and B.    That is, C = A x B.   A and B must be 3 element
  vectors.
  C = CROSS(A,B) returns the cross product of A and B along the
  first dimension of length 3.
  C = CROSS(A,B,DIM), where A and B are N-D arrays, returns the cross
  product of vectors in the dimension DIM of A and B. A and B must
  have the same size, and both SIZE(A,DIM) and SIZE(B,DIM) must be 3.
  See also DOT.
```

For example,

```
>> cross(a,b)
ans =

    -8
    -8
     8
>> cross(b,a)
ans =

     8
     8
    -8
```

We see that the order of factors in a cross product switches the sign of the cross product, akin to changing the sense of the unit normal $\hat{n}$.

A.2 Arrays: Simple Arrays, Cell Arrays, and Structures

Array manipulation is essential to data extraction from FEMLAB. FEMLAB has organized models conveniently (for its developers and programmers) around fem structures for multiphysics and xfem structures for extended multiphysics. Pruning structures and cell arrays to extract meaningful information is a useful way of interrogating FEMLAB models (and solutions).

Simple Arrays

Arrays have dimensions $(m \times n \times l \ ...)$.   A matrix is a two-dimensional array. Each dimension has a length. So two very important commands are size( ) and length( ).

```
>> a = [ 1 2 3 4; 5 6 7 8];
>> size(a)
ans =
     2    4
```

Size of an array is itself a row vector of length equal to the array dimensions.

```
>> length(a(1,:))
ans =
     4
```

The colon (:) placeholder in the second argument of a specifies the entire range of the second dimension, in this case elements 1:4, i.e. 1 thru 4.

```
>> length(a(:,3))
ans =
     2
>> a(1,2:4)
ans =
     2    3    4
```

In fact, the colon refers to subarrays of a lower dimension. a(1,:) is the first row; a(:,3) is the third column of a. a(1, 2:4) gives a subarray of elements 2 thru 4 of row 1. In higher dimensions, the subarrays extracted are more complicated. For instance

```
>>b=ones(2,2,2)
b(:,:,1) =
     1    1
     1    1
b(:,:,2) =
     1    1
     1    1
>> b(1,:)
ans =
     1    1    1    1
```

Here, the subarrays are matrices in the first two cases, but in the third case, the final two dimensions are rolled up into a single row vector.

FORTRAN programmers will probably feel more comfortable addressing single elements

```
>> a(1,3)
ans =
     3
```

rather than subarrays, perhaps by using looping structures.

Array Construction

Arrays can be automatically generated using colon notation, viz.

```
>> a=[0: 0.1: 1]*pi
a =
  Columns 1 through 8
0    0.3142    0.6283    0.9425    1.2566    1.5708    1.8850    2.1991
Columns 9 through 11
2.5133    2.8274    3.1416
```

which produces eleven values equally spaced between 0 and π. So does

```
a=linspace(0,pi,11)
a =
  Columns 1 through 8
0    0.3142    0.6283    0.9425    1.2566    1.5708    1.8850    2.1991
Columns 9 through 11
2.5133    2.8274    3.1416
```

linspace is a versatile command for automatic matrix generation, performing a role that is often done in looping constructs in older programming languages.

```
>> help linspace
 LINSPACE Linearly spaced vector.
    LINSPACE(X1, X2) generates a row vector of 100 linearly
    equally spaced points between X1 and X2.
    LINSPACE(X1, X2, N) generates N points between X1 and X2.
    For N < 2, LINSPACE returns X2.
    See also LOGSPACE, :.
```

logspace comes in handy as well.

```
>> help logspace
LOGSPACE Logarithmically spaced vector.
    LOGSPACE(X1, X2) generates a row vector of 50 logarithmically
    equally spaced points between decades 10^X1 and 10^X2.  If X2
    is pi, then the points are between 10^X1 and pi.
    LOGSPACE(X1, X2, N) generates N points.
    For N < 2, LOGSPACE returns 10^X2.
    See also LINSPACE, :.
```

Four other common array generators are zeros, ones, rand, and for matrices, eye. zeros initializes an array with zeros; ones with ones, rand with uniformly distributed random numbers (randn with normal deviates) and eye with the identity matrix.

```
>> help zeros
 ZEROS  Zeros array.
    ZEROS(N) is an N-by-N matrix of zeros.
    ZEROS(M,N) or ZEROS([M,N]) is an M-by-N matrix of zeros.
    ZEROS(M,N,P,...) or ZEROS([M N P ...]) is an M-by-N-by-P-by-...
    array of zeros.
    ZEROS(SIZE(A)) is the same size as A and all zeros.

>> help ones
ONES   Ones array.
    ONES(N) is an N-by-N matrix of ones.
    ONES(M,N) or ONES([M,N]) is an M-by-N matrix of ones.
    ONES(M,N,P,...) or ONES([M N P ...]) is an M-by-N-by-P-by-...
    array of ones.
    ONES(SIZE(A)) is the same size as A and all ones.
```

```
>> help rand
 RAND    Uniformly distributed random numbers.
    RAND(N) is an N-by-N matrix with random entries, chosen from
    a uniform distribution on the interval (0.0,1.0).
    RAND(M,N) and RAND([M,N]) are M-by-N matrices with random entries.
    RAND(M,N,P,...) or RAND([M,N,P,...]) generate random arrays.
    RAND with no arguments is a scalar whose value changes each time it
    is referenced.  RAND(SIZE(A)) is the same size as A.

>> help eye
 EYE Identity matrix.
    EYE(N) is the N-by-N identity matrix.
    EYE(M,N) or EYE([M,N]) is an M-by-N matrix with 1's on
    the diagonal and zeros elsewhere.

    EYE(SIZE(A)) is the same size as A.
```

Scalar – Array Math

Arithmetic of scalars acting on arrays threads across the array. For instance,

```
>> 3*a
ans =
Columns 1 through 8
0    0.9425    1.8850    2.8274    3.7699    4.7124    5.6549    6.5973
Columns 9 through 11
7.5398    8.4823    9.4248

>> 5+a
ans =
Columns 1 through 8
5.0000    5.3142    5.6283    5.9425    6.2566    6.5708    6.8850    7.1991
Columns 9 through 11
7.5133    7.8274    8.1416
```

Array – Array Element-Wise Math

Arithmetic of arrays on arrays is a tricky area. If the arrays are compatible sizes,
then dot-operators are applied element-wise:

```
>> b=linspace(1,11,11)
b =
    1    2    3    4    5    6    7    8    9    10    11
>> size(a)
ans =
    1    11
>> size(b)
ans =
    1    11
>> a.*b
ans =
Columns 1 through 8
0    0.6283    1.8850    3.7699    6.2832    9.4248    13.1947    17.5929
Columns 9 through 11
22.6195    28.2743    34.5575
```

Cell Arrays and Structures

You could write a whole chapter about cell arrays and structures. The important thing to note about both is that they are containers for heterogeneous mixtures of data types – floating point numbers, complex numbers, matrices, character strings, other cell arrays and structures. The cell array refers to each cell within the array by an array index. Cell arrays are directly defined with braces surrounding the list of array elements. The `cell` command will create an empty shell which can be assigned individual elements or subarrays.

```
>> ca = { 'every', 'good', 'boy', 'does', 'find', 3+3i, [0 1; -2 2] }
ca =
Columns 1 through 6
'every'     'good'     'boy'     'does'     'find'     [3.0000+ 3.0000i]
Column 7
[2x2 double]
```

Referencing can be done by array index, with parenthesis, returns the cell element.

```
>> ca(3)
ans =
    'boy'
```

Referencing by braces and index number returns the *contents* of the cell element.

```
>> ca{3}
ans =
boy
```

Perhaps this is clear with regard to the matrix cell constituent,

```
>> ca(7)
ans =
    [2x2 double]
>> ca{7}
ans =
    0     1
   -2     2
```

Structures are referenced by fields, which are named rather than enumerated, much as in the C programming language. The greatest utility in using a structure as averse to a cell array is that if you choose to alter the structure, addition or elimination of fields does not change the ordering of fields in a meaningful way. Elimination or addition of cell array elements, however, changes the numbering of cells or leaves "holes" in the array. The most common structure encountered by FEMLAB users is the fem structure, which is how FEMLAB organizes the complete set of data for its multiphysics models and their solutions. Exporting as fem structure to the MATLAB workspace from FEMLAB produces the following for our Benard convection model.

```
>> fem
fem =
             sdim: {'x'   'y'}
             appl: {[1x1 struct]   [1x1 struct]}
             draw: [1x1 struct]
         simplify: 'on'
             geom: [1x1 geom2]
              dim: {'u'   'v'   'p'   'T'}
             form: 'general'
              equ: [1x1 struct]
              bnd: [1x1 struct]
              pnt: [1x1 struct]
           border: 1
             expr: {1x0 cell}
              var: {}
           sshape: 2
          elemmph: {1x0 cell}
       eleminitmph: {1x0 cell}
             mesh: [1x1 struct]
          outform: 'general'
             diff: {'ga'   'g'   'f'   'r'   'expr'}
            shape: {'shlag(2,'u')'     'shlag(2,'v')'     'shlag(1,'p')'
'shlag(2,'T')'}
            rules: {1x0 cell}
              sol: [1x1 struct]
          version: [1x1 struct]
            xmesh: [1x1 struct]
            const: {'Ra'   [1710]   'Pr'   [1]}
```

The list is of fields in the structure fem above shows the description of the field contents. Each field can be addressed with the "dot" notation:

```
>> fem.sdim
ans =
    'x'        'y'
```

fem.sdim is a cell array with two cells; the cell array is small enough that its contents can be displayed. Since it is a cell array, the braces index reference will act on the contents of the cell element.

```
>> fem.sdim{1}
ans =
x
```

As we can see, the fem structure has cell arrays, other structures, characters, and numbers as its constituents. There is no reason why we cannot have cell arrays of fem structures, which is indeed the make up of the xfem structure used by FEMLAB for extended multiphysics, with one fem structure for each logical geometry. We have frequently had need of the postinterp command which acts on fem structures or xfem structures to produce values of solution variables interpolated at points within the domain discretized by finite elements:

```
[is,pe]=postinterp(xfem,xx);
[u]=postinterp(xfem,'u1',is);
```

Passing the whole of the xfem structure to the postinterp function gives it access to the complete description of the model and solution, for which it may have to execute different branches of commands given that specific structure. As users of FEMLAB, we need to know enough about the MATLAB data structure of a FEMLAB model and solution to extract relevant data if we have particular postprocessing or modeling requirements that are not built into the FEMLAB GUI.

A.3 Scalar and Vector Fields: MATLAB Function Representations

Physical properties of matter typically depend on position and sometimes time. At length scales observable to humans (by eye), most physical quantities are treated as a continuum – having values at every mathematical point. These quantities are called fields. Quantitites such as temperature and pressure that represent a single value are termed scalar fields. A scalar field is a single number, e.g.

$$\phi = \phi(\mathbf{x}) = \phi(x, y, z) \tag{A9}$$

A vector field in 3-D requires three components:

$$\mathbf{F}(\mathbf{x}) = \begin{bmatrix} F_1(x, y, z) \\ F_2(x, y, z) \\ F_3(x, y, z) \end{bmatrix} \tag{A10}$$

Each component of $\mathbf{F}$ is itself a scalar function of position.

Example. $\phi = x^2 + y^2$

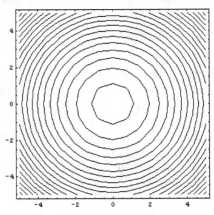

Figure A3. Contour lines for scalar function $\phi = x^2 + y^2 = C$ for 30 different values of C.

Figure A3 shows the contours of $\phi = x^2 + y^2 = C$ for several different values of the constant C. MATLA B has no data type to represent a field quantity. Rather, such quantities are represented as *functions*. There are three major ways of representing a function in MATLAB

(1) An inline function, defined only in the workspace

```
>> myfun = '1+log(r)';
>> myfuni=inline(myfun,'r')
myfuni =
     Inline function:
     myfuni(r) = 1+log(r)
>> a=feval(myfuni,1)
a =
     1
>> a=feval(myfuni,10)
a =
     3.3026
```

(2) An m-file function, which is stored as a disk file and can be called from either the workspace or an m-file script. For instance the m-file function temperat.m contains the following code and is stored in the MATLAB current directory.

```
function t=temperat(r)
%TEMPERAT evaluates T = 1 + ln r
%    T = temperat(r)
%
t=1+log(r);

>>a=temperat(10)
a =
     3.3026
```

(3) Interpolation functions. The values of the function at certain points are specified. Values at nearby points are estimated by assumption about how smooth the function is locally. Interpolation requires a series of MATLAB commands, but eventually results in a functional form. MATLAB has built-in functions for 1-D and 2-D data. The Rock Fracture Model in the FEMLAB Model Library (FEMLAB/Geophysics/rock_fracture) uses 2-D interpolation of a supplied dataset in a MATLAB mat-file, as an m-file function (flafun.m) that does 2-D interpolation:

```
function a=flafun(x,y)
%FLAFUN Interpolate aperture from sampled data.
%   A = FLAFUN(X,Y) interpolates rock fracture aperture from sampled
data.
%   X and Y are coordinates for node points in an unstructured grid.
%   A is the interpolated aperture in the the node points.
%
%   FLAFUN is a function used as diffusion coefficient in the
geophysics
%   ROCK_FRACTURE model in the Model Library.
```

```
%
%   Use FLAFUN as flafun(x,y)^3 in the diffusion coefficient in GUI.
%   This implements the cubic law for fracture conductivity in a
%   potential flow model.
%
%   The sampled data is stored in the file FLAPERTURE.MAT.
%
%   See also FLDOPING.

%   B. Sjodin 9-21-99.
%   Copyright (c) 1994-2000 by COMSOL AB
%   $Revision: 1.3 $   $Date: 2001/10/26 13:24:57 $

% Load the aperture data matrix.
load flaperture
% Create sample coordinates.
[m,n]=size(aperture);
dx=1;
dy=1;
[x1,y1]=meshgrid(0:dx:(m-1)*dx,0:dy:(n-1)*dy);
% Interpolate from rectangular grid to unstructured grid.
a=interp2(x1,y1,aperture,x,y);
```

Chapter three has a similar usage for using interpolant functions for representing velocity fields around a pellet. Chapter five represents a 1-D pressure field as an interpolant function in an m-file pinit.m:

```
function a=pinit(x)
presgrad=[
183.59
183.471
...
2.00851
0.0];
xlist=[0:0.1:10];
a=interp1(xlist, presgrad, x, 'spline');
```

We have judiciously abridged the pressure data set in presgrad. Here the cubic spline interpolation method is used forming a 1-D interpolant. The 2-D form above uses bilinear interpolation.

Typically FEMLAB field entry for coefficients and boundary data is done by in-line forms expressing the predefined independent, dependent, and derived variables. For instance, in general form with a single dependent variable u and independent variable x, expressions such as

$$u + 5 * x + \sin(3 * pi * x) + 3* u*ux$$

can be entered. But MATLAB m-file functions (including interpolants) can be used just as readily. An important point is that FEMLAB expects data entry as scalar components. If a vector or matrix is required, it is always through specification of scalar components, any of which can be (complex) functions.

FEMLAB represents its results in a FEM structure with the degrees of freedom specified in fem.sol for a mesh specified in fem.mesh (or fem.xmesh).

FEMLAB provides a special `postinterp` function to extract interpolated values from fem.sol for each dependent variable and derived variable. The book is littered with examples of using postinterp to represent functions. It can even be automated in an m-file function that calls the appropriate fem structure from a mat-file.

A.4 Differentiation in Multivariable Calculus

A.4.1 The gradient of a scalar field

If $\phi = \phi(x,y,z)$, then the vector

$$\nabla \phi = \frac{\partial \phi}{\partial x}\mathbf{i} + \frac{\partial \phi}{\partial y}\mathbf{j} + \frac{\partial \phi}{\partial z}\mathbf{k} = \begin{bmatrix} \dfrac{\partial \phi}{\partial x} \\[2mm] \dfrac{\partial \phi}{\partial y} \\[2mm] \dfrac{\partial \phi}{\partial z} \end{bmatrix} \qquad (A11)$$

is called the *gradient* of the scalar field ϕ, and is denoted as well by grad ϕ. The gradient operator ∇ (the nabla character) is the vector operator

$$\nabla = \mathbf{i}\frac{\partial}{\partial x} + \mathbf{j}\frac{\partial}{\partial y} + \mathbf{k}\frac{\partial}{\partial z} \qquad (A12)$$

in Cartesian coordinates in 3-D.

A FEMLAB example. Suppose $\phi = x^2 + y^2$, then $\nabla \phi = (2x, 2y, 0)$.

But MATLAB does not directly deal with such symbolic calculations, however its symbolic toolbox does. FEMLAB, however, routinely calculates the numerical approximation of the derivatives of a solution. So the gradient of a scalar field can be constructed by FEMLAB "primitive" operations. How do we easily access this information? Here's the recipe.

Model Navigator	2-D geom., PDE modes, general form (nonlin stat) independent variables: x,y dependent:phi
Options	Set Axes/Grid to [-1,1]×[-1,1]
Draw	Rectangular domain [-1,1]×[-1,1]
Boundary Mode/ Boundary Settings	Set all four domains to Neumann BCs
Subdomain Mode/ Subdomain Settings	In domain 1, set Γ = 0 0; da = 0; F = phi-x^2-y^2
Mesh Mode	Remesh using default (418 nodes, 774 elements)
Solve	Use default settings (nonlinear solver)
Post Process	Switch to arrow mode (automatically set to vectors of grad ϕ.) See Figure A4.

It should be noted that since no PDE is actually being solved, Neumann BCs amount to a neutral or non-condition on the boundaries. Otherwise, only if the boundary data are compatible with the condition 0 = phi-x^2-y^2 is a solution possible.

Now export the fem structure to MATLAB (file menu). We will use postinterp to get the approximate numerical value, along with MATLAB bilinear regression. The code below should look familiar to those who recall the porous catalyst (pellet) model of Chapter four.

```
>>x=0.5;y=0.5;
[xx,yy]=meshgrid(-1:0.01:1,-1:0.01:1);
xxx=[xx(:)'; yy(:)'];
phix=postinterp(fem,'phix',xxx);
phiy=postinterp(fem,'phiy',xxx);
uu=reshape(phix,size(xx));
vv=reshape(phiy,size(xx));
u=interp2(xx,yy,uu,x,y);
v=interp2(xx,yy,vv,x,y);
[u,v]

ans =
    1.0000    1.0000
```

X	y	phix	phiy
0.5	0.5	1.0000	1.0000
-0.25	0.75	-0.5000	1.5000
0.75	-0.5	1.5000	-1.0000
0.25	-0.75	0.5000	-1.5000

Table A1. Numerical estimates of grad ϕ using FEMLAB model.

By any accounting method, the use of FEM for finding first derivatives is fairly accurate. The global error of $O(10^{-16})$ as reported in the convergence criteria leads to a minimum of four decimal places in the estimated gradients.

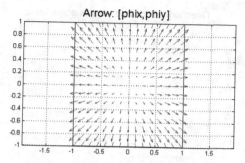

Figure A4. Arrow plot of vectors of grad ϕ.

The Directional Derivative

The directional derivative of ϕ is the rate of change of ϕ (x,y,z) along a given direction. If $\hat{n}$ is the unit vector in that direction, then the directional derivative is given by

$$\frac{\partial \phi}{\partial n} = \hat{n} \cdot \nabla \phi \tag{A13}$$

The coordinate directions are the easiest to compute, e.g.

$$\underline{i} \cdot \nabla \phi = \frac{\partial \phi}{\partial x} \tag{A14}$$

We used directional derivatives in the ECT models of Chapter seven to directly compute the normal derivatives of the electric potential (see §7.3.2 and equation (7.5)). Clearly, directional derivatives are intimately related to the concept of flux. The total flux across a material surface for a "linear" property (Fick's Law, Fourier's Law, etc.) is proportional to the integral of the normal derivative along that surface. The local flux is proportional to the normal derivative at a point.

At this point in most vector calculus texts, it is demonstrated that the direction in which the rate of change of ϕ is greatest is the direction of grad ϕ, and that | grad ϕ | is the rate of change in that direction. We can show this at say the point $(x,y)=(0.25,-0.75)$ by stepping through the angles $\theta=[0,\pi]$ and plotting the (scalar value) $\frac{\partial \phi}{\partial n}$. MATLAB code that achieves this is written below.

```
>> theta=linspace(0, pi, 100);
dirder = zeros(size(theta));
for k=1:length(theta)
dirder(k)=cos(theta(k))*u+sin(theta(k))*v;
end
plot(theta, dirder)
```

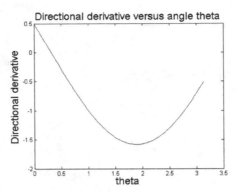

Figure A5. Directional derivative versus direction (angle θ) in radians. Note the presence of a minimum in directional derivative – the direction of steepest descent, which corresponds to the gradient direction.

Of course I refuse to apologize for my FORTRAN-ish programming bias which is revealed in the looping structure above. Were I in a more MATLAB-ish mode, then judicious use of threading achieves the same results without the loop:

```
>>dirder = u*cos(theta)+v*sin(theta);
plot(theta,dirder)
```

`cos` and `sin` functions thread across each element of the vector theta, producing an output vector of the same length.

Level Sets/Level Surfaces

Note that the directional derivative (dirder) crosses the x-axis, i.e. there is a direction for which the directional derivative is zero – no rate of change at all in that direction. It can be shown that the direction $\hat{n}$ for which $\dfrac{\partial \phi}{\partial n} = 0$ is perpendicular to the gradient direction. So in this direction, ϕ=constant locally. Tracing out the curve (in 2D) or surface (in 3D) of each constant identifies a family of curves (surfaces) called level sets of ϕ (see Chapter eight). In 2D, level sets are also called contours. The terminology of the directional derivative is analogous to survey maps, where ϕ is the elevation of land. The contours all have the same height above sea level (level sets); the directional derivative $\hat{n} \cdot \nabla \phi$ is the rate of climb in the direction $\hat{n}$, and the gradient is in the direction of steepest climb (or descent) and the rate of climb is | grad ϕ |. In fluid dynamics, the quantity that is most often represented by a contour plot is the streamfunction, with contours all being streamlines (particle paths in steady flow) tangent to the velocity field. In Chapter three, the buoyant convection example shows how to compute streamfunction (see equation (3.3)).

A.4.2 Derivatives of vector fields

The vector differential operator ∇ may be applied to a vector field $\boldsymbol{F(x)}$ in two ways: (1) the scalar product $\nabla \cdot \boldsymbol{F}$ called the divergence, (2) the vector product $\nabla \times \boldsymbol{F}$, called the curl.

The divergence is given by

$$\operatorname{div} F = \nabla \cdot F = \sum_{k=1}^{3} \frac{\partial F_k}{\partial x_k}$$

$$= \left(\frac{\partial}{\partial x}, \frac{\partial}{\partial y}, \frac{\partial}{\partial z} \right) \cdot (F_1, F_2, F_3) \qquad (A15)$$

$$= \frac{\partial F_1}{\partial x} + \frac{\partial F_2}{\partial y} + \frac{\partial F_3}{\partial z}$$

The curl is given by

$$\operatorname{curl} F = \nabla \times F = \sum_{i=1}^{3}\sum_{j=1}^{3}\sum_{k=1}^{3} \varepsilon_{ijk} \frac{\partial F_k}{\partial x_j} \hat{e}_i$$

$$= \begin{vmatrix} i & j & k \\ \dfrac{\partial}{\partial x} & \dfrac{\partial}{\partial y} & \dfrac{\partial}{\partial z} \\ F_1 & F_2 & F_3 \end{vmatrix} \qquad (A16)$$

$$= \left(\frac{\partial F_3}{\partial y} - \frac{\partial F_2}{\partial z}, \frac{\partial F_1}{\partial z} - \frac{\partial F_3}{\partial x}, \frac{\partial F_2}{\partial x} - \frac{\partial F_1}{\partial y} \right)$$

ε_{ijk} is the permutation tensor introduced earlier. Of course one can see readily that div **F** is a scalar, while curl **F** is a vector.

The operator $\boldsymbol{F} \cdot \nabla$ is often seen in advection terms in heat or mass transport equations. Clearly, it is *not* the divergence, since

$$F \cdot \nabla = F_1 \frac{\partial}{\partial x} + F_2 \frac{\partial_2}{\partial y} + F_3 \frac{\partial}{\partial z} \qquad (A17)$$

which is still an operator, in contrast to (A15), which is a scalar.

As we saw before, the numerical approximation of derivatives is a "primitive" of FEMLAB, so we should be able to compute approximations to both div and curl.

<u>A FEMLAB example</u>. Suppose $F = \left(x^2, 3xy, x^3 \right)$. Here's the recipe.

Model Navigator	3-D geom., PDE modes, general form (nonlin stat) independent variables: x,y z; 3dependent: u1, u2, u3
Options	Set Axes/Grid to [0,1]×[0,1]×[0,1]
Draw	Block BLK1= [0,1]×[0,1]×[0,1]
Boundary Mode/ Boundary Settings	Set all four domains to Neumann BCs
Subdomain Mode/ Subdomain Settings	set Γ1 = 0 0 0; da1 = 0 0 0; F1 = u1-x^2 set Γ2 = 0 0 0; da2 = 0 0 0; F2 = u2-3*x*y set Γ3 = 0 0 0; da3 = 0 0 0; F3 = u3-x^3
Mesh Mode	Remesh using mesh scaling factor 3 (201 nodes, 719 elements)
Solve	Use default settings (nonlinear solver)
Post Process	1. Color plot of u1x+u2y+u3z for the divergence 2. Arrow plot for the curl of (u3y-u2z,u1z-u3x,u2x-u1y)

Again, it should be noted that since no PDE is actually being solved, Neumann BCs amount to a neutral or non-condition on the boundaries. Otherwise, only if the boundary data are compatible with the conditions 0 = u1-x^2, 0 = u2-3*x*y, 0 = u3-x^3 is a solution possible.

Symbolically, it is straightforward to compute

$$\nabla \cdot \mathbf{F} = 5x$$

$$\nabla \times \mathbf{F} = \left(0, -3x^2, 3y \right)$$

So how good is the numerical approximation? Try the divergence:

```
>> xxx=[0.42; 0.57; 0.33];postinterp(fem,'u1x+u2y+u3z',xxx)
ans =    2.1137
>> 5*0.42
ans =    2.1000
```

Clearly, for such a coarse mesh, half a percent error is not a bad result. Now for the curl.

```
>>xxx=[0.42;0.57;0.33];[postinterp(fem,'u3y-u2z',xxx);
postinterp(fem,' u1z-u3x',xxx);postinterp(fem,'u2x-u1y',xxx)]
ans =
    0.0043
   -0.5319
    1.7100
>> [0; -3*0.42^2; 3*0.57]
ans =
         0
   -0.5292
    1.7100
```

The worst error here is again half a percent.

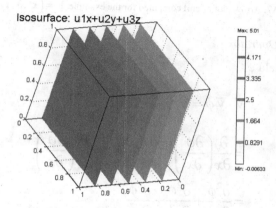

Figure A6. Isosurfaces of divergence computed for the example $\mathbf{F} = \left(x^2, 3xy, x^3 \right)$.

Figure A6 shows the numerical approximation by FEM to the divergence, which qualitatively shows isosurfaces consistent with $\nabla \cdot \mathbf{F} = 5x$. Figure A7 shows the arrow plot of curl F. Since most of us have little feel for three-dimensional vector plots, determining whether the plot is consistent with the closed form calculation is beyond our visual capacity for numeracy. Nevertheless, the FEM solution shows the very important feature of numerical solutions – visualization of solutions. Does anyone have a feel for the analytic solution $\nabla \times \mathbf{F} = \left(0, -3x^2, 3y \right)$ either?

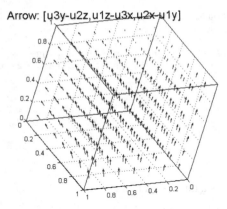

Figure A7. Arrow plot of curl computed for the example $\mathbf{F} = \left(x^2, 3xy, x^3 \right)$.

The Laplacian Operator

For the scalar field ϕ

$$
\begin{aligned}
\nabla \cdot \nabla \phi &= \nabla \cdot \left(\frac{\partial \phi}{\partial x}, \frac{\partial \phi}{\partial y}, \frac{\partial \phi}{\partial z} \right) \\
&= \frac{\partial}{\partial x} \left(\frac{\partial \phi}{\partial x} \right) + \frac{\partial}{\partial y} \left(\frac{\partial \phi}{\partial y} \right) + \frac{\partial}{\partial z} \left(\frac{\partial \phi}{\partial z} \right) \\
&= \frac{\partial^2 \phi}{\partial x^2} + \frac{\partial^2 \phi}{\partial y^2} + \frac{\partial^2 \phi}{\partial z^2}
\end{aligned}
\tag{A18}
$$

By parallel with other nabla operators,

$$
\begin{aligned}
\text{div}\left(\text{grad}\right) &= \nabla \cdot \nabla \\
&= \left(\frac{\partial}{\partial x}, \frac{\partial}{\partial y}, \frac{\partial}{\partial z} \right) \cdot \left(\frac{\partial}{\partial x}, \frac{\partial}{\partial y}, \frac{\partial}{\partial z} \right) \\
&= \frac{\partial^2}{\partial x^2} + \frac{\partial^2}{\partial y^2} + \frac{\partial^2}{\partial z^2}
\end{aligned}
\tag{A19}
$$

So for shorthand, the operator div(grad) is called the Laplacian and denoted ∇^2. Typically, the Laplacian is used in differential equations, rather than computed directly. For instance, Laplace's equation

$$\nabla^2\phi = 0 \tag{A20}$$

is an example where the Laplacian is known (zero) but the function ϕ is to be found. FEMLAB routinely computes the first derivatives of a dependent variable, but not necessarily the second, directly. But as we already know how to compute both div and grad separately, computing div(grad) is a matter of using auxiliary dependent variables v1,v2,v3 that are assigned values in the last example of

$$F1=v1-ux; \quad F2=v2-uy; \quad F3=v3-uz$$

so that

$$\nabla^2 u = v1x + v2y + v3z \tag{A21}$$

Scalar and Vector Potentials

Quite a lot of space in vector calculus books is devoted to the topics of scalar and vector potentials.

A scalar potential ϕ for a vector field $\mathbf{F}$ is a scalar function for which $\nabla\phi = \mathbf{F}$. The textbooks show that this is only possible if, and only if, curl $\mathbf{F}=0$.

Similarly, a vector potential $\mathbf{A}$ for a vector field $\mathbf{F}$ is a vector function for which $\mathbf{F}$=curl $\mathbf{A}$. Again, the textbooks show that this is only possible, if and only if, div $\mathbf{F}=0$.

Scalar and vector potentials are useful for simplifying pde systems that are either irrotational or divergence free (solenoidal). In the case of fluid flow, either inviscid or completely viscous flow are simplified dramatically by such potentials. One might ask, can FEMLAB help in the task of identifying these potentials? In the case of 2-D flows, we already saw that the streamfunction acts like a vector potential (3.3), so the answer is a qualified yes. For many years in both electrodynamics and hydrodynamics, the hunt for vector potentials or scalar potentials to simplify calculations was paramount – many analyses end is solving, even approximately, for such a potential. Yet whether sufficient symmetries exist in a given modeling situation to use scalar and vector potentials to simplify the calculations is now almost a moot point. General pde engines like FEMLAB can compute numerical approximations to the primitive variables in the most general cases, limited only by their CPU requirements. So the virtue of finding such simplifications is a reduction of CPU usage, for which we must still pay the price of numerical differentiation to arrive at the primitive variables (using our grad and curl recipes) if detailed solutions are required.

It is perhaps a sobering note to end our Appendix on that general purpose numerical solvers like FEMLAB limit the need for many of the complexities of vector calculus, since much of the higher theory was developed to treat intentionally idealized models. Nevertheless, the basics of vector calculus are necessary to understand what such pde engines do, and how they do it. Theory simultaneously becomes more important in some aspects – dealing with complexities that are still beyond computability, proposing physical models that are amenable to numerical computation – but also less necessary for "run-of-the-mill" applications. Theorists should be challenged that they must remain ahead of the game to still be relevant practioners due to the advent of general purpose solvers like FEMLAB.

Epilogue (WBJZ)

As an epilogue to this book, I have found that MATLAB programming is not really essential to modeling with FEMLAB. The GUI, with experience, serves for most purposes rather well. I believe that many experienced FEMLAB users are amazed at the flexibility I have teased out of the FEMLAB GUI. So MATLAB programming is really essential in only a few cases: (1) Massively parallel parameter space studies; (2) Non-PDE models of nonlocal, discrete coupling; (3) Moving boundaries; (4) Getting the detailed data out for post-processing. This list is probably not complete. I am not sure that FEMLAB developers envisaged people would do (1)-(3) with their tools. (4) should be addressed in later editions of FEMLAB. An I/O wizard of some sort to read in data into initial conditions or functions and to write output into standard formats would go a long way in removing the need to write your own MATLAB code. Until then, information concentrated in this Appendix on the MATLAB/FEMLAB interface, and sprinkled throughout the book in worked examples, will prove invaluable to the budding expert user.

A.5 End Note: Platform Dependence of Meshes

I could not think of a good place to put this note. As you might have noticed, the book was developed under both FEMLAB 2.2 and FEMLAB 2.3LCS/2.3/2.3a/2.3b, with either MATLAB 5.3/6.1/6.5, while using both Microsoft Windows and linux operating systems. This is bad news in that your MAT-files are not necessarily compatible across versions and platforms. So the advice here is to save your models as m-files for cross platform transfer. But in many cases, it is the results of the computations that are required to cross platforms. For instance, for us it was useful to do long computations on our best linux workstation as background jobs, then save the solution (the whole fem or xfem structure) and read it into the MATLAB workspace, and upload the relevant parts into the FEMLAB GUI (see all those import options on the file menu). For several of our models, we found that the same sequences of m-file commands generated different meshes (node and element numbers and places) so that the solution formats (degrees of freedom in fem.sol) were incompatible across platforms, FEMLAB and MATLAB versions. The workaround we found was to save the solutions (fem.sol) and the mesh on the compute server, and upload these through MATLAB into the FEMLAB GUI as imports on the file menu. The best advice is to always maintain a consistent environment, but if not, then carry both solutions and meshes in MATLAB MAT-files and do your model set up in an m-file – it gets the closest to portability!

INDEX

Printed in the United States
By Bookmasters